Process Selection
From design to manufacture

Second edition

K. G. Swift

Department of Engineering, University of Hull, UK

and

J. D. Booker

Department of Mechanical Engineering, University of Bristol, UK

OXFORD AMSTERDAM BOSTON LONDON NEW YORK PARIS
SAN DIEGO SAN FRANCISCO SINGAPORE SYDNEY TOKYO

Butterworth-Heinemann
An imprint of Elsevier
Linacre House, Jordan Hill, Oxford OX2 8DP
200 Wheeler Road, Burlington MA 01803

First published in 1997 by Edward Arnold
Second edition 2003

British Library Cataloguing in Publication Data
A catalogue record for this book is available from the British Library

Library of Congress Cataloguing in Publication Data
A catalogue record for this book is available from the Library of Congress

ISBN 0 7506 5437 6

For information on all Butterworth-Heinemann publications visit our website at www.bh.com

Typeset by Integra Software Services Pvt. Ltd, Pondicherry 605 005, India
www.integra-india.com
Printed and bound in Great Britain by Biddles Ltd, *www.biddles.co.uk*

Contents

Preface to second edition

Recent experiences from carrying out Design for Assembly (DFA) and Design for Manufacture (DFM) product studies in industry have reinforced the authors' belief that consideration of manufacturing problems at the design stage is the major means available for improving product quality, reducing manufacturing costs and increasing productivity. In the second edition, as well as providing further information to help select processes for components and the joining of components, we have included data on assembly process selection and costing. This can all be used to support DFA/DFM projects and associated activities.

The inclusion of assembly is very conscious, in that assembly issues are often neglected in product engineering. Through consideration of assembly, many strategically important issues can be addressed. For example, DFA impacts much more than assembly itself. In addition to reducing component assembly and handling costs, DFA encourages part-count optimization, variety reduction and standardization.

The authors wish to acknowledge the further support of individuals at CSC and Richard Batchelor of TRW. Special thanks are given to Bob Swain for help in the preparation of the figures and to Nathan Brown for research into joining process selection, both at the University of Hull. Thanks are also due to EPSRC for continued support of research under the Designers' Sandpit Project (GR/M53103 and GR/M55145).

K. G. Swift and J. D. Booker
October 2002

Preface to first edition

In order to facilitate the achievement of the required quality and cost objectives for the manufacture of a component design solution it is necessary to carry out the interrelated activities of selecting candidate processes and tuning a design to get the best out of a chosen manufacturing route. These are difficult decision-making tasks that few experts do well, particularly in the situation of new product introduction.

Failure to get this right often results in late engineering change, with its associated problems of high cost and lead time protraction, or having to live with components that are of poor quality and/or expensive to make.

There is a need for specialist knowledge across a range of manufacturing technologies to enable the correct design decisions to be made from the breadth of possibilities. The difficulties faced by businesses in this area are frequently due to a lack of the necessary detailed knowledge and the absence of process selection methods.

The main motivation behind the text is the provision of technological and economic data on a range of important manufacturing processes. Manufacturing **PR**ocess **I**nformation **MA**ps (PRIMAs) provide detailed data on the characteristics and capabilities of each process in a standard format under headings including: material suitability, design considerations, quality issues, general economics and process fundamentals and variations. A distinctive feature is the inclusion of process tolerance capability charts for processing key material types.

Another distinctive feature of the book is the inclusion of a method for estimating component costs, based on both design characteristics and manufacturing process routes. The cost associated with processing a design is based on the notion of a design independent basic processing cost and a set of relative cost coefficients for taking account of the design application including geometry, tolerances, etc. The overall component cost is logically based on the sum of the material processing and material purchase cost elements. While the method was primarily designed for use with company specific data, approximate data on a sample of common manufacturing processes and material groups is included to illustrate the design costing process and quantify the effect of design choices and alternative process routes on manufacturing cost.

The work is presented in three main parts. Part I addresses the background to the problem and puts process selection and costing into the context of modern product introduction processes and the application of techniques in design for manufacture. Part II presents the manufacturing process information maps (PRIMAs) and their selection. Part III is concerned with methods and data for costing design solutions.

The book is primarily intended to be useful to engineering businesses as an aid to the problem of selecting processes and costing design alternatives in the context of concurrent engineering. The work will also be useful as an introduction to manufacturing processes and their selection for all students of design, technology and management.

The authors are very grateful to Liz Davidson of CMB Ltd for her efforts in collecting data on many of the processes included, and to Robert Braund of T&N Ltd for his contribution to extending the data sheets and particularly for his work on the effects of component section thickness and size on process selection and costing. The authors are also greatly indebted to Adrian Allen for his valuable contribution to the research concerned with methodologies for manufacturing process selection and costing.

Thanks are also due to Phil Baker, Graham Hird, Duncan Law and Brian Miles of CSC Manufacturing Ltd (formerly Lucas Engineering & Systems Ltd) for their encouragement and enthusiastic support, and to Bob Swain of the University of Hull for help with manuscript preparation.

The Engineering and Physical Sciences Research Council (EPSRC) of the UK is gratefully acknowledged for support (under research grant number GR/J97922) for research concerned with process capability and design costing.

K. G. Swift and J. D. Booker

Notation used

List of terms

A = total average cost of setting up and operating a specific process, including plant, labor, supervision and overheads, per second in the chosen country

B = average annual cost of tooling for processing an ideal component, including maintenance

A_h = basic handling index for an ideal design using a given handling process

A_f = basic fitting index for an ideal design using a given assembly process

C_c = relative cost associated with producing components of different geometrical complexity

C_f = relative cost associated with obtaining a specified surface finish

C_{ft} = value of C_t or C_f (whichever is greatest)

C_l = labor rate

C_{ma} = total cost of manual assembly

C_{mp} = relative cost associated with material-process suitability

C_{mt} = cost of the material per unit volume in the required form

C_s = relative cost associated with size considerations and achieving component section reductions/thickness

C_t = relative cost associated with obtaining a specified tolerance

F = component fitting index

H = component handling index

M_c = material cost

M_i = manufacturing cost (pence)

n = number of operations required to achieve the finished component

N = total production quantity per annum

P_a = penalty for additional assembly processes on parts in place

P_c = basic processing cost for an ideal design of component by a specific process

P_f = insertion penalty for the component design

P_g = general handling property penalty

P_o = orientation penalty for the component design

Ra = roughness average (surface finish)

R_c = relative cost coefficient assigned to a component design

T = process time in seconds for processing an ideal design of component by a specific process

V = volume of material required in order to produce the component

V_f = finished volume of the component

W_c = waste coefficient

α = cost of setting up and operating a specific process, including plant, labor, supervision and overheads, per second

β = process specific total tooling cost for an ideal design

■ Units

m	=	meter
μm	=	micron/micrometer
mm	=	millimeter
t	=	ton (metric)
kg	=	kilogram
g	=	gram
h	=	hour
min	=	minute
s	=	second
rpm	=	revolutions per minute

■ Acronyms – general

CA	Conformability Analysis
CAD	Computer-aided Design
DFA	Design for Assembly
DFM	Design for Manufacture
DOE	Design of Experiments
FMEA	Failure Mode and Effects Analysis
PDS	Product Design Specification
PIM	Product Introduction Management
PRIMA	Process Information Map
QFD	Quality Function Deployment

■ Acronyms – manufacturing processes

AJM	Abrasive Jet Machining
ATB	Automated Torch Brazing
ATS	Automated Torch Soldering
CM	Chemical Machining
CNC	Computer Numerical Control
CW	Cold Welding
DB	Dip Brazing
DS	Dip Soldering
DFW	Diffusion bonding (Welding)
DFB	Diffusion Brazing
EBM	Electron Beam Machining
EBW	Electron Beam Welding
ECG	Electrochemical Grinding
ECM	Electrochemical Machining
EDG	Electrical Discharge Grinding

EDM   Electrical Discharge Machining
EGW   Electrogas Welding
ESW   Electroslag Welding
EXW   Explosive Welding
FB   Furnace Brazing
FS   Furnace Soldering
FCAW   Flux Cored Arc Welding
FRW   Friction Welding
FW   Flash Welding
GW   Gas Welding
IB   Induction Brazing
INS   Iron Soldering
IRB   Infrared Brazing
IRS   Infrared Soldering
IS   Induction Soldering
LBM   Laser Beam Machining
LBW   Laser Beam Welding
MIG   Metal Inert-gas Welding
MMA   Manual Metal Arc Welding
NDT   Non-Destructive Testing
NTM   Non-Traditional Machining
PAW   Plasma Arc Welding
RB   Resistance Brazing
RPW   Resistance Projection Welding
RS   Resistance Soldering
RSEW   Resistance Seam Welding
RSW   Resistance Spot Welding
SAW   Submerged Arc Welding
SW   Stud Arc Welding
TB   manual Torch Brazing
TIG   Tungsten Inert-gas Welding
TS   Manual Torch Soldering
TW   Thermit Welding
USM   Ultrasonic Machining
USW   Ultrasonic Welding
USEW   Ultrasonic Seam Welding
WS   Wave Soldering

▓ Manufacturing process key (for Part III)

AM   Automatic Machining
CCEM   Cold Continuous Extrusion (Metals)
CDF   Closed Die Forging
CEP   Continuous Extrusion (Plastics)
CF   Cold Forming
CH   Cold Heading
CM2.5   Chemical Milling (2.5 mm depth)

CM5   Chemical Milling (5 mm depth)
CMC   Ceramic Mold Casting
CNC   Computer Numerical Controlled Machining
CPM   Compression Molding
GDC   Gravity Die Casting
HCEM   Hot Continuous Extrusion (Metals)
IC   Investment Casting
IM   Injection Molding
MM   Manual Machining
OEM   Original Equipment Manufacturing
PDC   Pressure Die Casting
PM   Powder Metallurgy
SM   Shell Molding
SC   Sand Casting
SMW   Sheet-Metal Work
VF   Vacuum Forming

■ Materials key (for plastics processing)

ABS   Acrylonitrile Butadiene Styrene
CA   Cellulose Acetate
CP   Cellulose Propionate
PF   Phenolic
PA   Polyamide
PBTP   Polybutylene Terephthalate
PC   Polycarbonate
PCTFE Polychlorotrifluoroethylene
PE   Polyethylene
PESU   Polyethersulfone
PETP   Polyethyleneterephthalate
PMMA   Polymethylmethacrylate
POM   Polyoxymethylene
PPS   Polyphenylene Sulphide
PP   Polypropylene
PS   Polystyrene
PSU   Polysulfone
PVC-U Polyvinylchloride – Unplasticized
SAN   Styrene Acrylonitrile
UP   Polyester
SMC   Sheet Molding Compounds
BMC   Bulk Molding Compounds

Part I

A strategic view

Some background to the problem and placing process selection and costing into the context of modern product introduction and the application of techniques in design for manufacture and assembly.

1.1 Problems

In today's environment, manufacturing businesses are facing fierce competition and operating in changing markets. Customer demands for higher quality products at lower costs and shorter product lifecycles are putting extra pressure on the product introduction process. Cost and quality are essentially designed into products in the early stages of this process. The designer has the great responsibility of ensuring that the product will conform to customer requirements, comply with specification, and ensuring quality in every aspect of the product, including its manufacture and assembly, all within compressed time-scales. The company that waits until the product is at the end of the line to measure its conformity, performance and cost will not be competitive. The need to understand and quantify the consequences of design decisions on product manufacture and quality has never been greater.

There is extensive evidence to show that products are being designed with far too many parts and with many complex assembly and manufacturing requirements. It has been found that more than 30 per cent of product development effort can be wasted on rework (1.1)* and it is not uncommon for manufacturing operations to have a 'cost of quality' equal to 25 per cent of total sales revenues (1.2). Even Fortune 500 quality leaders face intimidating quality losses (1.3).

Why do businesses continually face such difficulties? The costs 'fixed' at the planning and design stages in product development are between 60 and 85 per cent, while the costs actually incurred at that stage range from only 5 to 7 per cent. Therefore, the more the problems prevented early on, through careful design, the fewer the problems that have to be corrected later when they are difficult and expensive to change. However, to achieve this, it is necessary to reduce the 'knowledge gap' between design and manufacture as shown in Figure 1.1.

Some designers have practical experience of production, and understand the limitations and capabilities they must work within. Unfortunately, there are many who do not. Furthermore, the effects of assigning tolerances and specifying geometry and materials in design have far reaching implications on manufacturing operations and service life, and the associated risks are not properly understood. Understanding the effects of variability and the severity/cost of failure is crucial to risk assessment and its management.

---

* Numbers in parentheses indicate References. These are found on p. 309.

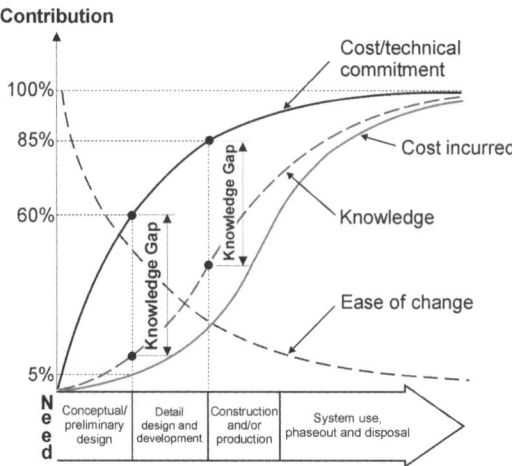

Fig. 1.1 Commitment and incursion of costs during product development (or the 'knowledge gap' principle) (after (1.4)).

■ 1.2 Manufacturing information for design

The need to provide the design activity with information regarding manufacturing process capabilities and costs has been recognized for many years, and some of the work that has been done to address this problem will be touched on. However, there is relatively little published work in this area. The texts on design rarely include relevant data and while a few of the volumes on manufacturing processes do provide some aid in terms of process selection and costs (1.5–1.10), the information is seldom sufficiently detailed and systematically presented to do more than indicate the apparent enormity of the problem. Typically, the facts tend to be process specific and described in different formats in each case, making the engineer's task more difficult. There is a considerable amount of data available but precious little knowledge of how it can be applied to the problem of manufacturing process selection. The available information tends to be inconsistent; some processes are described in great detail, whilst others are perhaps neglected. This may give a disproportionate impression of the processes and their availability.

Information in manufacturing texts can also be found displayed in a tabulated and comparative form on the basis of specific process criteria. While useful, the design related data tends to be somewhat limited in scope and detail. Such forms may be adequate if the designer has expertise in the respective processes, but otherwise gaps in the detail leave room for misconceptions and may be a poor foundation for decision-making. Manufacturing catalogues and information can be helpful, however, they tend to be sales oriented and again, data is presented in different formats and at various levels of detail. Suppliers rarely provide much on design considerations or information on process capability. In addition, there are often differences in language between the process experts and the users.

In recent years, a number of research groups have concentrated specifically on the design/manufacture interface. Processes and systems for cost estimation have been under development in areas such as machining, powder metallurgy, die casting and plastic molding and on broader techniques with the goal of providing Design for Manufacture (DFM) and cost related information for the designer (1.11–1.20). A review of cost estimation techniques for the early stages of design and a method for relating product cost to material cost, total batch size and level of underlying technology can be found in References 1.21–1.24.

Companies recognizing the importance of design for manufacture have also searched for many years for a solution to this problem with most opting for some kind of product 'team' approach, involving a multitude of persons supposedly providing the necessary breadth of experience in order to obtain 'production friendly products'. While sometimes obtaining reasonable results, this approach often faces a number of obstacles, such as: assembling the persons with the relevant experience; lack of formal structure (typically such meetings tend to be unstructured and often *ad hoc* attacks on various 'pet' themes); and the location of the persons required in the team (not only can designers and production engineers be found in different functional departments, but they can frequently be on different sites and are in the case of sub-contractors in different companies). In addition, the chances are that the expertise in the team will only cover the primary activities of the business, and hence opportunity to exploit any benefits from alternative processes may be lost.

The greatest opportunity in design for manufacture occurs at the initial design stage, for while there are also possibilities for a product in production to be modified, there are many additional constraints. This is also illustrated in Figure 1.1. On top of the problems of tooling and equipment, it is not uncommon to find that the 'ownership' of a design changes many times. Consequently, the logic behind a design can become clouded, with the result that subsequent 'owners' tend to assume that existing features must be for good reason and resist change, even though in fact there may be great opportunities for cost reductions. Some companies do have a structured and formal approach to design ownership and alteration. However, these are not always sufficiently annotated. The problems associated with the traditional, functionally organized product introduction process are summarized in Figure 1.2 (1.25).

The traditional, functionally organised product introduction process is incapable of meeting the new pressures placed upon it. Its problems are summarised below:

- Sequential activity results in protracted lead times

- Customer requirements, product design and method of manufacture are inextricably linked with many tradeoffs: they cannot be addressed independently by marketing, engineering and manufacturing functions

- Scarce design resources are wasted on interdepartmental communications, progress chasing and non-value added activities correcting designs that prove difficult to make or do not fully meet customer's requirements

- Manufacturability issues are discovered too late and are the subject of quick-fix solutions and compromises

- All design activity is pushed through a single, ill-defined activity

- Products are designed with an excessive number of component parts which in addition to the cost of these parts adds to the cost of supply and stock control.

Fig. 1.2 Problems with the traditional approach to product introduction.

■ 1.3 Competitive product introduction processes

Faced with the above issues, some companies are currently making dramatic changes to the way in which new products are brought to market. The traditional engineering function led sequential product introduction process is being replaced by a faster and far more effective team based simultaneous engineering approach (1.25). For example, the need for change has been recognized in TRW (formerly Lucas Varity) and has led to the development of a Product Introduction Management (PIM) process (1.26, 1.27) for use in all TRW operating businesses with the declared targets of reducing:

- Time to market by 30 per cent
- Product cost by 20 per cent
- Project cost by 30 per cent.

The generic process is characterized by five phases and nine reviews as indicated in Figure 1.3. Each review has a relevant set of commercial, technical and project criteria for sign off and hand over to the next stage. (The TRW PIM process effectively replaces the more conventional design methodology and provides a more business process orientated approach to product development.)

The process defines what the enterprise has to deliver. The phases, the review points, and the technical and commercial deliverables are clearly defined, and the process aims to take account of market, product design, and manufacturing and financial aspects during each process stage. The skill requirements are defined, together with the necessary supporting tools and techniques. The process runs across the functional structure and includes customer and supplier representation. The PIM process is owned by a senior manager and each product introduction project is also owned by a senior member of staff.

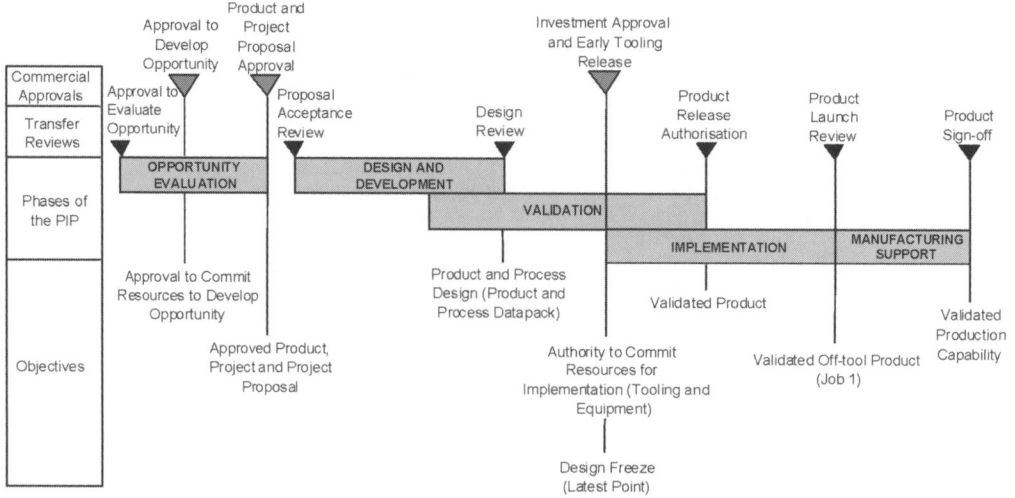

Fig. 1.3 The TRW PIM process.

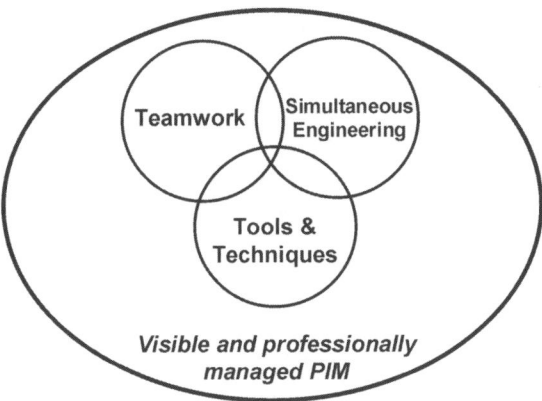

Fig. 1.4 Key elements of successful PIM (after 1.28).

In essence the product introduction process requires the collaborative use of:

- *Teamwork* – Product development undertaken by a full time co-located team with representation from Marketing, Product Development Engineering, Manufacturing Systems Engineering, Manufacturing, Suppliers and Customers formed at the requirements definition stage and selected for team-working and technical skills.
- *Simultaneous Engineering* – The simultaneous design of product, its method of manufacture and, the manufacturing system, against clear customer requirements at equal levels of product and process definition.
- *Project Management* – The professional management of every product introduction project against clearly defined and agreed cost, quality and delivery targets specified to achieve complete customer satisfaction and business profitability.
- *Tools and Techniques* – The routine use of concurrent engineering tools to structure the team's activity, thereby improving the productivity of the team and quality of their output.

The linkage between the above elements is represented diagrammatically in Figure 1.4. Design for Assembly (DFA) is one of the main tools and techniques prescribed by the PIM process. Other main tools and techniques currently specified include: Quality Function Deployment (QFD) (1.29), Failure Mode and Effects Analysis (FMEA) (1.30), Design of Experiments (DOE) (1.31) and Conformability Analysis (CA) (1.32).

▪ 1.4 Techniques in design for manufacture and assembly

The application of tools and techniques that quantify manufacturing and assembly problems and identify opportunities for redesign is the major means available for bridging the knowledge gap. It has been found that DFM/DFA analysis leads to innovative design solutions where considerable benefits accrue, including functional performance and large savings in manufacturing and assembly cost. DFA is particularly powerful in this connection and is one of the most valuable product introduction techniques. Although the use of design for manufacture and assembly techniques requires additional up-front effort when compared with the more conventional design activity, overall the effect is to reduce the time-to-market quite considerably. This is primarily due to fewer engineering changes, fewer parts to detail, document and plan, and a less complex product with good assembly and manufacturing characteristics. An illustration of the business benefits of reducing time-to-market is given in Figure 1.5 (1.33).

Very substantial reductions in part-count and component manufacture and assembly costs have resulted from using DFA techniques in product development teams. Figures 1.6 and 1.7 give examples of what can be achieved in terms of product rationalization. The contractor assembly DFA study shown in Figure 1.6 resulted in a 66% reduction in part-count. Figure 1.7 shows the overall results of a study on an assembly test machine and a redesign of part of the system, a pump stand, where 14 parts were replaced by a single casting.

The results of 60 documented applications, carried out recently in a wide variety of industries, show that the average part-count reduction was almost 48 per cent and the assembly cost saving was 45 per cent (see Figure 1.8). It is interesting to note that there proved to be little difference, in terms of means and standard deviations, across the aerospace/defence, automotive and industrial equipment business sectors. This indicates that the applicability of the methods is not particularly sensitive to product demand levels or technology. Indeed the largest single benefit achieved resulted from the redesign of a range of assembly and test machines.

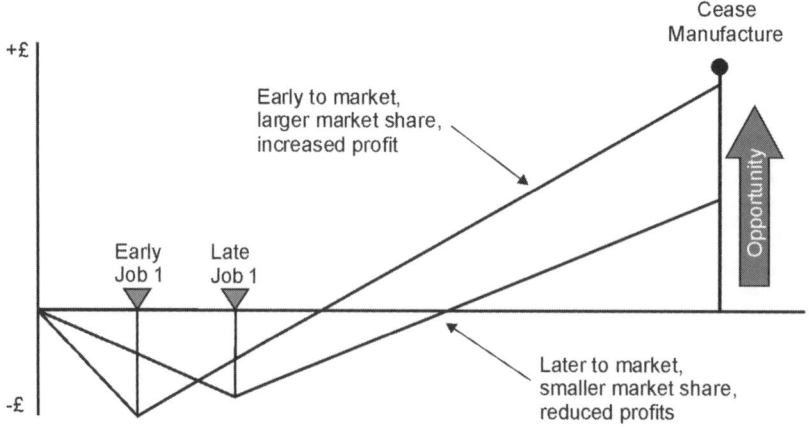

Fig. 1.5 Benefits of reducing time-to-market (after 1.33).

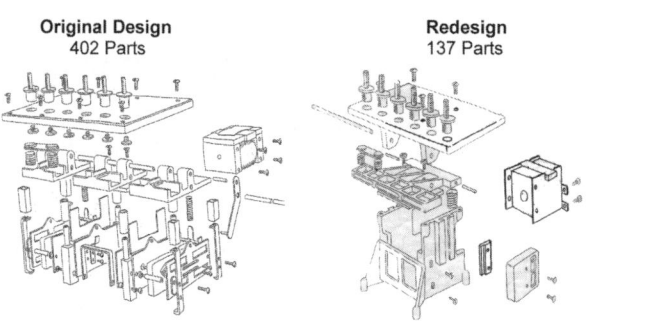

Fig. 1.6 Contactor assembly.

Overall Results

- 70% fewer parts
- 15% reduction in total cost
- 35% shorter build time
- Simpler Manufacturing system
- Improved reliability and performance

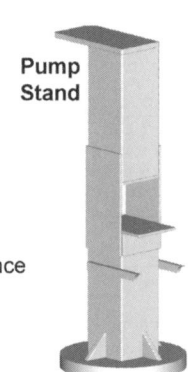

Pump Stand

Original Design
- 60 off
- 14 parts
- Mild steel
- Fabricated (welded)
- Normalised
- No machining (as supplied)
- £295 per unit

Redesign
- Single piece casting
- Manufacturing estimate £30 per unit
- Quote £29 per unit
- Pattern £500
- Total cost £36 per unit

Fig. 1.7 Pump assembly and test machine overall assembly results.

Business Sector	Automotive	Aerospace/ Defence	Industrial Equipment	All Sectors
Average Part-count Reduction	44%	52%	51%	48%
Average Assembly Cost Saving	43%	47%	48%	45%

Fig. 1.8 Results from 60 product studies.

Similar savings have been reported by others involved with the application of techniques in design for manufacture and assembly (1.34). It is also worth commenting that the designs coming out of the process tend to be more reliable and easier to manufacture.

As can be seen from the above results, DFA techniques (1.35–1.38) when used in industry are highly effective in realizing part-count reduction and taking costs out of manufacture and assembly. The analysis metrics associated with part-count and potential costs are inputs to concept design and development. As part of the DFA process, the product development team needs to generate improved product design solutions, with better DFA metrics, by simplifying the product structure, reducing part-count and simplifying component assembly operations. DFA is particularly interesting in the context of this book, since its main benefits result from systematically reviewing functional requirements, and replacing component clusters by single integrated pieces and selecting alternative joining processes (1.34)(1.38). Invariably the proposed design solutions rely heavily on the viability of adopting different processes and/or materials as shown in two part-count reduction examples in Figure 1.9. A number of guidelines for assembly-orientated design are provided in Appendix A for the reader.

DFM further involves the simultaneous consideration of design goals and manufacturing constraints in order to identify and alleviate manufacturing problems while the product is being designed, thereby reducing the lead time and improving product quality. This includes an understanding of the technical capabilities and limitations of the manufacturing processes chosen by invoking a series of guidelines, principles and recommendations, commonly termed 'producibility' guidelines, to modify component designs for subsequent manufacture. The use

Example 1

After (1.34)

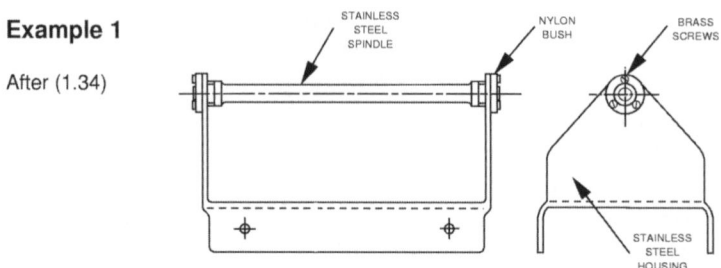

This spindle/housing assembly (sheet metal housing) has ten separate parts and an assembly efficiency rating of 7%.

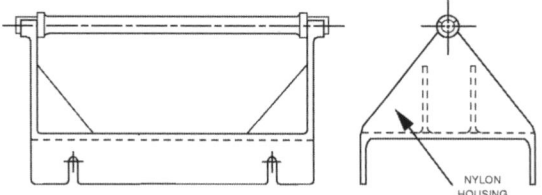

The two part design, utilising an injection-moulded nylon housing, has an assembly efficiency rating of 93%.

Example 2

After (1.38)

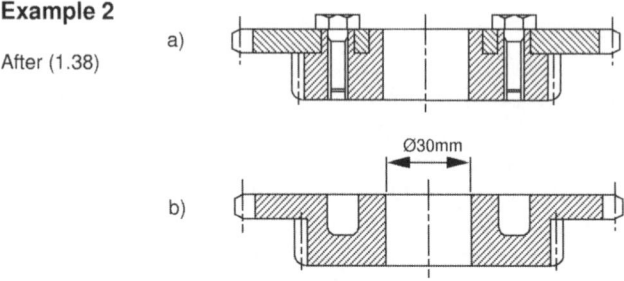

Two different types of sprocket and gear wheel: (a) includes assembly: this is made of steel and produced by machining. The individual teeth are cut. It is necessary to divide it in to two elements due to production technique reasons. (b) does not include assembly: this is produced from sintered metal, the teeth being sintered in accordance with the required tolerance and surface quality; the advantages being no waste material, short processing time and no assembly.

Fig. 1.9 Examples of part-count reduction (after 1.34, 1.38).

of techniques to assist costing of component designs also aids the process of cost optimization. Since few formal DFM methods exist, unlike DFA, implementing a strategy is not straightforward, and companies tend to develop DFM guidelines in-house. This takes the focus away from quality to a large extent because of the difficulties in establishing the methods to verify it in the first place.

$\rightarrow$ Identify critical characteristics (tolerances, surface finishes)
$\rightarrow$ Identify factors that influence the manufacture of critical characteristics
$\rightarrow$ Estimate manufacturing costs
$\rightarrow$ Minimise component cost
$\rightarrow$ Establish maximum tolerances for each characteristic
$\rightarrow$ Determine process capability of characteristics early
$\rightarrow$ Avoid tight tolerances
$\rightarrow$ Design the part to be easily inspectable
$\rightarrow$ Minimise number of machined surfaces
$\rightarrow$ Minimise number re-orientations during manufacture
$\rightarrow$ Use standard manufacturing processes where possible
$\rightarrow$ Use generous radii/fillets on castings, mouldings and machined parts
$\rightarrow$ Avoid secondary processes
$\rightarrow$ Design parts for easy tooling/jigging using standard systems
$\rightarrow$ Utilise special characteristics of processes (moulded-inserts, colours)
$\rightarrow$ Use good detail design for manufacture and conform to drafting standards

Fig. 1.10 DFM rules and objectives.

A number of general rules have been developed to aid designers when thinking about the manufacture of the product:

- Holes in machined, cast, molded, or stamped parts should be spaced such that they can be made in one operation without tooling weakness. This means that there is a limit on how close holes may be spaced due to strength in the thin section between holes.
- Generalized statements on drawings should be avoided, like 'polish this surface' or 'tool-marks not permitted', which are difficult for manufacturing personnel to interpret. Notes on engineering drawings must be specific and unambiguous.
- Dimensions should be made from specific surfaces or points on the part, not from points in space. This greatly facilitates the making of gauges and fixtures.
- Dimensions should all be from a single datum line rather than from a variety of points to avoid overlap of tolerances.
- The design should aim for minimum weight consistent with strength and stiffness requirements. While material costs are minimized by this criterion, there also will usually be a reduction in labor and tooling costs.
- Wherever possible, design to use general-purpose tooling rather than special dies, form cutters, etc. An exception is high-volume production, where special tooling may be more cost-effective.
- Generous fillets and radii on castings, molded, formed, and machined parts should be used.
- Parts should be designed so that as many operations as possible can be performed without requiring repositioning. This promotes accuracy and minimizes handling.

Figure 1.10 provides a number of specific design rules and objectives associated with effective DFM.

As mentioned previously, selecting the right manufacturing process is not always simple and obvious. In most cases, there are several processes that can be used for a component, and selection depends on a large number of factors. Some of the main process selection drivers are shown in Figure 1.11. The intention is not to infer that these are necessarily of equal importance or occur in this fixed sequence.

The problem is compounded by the range of manufacturing processes and wide variety of material types commonly in use. Figures 1.12–1.16 provide a general classification and guide to the range of materials and processes (component manufacturing, assembly, joining and, bulk and surface engineering, respectively) that are widely available. (All, except the latter, of

$\rightarrow$ Product quantity
$\rightarrow$ Equipment costs
$\rightarrow$ Tooling costs
$\rightarrow$ Processing times
$\rightarrow$ Labour intensity and work patterns
$\rightarrow$ Process supervision
$\rightarrow$ Maintenance
$\rightarrow$ Energy consumption and other overhead costs
$\rightarrow$ Material costs and availability
$\rightarrow$ Material to process compatibility
$\rightarrow$ Component form and dimensions
$\rightarrow$ Tolerance requirements
$\rightarrow$ Surface finish needs
$\rightarrow$ Bulk treatment and surface engineering
$\rightarrow$ Process to component variability
$\rightarrow$ Process waste
$\rightarrow$ Component recycling

Fig. 1.11 Key process selection drivers.

these processes are discussed in detail in Part II of the book.) To be competitive, the identification of technologically and economically feasible process and material combinations is crucial. The benefits of picking the right process can be enormous, as shown in Figure 1.17 for a number of components and processing routes.

The placing in the product design cycle of process selection in the context of engineering for manufacture and assembly is illustrated in Figure 1.18. The selection of an appropriate set of processes for a product is very difficult to perform effectively without a sound Product Design Specification (PDS). A well-constructed PDS lists all the needs of the customers, end users and the business to be satisfied. It should be written and used by the Product Team and provide a reference point for any emerging design or prototype. Any conflict between customer needs and product functionality should be referred back to the PDS.

The first step in the process is to analyze the design or prototype with the aim of simplifying the product structure and optimizing part-count. As shown earlier, without proper analysis design solutions invariably tend to have too many parts. Therefore, it is important to identify components that are candidates for elimination or integration with mating parts. (Every component part must be there for a reason and the reason must be in the PDS.) This must be done with due regard for the feasibility of material process combinations and joining technology. A number of useful approaches are available for material selection in engineering design. For more information see references (1.10), (1.39) and (1.40).

The next steps give consideration to the problems of component handling and fitting processes, the selection of appropriate manufacturing processes and ensuring that components are tuned to the manufacturing technology selected. Estimation of component manufacture and assembly costs during the design process is important for both assessing a design against target costs and in trade-off analysis. Overall, the left-hand side of Figure 1.18 is closely related to DFA, while the right-hand side is essentially material/process selection and component design for processing, or consideration in DFM. A reader interested in more background information on DFA/DFM and materials and process selection in product development is directed to references (1.40–1.45).

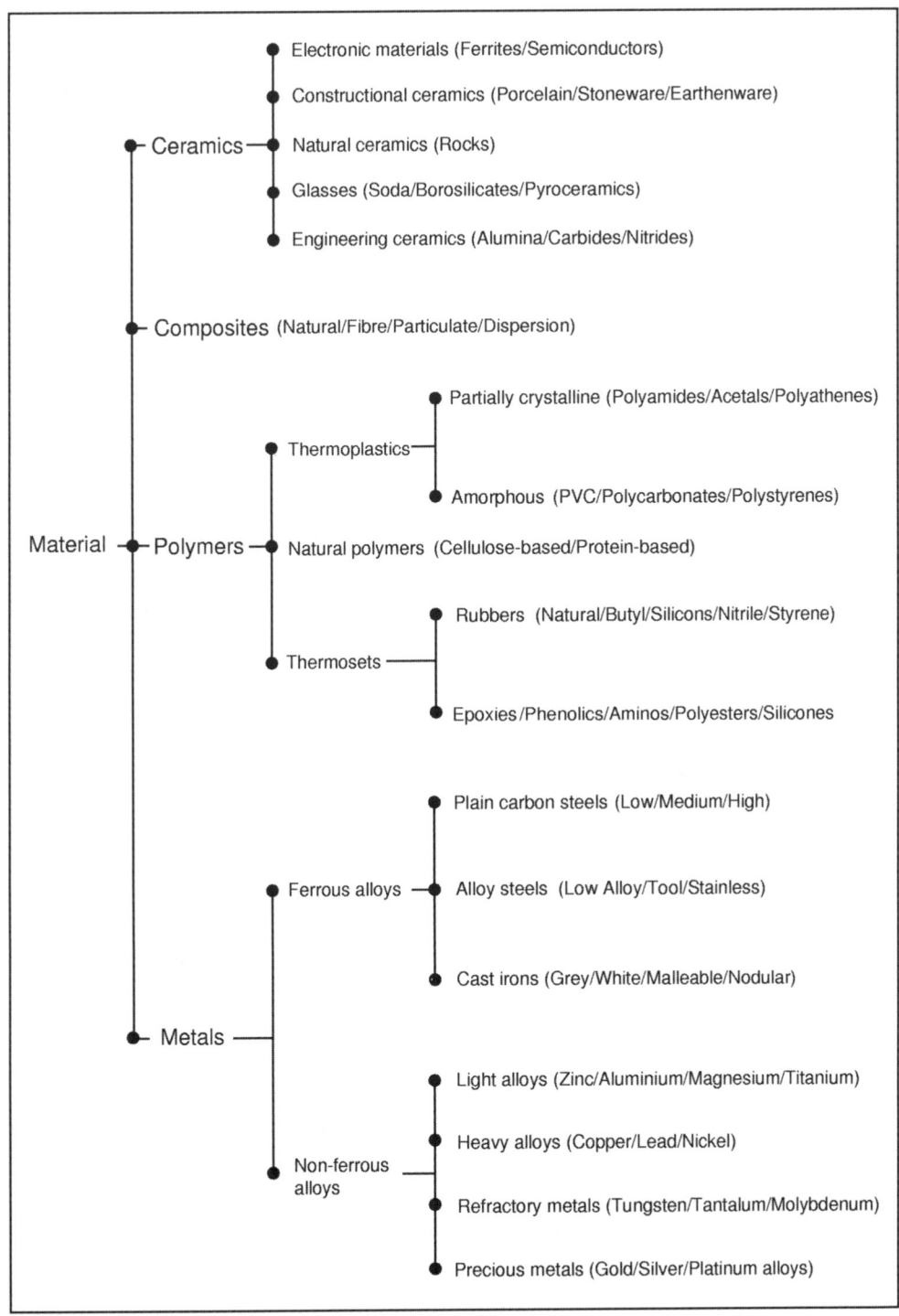

Fig. 1.12 General classification of materials.

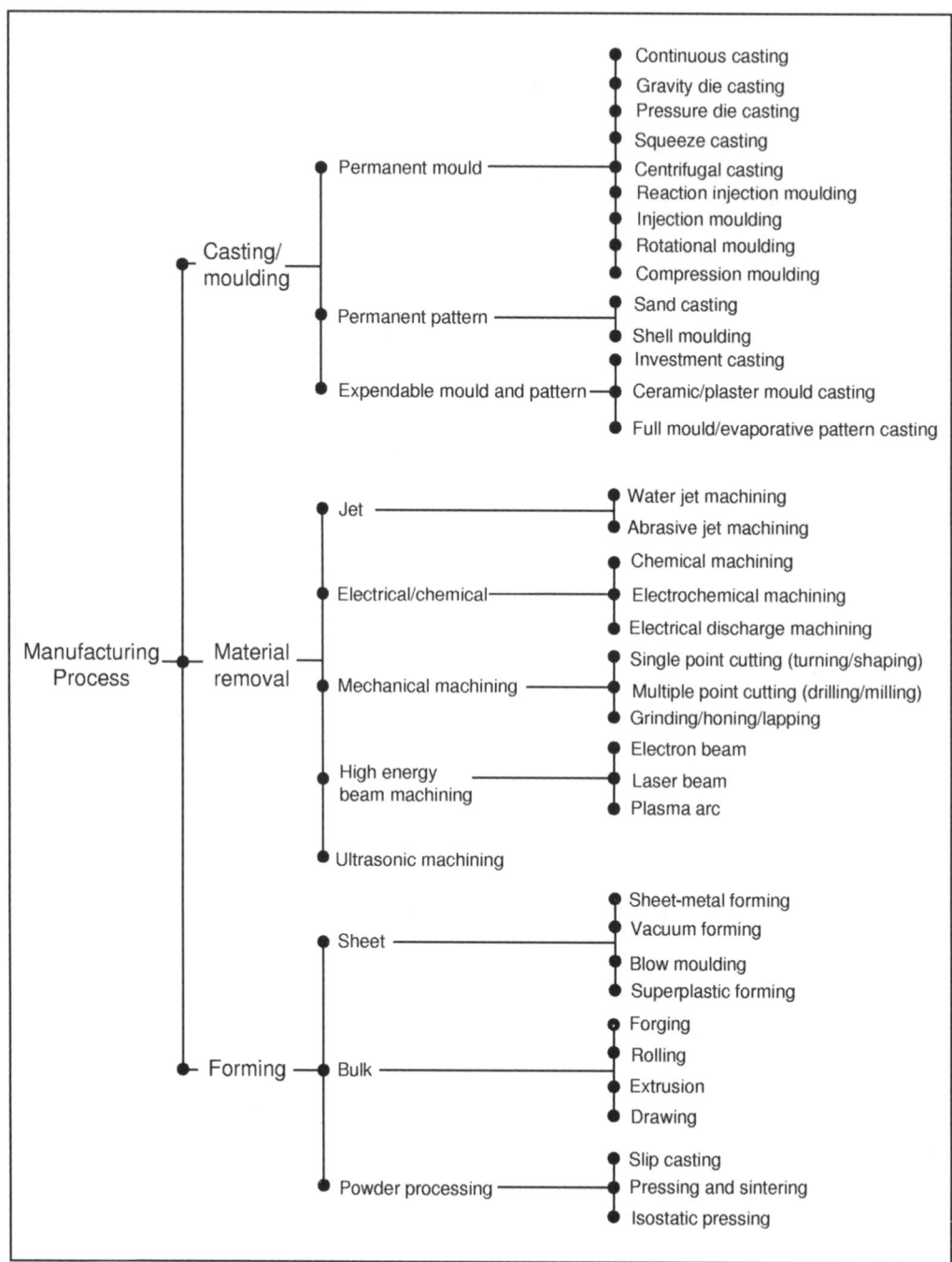

Fig. 1.13 General classification of manufacturing processes.

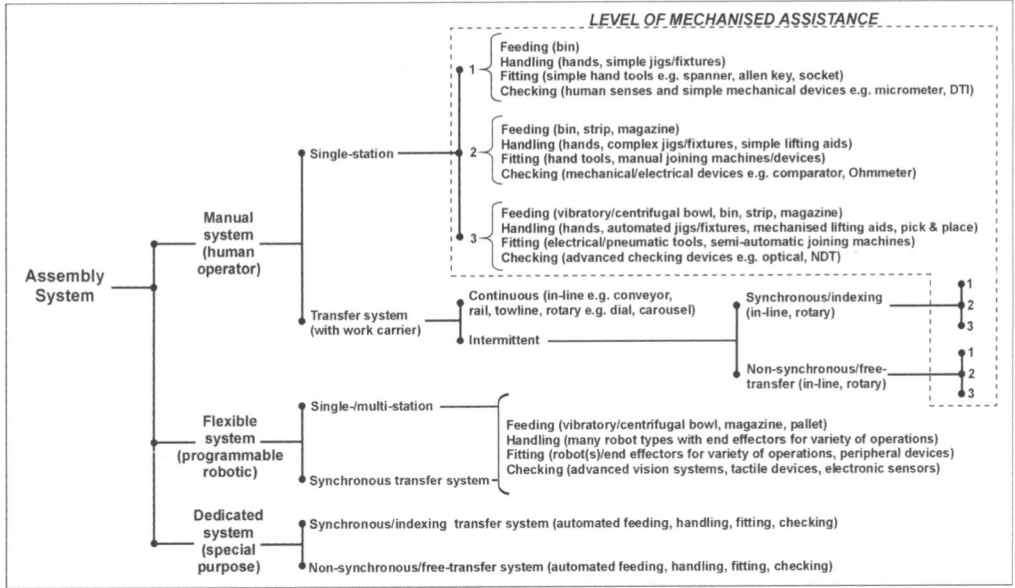

Fig. 1.14 General classification of assembly systems.

▨ 1.5 Process selection strategy

In considering alternative design solutions for cost and quality, it is necessary to explore candidate materials, geometries and tolerances, etc., against possible manufacturing routes. This requires some means of selecting appropriate processes and estimating the costs of manufacture early on in product development, across a whole range of options. In addition, the costs of non-conformance (1.46) need to be understood, that is appraisal (inspection and testing) and failure, both internal (rework, scrap, design changes) and external (warranty claims, liability claims and product recall). Therefore, we also need a way of exploring conformance levels before a process is selected. For more information on this important aspect of design, the reader is directed to Reference 1.32.

The primary objective of the text is to provide support for manufacturing process selection in terms of technological feasibility, quality of conformance and manufacturing cost. The satisfaction of this objective is through:

- The provision of data on the characteristics and capabilities of a range of important manufacturing, joining and assembly processes. The intention is to promote the generation of design ideas and facilitate the matching and tuning of a design to a process, and
- The provision of methods and data to enable the exploration of design solutions for component manufacturing and assembly costs in the early stages of the design and development process.

To provide for the first point, a set of so-called manufacturing **PR**ocess Information **MA**ps (PRIMAs) have been developed. In a standard format for each process, the PRIMAs present knowledge and data on areas including material suitability, design considerations, quality issues, economics and process fundamentals and process variants. The information includes

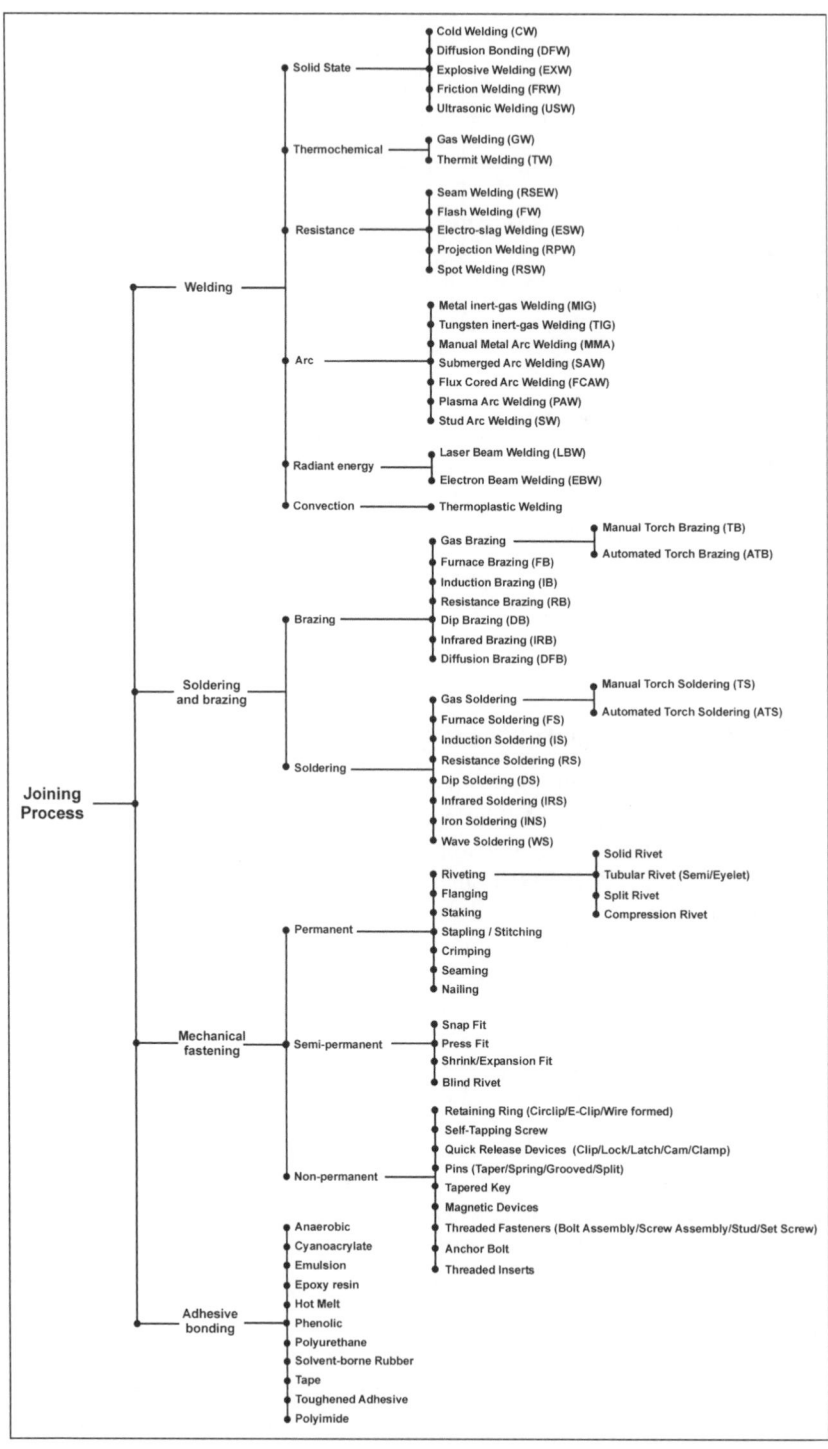

Fig. 1.15 General classification of joining processes.

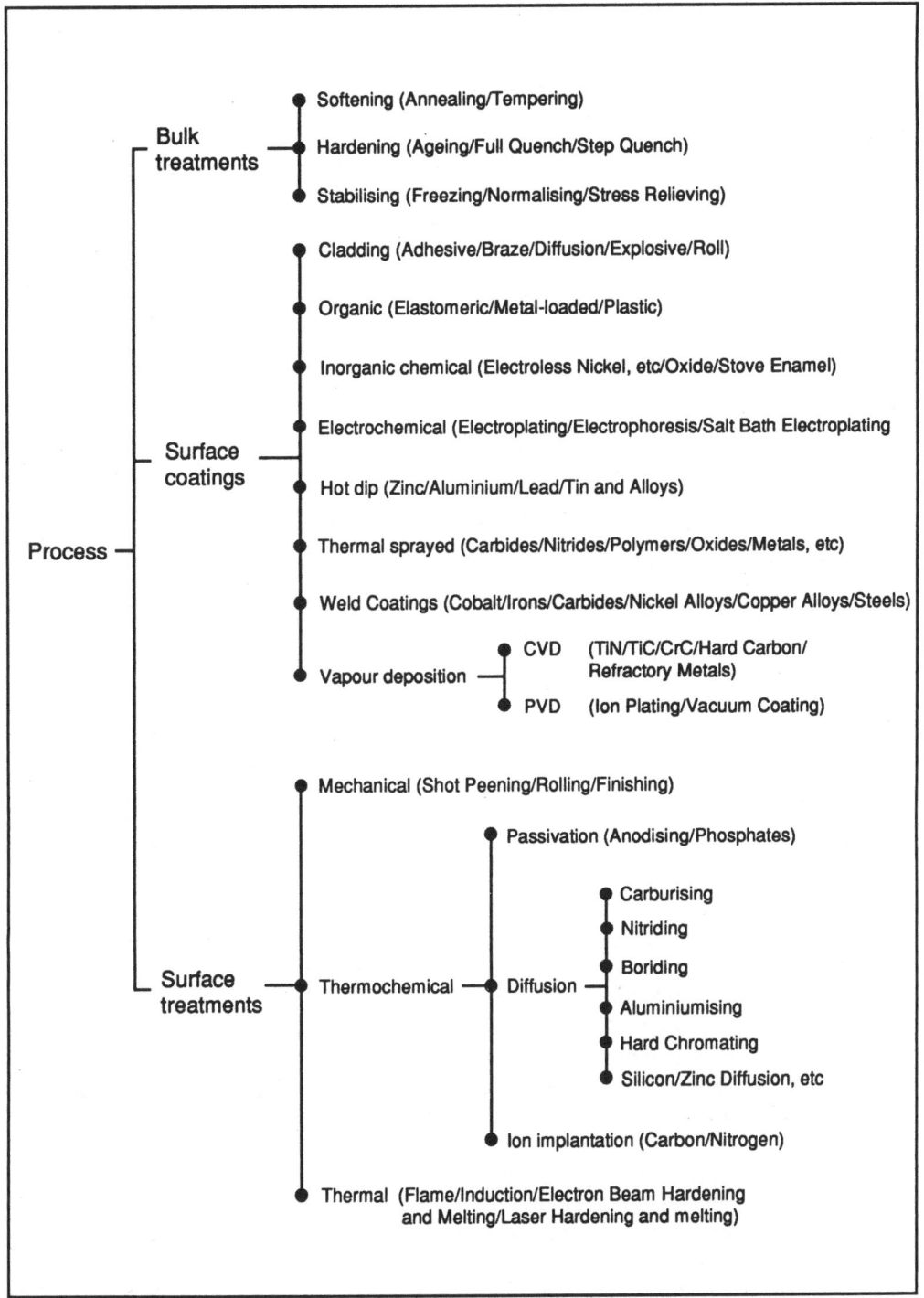

Fig. 1.16 General classification of bulk and surface engineering processes.

Component	Material	Number Per Annum	Manufacturing Process	Relative Economic and Technical Merits	Relative Cost
Plug Body	Low Carbon Steel	1,000,000	Machining	High waste Low to medium production rates Poor strength	3.9
			Cold Forming	Little waste Very high production rates High strength	1
Plain Bearing	Bronze	50,000	Machining	High waste Low to medium production rates Non-porous properties	2.2
			Powder Metal Sintering	No waste High production rates Porous product	1
Cover	Alum. Alloy	5,000	Spinning	High labour costs Low production rates Limited detail and accuracy	1.8
			Deep Drawing	Low labour costs High production rates High detail and accuracy	1
Connecting Rod (after (1.6))	Medium Carbon Steel	100,000	Closed Die Forging	Long lead times High tooling costs High equipment costs	1.3
			Sand Casting	Short lead times Low tooling costs Low equipment costs	1
Pump Gear	Low Carbon Steel	5,000	Machining	High waste Low to medium production rates Poor strength	2.6
			Cold Extrusion	Little waste Very high production rates High strength	1

Fig. 1.17 Contrast in component cost for different processing routes.

not only design considerations relevant to the respective processes, but quite purposefully, an overview of the functional characteristics of each process, so that a greater overall understanding may be achieved. Within the standard format, a similar level of detail is provided on each of the processes included. The format is very deliberate. Firstly, an outline of the process itself – how it works and under what conditions it functions best. Secondly, a summary of what it can do – limitations and opportunities it presents – and finally an overview of quality considerations including process capability charts for relating tolerances to characteristic dimensions.

To provide for the second point, techniques are put forward that can be used to estimate the costs of component manufacture and assembly for concept designs. It enables the effects of product structure, design geometry and materials to be explored against various manufacturing and assembly routes. A sample data set is included, which enables the techniques to be used to predict component manufacturing and assembly costs for a range of processes and materials. The process of cost estimation is illustrated through a number of case studies, and the scope for and importance of application with company specific data is discussed.

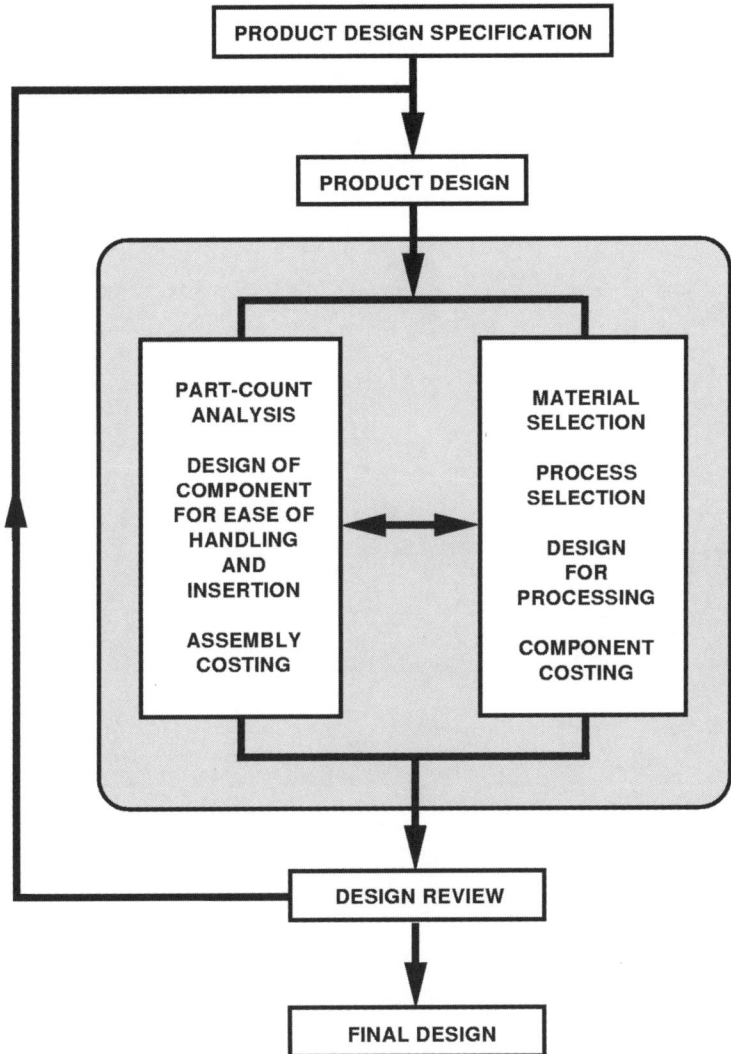

Fig. 1.18 Outline process for design for manufacture and assembly.

Part II begins with the strategies employed for PRIMA selection, where attention is focused on identification of candidate processes based on strategic criteria such as material, process technology and production quantity. Having identified the possible targets, the data in the PRIMAs are used to do the main work of selection. The PRIMAs include the main five manufacturing process groups: casting, plastic and composite processing methods, forming, machining and non-traditional processes. In addition, the main assembly technologies and the majority of commercially available joining processes are covered. In all, sixty-five PRIMAs are presented, giving reference to over one hundred manufacturing, assembly and joining processes.

Part III of the text concentrates on the cost estimation methodologies for components and assemblies, their background, theoretical development and industrial application. In practice, Part II of the work can be used to help select the candidate processes for a design from the whole range of possibilities. Part III is concerned with getting a feel for the manufacturing and assembly costs of the alternatives. The book finishes with a statement of conclusions and a list of areas where future work might be usefully directed.

Part II

Selecting candidate processes

Strategies and data relevant to selecting candidate processes for design solutions.

■ 2.1 Introduction

Selecting the right process and optimizing the design to suit the process selected involves a series of decisions which exert considerable influence on the quality and cost of components and assemblies. Such decisions can significantly effect the success of a product in the market place. As mentioned previously, in selecting processes and tuning designs for processing many factors need to be taken into consideration. The manufacturing PRIMAs presented in this part of the book attempt to provide the knowledge and data required to underpin this decision making process provided by the various process selection strategies. However, it is the PRIMAs that provide the means of making more detailed assessments regarding the technological and economic feasibility of a process.

Design considerations are provided to enable the designer to understand more about the technical feasibility of the design decisions made. The process quality considerations give valuable information on process conformance, including data on process tolerance capability associated with characteristic dimensions. A good proportion of the PRIMAs is taken up with quality considerations. No excuse is made for this. Non-conformance often represents a large quality cost in a business. As touched on earlier, such losses result from rework, order exchange, warranty claims, legal actions and recall. In many businesses, these losses account for more than 10 per cent of turnover (2.1). The goal is to provide data which enables the selection of processes that have the capability to satisfy the engineering needs of the application, including those associated with conformance to quality requirements.

■ 2.2 PRIMAs (Process Information Maps)

Each PRIMA is divided into seven categories, as listed and defined below, covering the characteristics and capabilities of the process:

- *Process description*: an explanation of the fundamentals of the process together with a diagrammatic representation of its operation and a finished part.
- *Materials*: a description of the materials currently suitable for the given process.

- *Process variations*: a description of any variations of the basic process and any special points related to those variations.
- *Economic considerations*: a list of several important points including production rate, minimum production quantity, tooling costs, labor costs, lead times, and any other points which may be of specific relevance to the process.
- *Typical applications*: a list of components or assemblies that have been successfully manufactured or fabricated using the process.
- *Design aspects*: any points, opportunities or limitations that are relevant to the design of the part as well as standard information on minimum section, size range and general configuration.
- *Quality issues*: standard information includes a process capability chart (where relevant), typical surface roughness and detail, as well as any information on common process faults.

A key feature of the PRIMAs is the inclusion of process capability charts for the majority of the manufacturing processes. Tolerances tend to be dependent on the overall dimension of the component characteristic, and the relationship is specific and largely non-linear. The charts have been developed to provide a simple means of understanding the influence of dimension on tolerance capability. The regions of the charts are divided by two contours. The region bounded by these two contours represents a spectrum of tolerance-dimension combinations where $C_{pk} \geq 1.33$* is achievable. Below this region, tolerance-dimension combinations are likely to require special control or secondary processing if $C_{pk} = 1.33$ is to be realized.

In the preparation of the process capability charts it has been assumed that the geometry is well suited to the process and that all operational requirements are satisfied. Where the material under consideration is not mentioned on the charts, care should be taken. Any adverse effects due to this or geometrically driven component variation should be taken into consideration. For more information the reader is referred to reference (1.32). The data used in the charts has been compiled from contacts in industry and from published work. Although attempts have been made to standardize the data as far as possible, difficulties were faced in this connection, since it was not always easy to obtain a consensus view. Consequently, as many as twenty different data sources have been used in the compilation of the individual process capability charts to provide an understanding of the general tolerance capability range offered by each manufacturing process.

■ 2.3 PRIMA selection strategies

Different manufacturing technologies such as primary shape generating processes, joining techniques, assembly systems and surface engineering processes require that selection takes place based on the factors relevant to that particular technology. For example, the selection of a joining technique may be heavily reliant on the ability of the process to join dissimilar

---

* C_{pk} – process capability index. If the process characteristic is a normal distribution, C_{pk} can be related to a parts-per-million (ppm) defect rate. $C_{pk} = 1.33$ equates to a defect rate of 30 ppm at the nearest limit. At $C_{pk} = 1$, the defect rate equates to approximately 1350 ppm (see reference 2.2 for more information about process capability indices).

materials and materials of different thickness. This is a particular requirement not necessarily defined by the PDS, but one that has been arrived at through previous design decisions, perhaps based on spatial or functional requirements. Whereas assembly system selection may simply be dictated by a low labor rate in the country of manufacture and therefore manual assembly becomes viable for even relatively large production volumes.

Although there may be many important selection drivers with respect to each process technology, a simple and effective strategy for selection must be sought for the general situation and for usability. Selection strategies can be developed by concentrating on several key economic and technical factors which are easily interpreted from the PDS or other requirements. Put in a wider context, the selection strategies, together with the information provided in the PRIMAs, must complement business strategy and the costing of designs, in order to provide a procedure that fully justifies the final selection. A flowchart is shown in Figure 2.1, relating all the factors relevant to the process selection strategies discussed in detail.

2.3.1 Manufacturing process selection

Manufacturing processes represent the main shape generating methods such as casting, molding, forming and material removal processes. The individual processes specific to this section are classified in Figure 1.13. The purpose of this section is to provide a guide for the selection of the manufacturing processes that may be suitable candidates for a component.

The manufacturing process selection strategy is given below, but points 4, 5 and 6 apply to all selection strategies:

1 Obtain an estimate of the annual production quantity.
2 Choose a material type to satisfy the PDS.
3 Refer to Figure 2.2 to select candidate PRIMAs.
4 Consider each PRIMA against the engineering and economic requirements such as:

  - Understand the process and its variations
  - Consider the material compatibility
  - Assess conformance of component concept with design rules
  - Compare tolerance and surface finish requirements with process capability data.

5 Consider the economic positioning of the process and obtain component cost estimates for alternatives.
6 Review the selected manufacturing process against business requirements.

The principal intention is that the candidate processes are selected before the component design is finalized, so that any specific constraints and/or opportunities may be borne in mind. To this end, the manufacturing process PRIMA selection matrix (see Figure 2.2) has been devised based on two basic variables:

- *Material type* – Accounts for the compatibility of the parent material with the manufacturing process, and is therefore a key technical selection factor. A large proportion of the materials used in engineering manufacture have been included in the selection methodology, from ferrous alloys to precious metals, as classified in Figure 1.12.

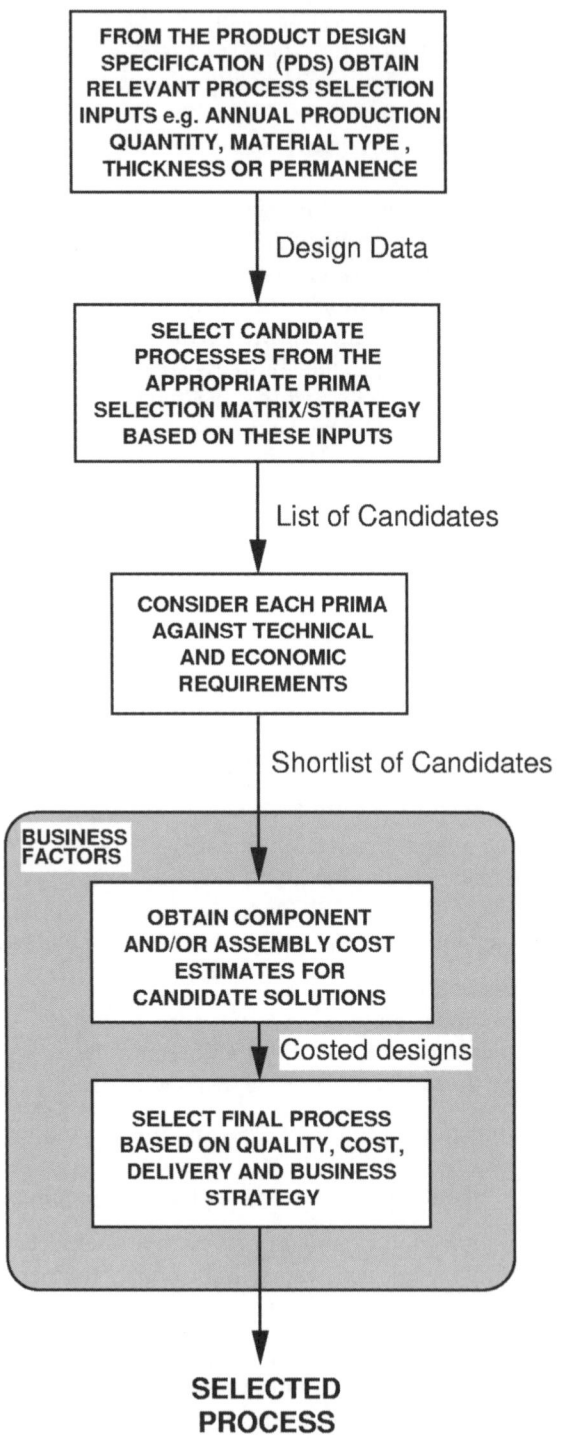

Fig. 2.1 General process selection flowchart.

Note - The PRIMA selection matrix cannot be regarded as comprehensive and should not be taken as such. It represents the main common industrial practice but there will always be exceptions at this level of detail. Also, the order in which the PRIMAs are listed in the nodes of the matrix has no significance in terms of preference.

MATERIAL / QUANTITY	IRONS	STEEL (carbon)	STEEL (tool, alloy)	STAINLESS STEEL	COPPER & ALLOYS	ALUMINIUM & ALLOYS	MAGNESIUM & ALLOYS	ZINC & ALLOYS	TIN & ALLOYS	LEAD & ALLOYS	NICKEL & ALLOYS	TITANIUM & ALLOYS	THERMOPLASTICS	THERMOSETS	FR COMPOSITES	CERAMICS	REFRACTORY METALS	PRECIOUS METALS
VERY LOW 1 TO 100	[1.2][1.5] [1.6][1.7] [4.M] [5.3][5.4]	[1.5][1.7] [3.10][4.M] [5.1][5.5] [5.6]	[1.1][1.5][1.7] [3.10][4.M] [5.1][5.5] [5.6][5.7]	[1.5][1.7][3.7] [3.10][4.M] [5.1][5.5][5.6]	[1.5][1.7] [3.10][4.M] [5.1]	[1.5][1.7] [3.7][3.10] [4.M][5.5]	[1.6][1.7] [3.10][4.M] [5.1][5.5]	[1.1][1.7] [3.10][4.M] [5.5]	[1.1][1.7] [3.10][4.M] [5.5]	[1.1][3.10] [4.M][5.5]	[1.5][1.7] [3.10][4.M] [5.1][5.5][5.6]	[1.1][1.6] [3.7][3.10] [4.M][5.1] [5.5][5.6][5.7]	[2.5] [2.7]	[2.5] [5.7]	[2.2] [2.8] [5.7]	[1.5] [5.1] [5.6] [5.7]	[1.1] [5.7]	[5.5]
LOW 100 TO 1,000	[1.2][1.5] [1.6][1.7] [4.M] [5.3][5.4]	[1.2][1.5] [1.7][3.10] [4.M][5.1] [5.3][5.4][5.5]	[1.1][1.2][1.7] [4.M][5.1] [5.3][5.4] [5.5][5.6][5.7]	[1.2][1.7] [3.7][3.10] [4.M][5.1] [5.3][5.4][5.5]	[1.2][1.5] [1.7][1.8][3.5] [3.10][4.M][5.1] [5.3][5.4]	[1.2][1.5][1.7] [1.8][3.7][3.10] [4.M][5.3] [5.4][5.5]	[1.6][1.7] [1.8][3.10] [4.M][5.5]	[1.1][1.7] [1.8][3.10] [4.M][5.5]	[1.1][1.7] [1.8][3.10] [4.M][5.5]	[1.1][1.8] [3.10][4.M] [5.5]	[1.2][1.5][1.7] [3.10][4.M] [5.4][5.5]	[1.1][1.6][3.7] [3.10][4.M][5.1] [5.3][5.4][5.5] [5.6][5.7]	[2.3] [2.5] [2.7]	[2.2] [2.3]	[2.2] [2.3] [2.8] [5.7]	[5.1] [5.3] [5.6] [5.7]	[5.7]	[5.5]
LOW TO MEDIUM 1,000 TO 10,000	[1.2][1.5][1.6] [1.7][3.11] [4.A][5.2]	[1.2][1.5][1.6] [1.7][3.10][3.11] [4.A][5.2][5.3] [5.4][5.5]	[1.2][1.5][1.7] [3.1][3.4][5.2] [3.11][4.A][5.3] [5.4][5.5]	[1.2][1.5][1.7] [3.1][3.11] [3.10][4.A] [5.4][5.5]	[1.2][1.3][1.5] [1.8][3.10][3.11] [3.10][3.11][4.A] [3.4][5.2][5.3]	[1.2][1.3][1.5] [1.8][3.10][3.11] [3.7][3.10][3.11] [4.A][5.2]	[1.3][1.6] [1.8][3.1] [3.3][3.4][3.10] [4.A][5.5]	[1.3][1.8] [3.3][3.10] [4.M][5.5]	[1.3][1.8] [3.3][3.10]	[1.3][1.8] [3.3][3.10]	[1.2][1.5][1.6] [1.7][3.1][3.3] [3.11][4.A][5.2] [3.10]	[3.1][3.7] [3.10][3.11] [4.A][5.2] [5.3][5.4][5.5]	[2.3] [2.6] [2.7]	[2.2] [2.3] [2.4]	[2.1] [2.2] [2.3]	[5.2] [5.3] [5.4] [5.5]	[5.7]	[5.5]
MEDIUM TO HIGH 10,000 TO 100,000	[1.2][1.3] [3.11][4.A]	[1.9][3.1][3.3] [3.4][3.5] [3.11][3.12] [4.A][5.2][5.5]	[3.1][3.4][3.5] [3.11][3.12] [4.A][5.2]	[1.9][3.1] [3.3][3.4] [3.5][3.11] [3.12][4.A]	[1.2][1.4][1.9] [3.1][3.3] [3.4][3.5][3.11] [3.12][4.A][5.5]	[1.2][1.3][1.4] [1.9][3.1][3.3] [3.4][3.5][3.11] [3.12][4.A][5.5]	[1.3][1.4] [3.1][3.3] [3.4][3.5] [3.12][4.A]	[1.3][1.4] [3.3][3.4] [3.5][3.12] [4.A]	[1.3][1.4] [3.3][3.4] [3.12]	[1.3][1.4] [3.3][3.4] [3.5][3.12] [4.A]	[3.1][3.3][3.5] [3.4][3.11] [3.12][4.A] [5.2][5.5]	[3.1][3.4] [3.11][3.12] [4.A][5.2][5.5]	[2.1] [2.3] [2.5] [2.6] [2.9]	[2.1] [2.3] [2.4] [2.9]	[2.1] [2.3]	[3.11][3.12] [3.5]	[3.12][3.11]	[3.5]
HIGH 100,000+	[1.2][1.3] [3.11] [4.A]	[1.9][3.1] [3.2][3.3] [3.4][3.5] [3.12][4.A]	[4.A]	[1.9][3.2] [3.3][4.A]	[1.2][1.9][3.1] [3.2][3.3][3.4] [3.5][3.7][3.8] [3.11][3.12][4.A]	[1.2][1.9][3.1] [3.2][3.3][3.4] [3.5][3.7][3.8] [3.8][3.12][4.A]	[1.3][1.4] [3.1][3.3][3.4] [3.8][3.12][4.A]	[1.4][3.2] [3.3][4.A] [3.5][4.A]	[1.4][3.3] [3.4][4.A]	[1.4][3.2] [3.3][3.4] [4.A]	[3.2][3.3] [4.A]	[4.A]	[2.1] [2.6] [2.9]	[2.1] [2.4] [2.9]	[2.1]	[3.7] [3.11]		[3.5]
ALL QUANTITIES	[1.1]	[1.1][1.6] [3.6][3.8] [3.9]	[1.6][3.6]	[1.1][1.6] [3.6][3.8][3.9]	[1.1][1.6] [3.6][3.8] [3.9][5.5]	[1.1][1.6] [3.6][3.8] [3.9]	[1.1][3.6] [3.8][3.9]	[3.6][3.8] [3.9]		[3.6]	[1.1][1.6] [3.6][3.8] [3.9]	[3.8][3.9]				[5.5]	[1.6]	[1.6]

KEY TO MANUFACTURING PROCESS PRIMA SELECTION MATRIX:

CASTING PROCESSES
[1.1] SAND CASTING
[1.2] SHELL MOULDING
[1.3] GRAVITY DIE CASTING
[1.4] PRESSURE DIE CASTING
[1.5] CENTRIFUGAL CASTING
[1.6] INVESTMENT CASTING
[1.7] CERAMIC MOULD CASTING
[1.8] PLASTER MOULD CASTING
[1.9] SQUEEZE CASTING

PLASTIC & COMPOSITE PROCESSING
[2.1] INJECTION MOULDING
[2.2] REACTION INJECTION MOULDING
[2.3] COMPRESSION MOULDING
[2.4] TRANSFER MOULDING
[2.5] VACUUM FORMING
[2.6] BLOW MOULDING
[2.7] ROTATIONAL MOULDING
[2.8] CONTACT MOULDING
[2.9] CONTINUOUS EXTRUSION (PLASTICS)

FORMING PROCESSES
[3.1] CLOSED DIE FORGING
[3.2] ROLLING
[3.3] DRAWING
[3.4] COLD FORMING
[3.5] COLD HEADING
[3.6] SWAGING
[3.7] SUPERPLASTIC FORMING
[3.8] SHEET-METAL SHEARING
[3.9] SHEET-METAL FORMING
[3.10] SPINNING
[3.11] POWDER METALLURGY
[3.12] CONTINUOUS EXTRUSION (METALS)

MACHINING PROCESSES
[4.A] AUTOMATIC MACHINING
[4.M] MANUAL MACHINING

(THE ABOVE HEADINGS COVER A BROAD RANGE OF MACHINING PROCESSES AND LEVELS OF CONTROL TECHNOLOGY. FOR MORE DETAIL, THE READER IS REFERRED TO THE INDIVIDUAL PROCESSES.)

NTM PROCESSES
[5.1] ELECTRICAL DISCHARGE MACHINING (EDM)
[5.2] ELECTROCHEMICAL MACHINING (ECM)
[5.3] ELECTRON BEAM MACHINING (EBM)
[5.4] LASER BEAM MACHINING (LBM)
[5.5] CHEMICAL MACHINING (CM)
[5.6] ULTRASONIC MACHINING (USM)
[5.7] ABRASIVE JET MACHINING (AJM)

Fig. 2.2 Manufacturing process PRIMA selection matrix.

- *Production quantity per annum* – The number of components to be produced to account for the economic feasibility of the manufacturing process. The quantities specified for selection purposes are in the ranges:

Very low volume	= 1–100
Low volume	= 100–1000
Medium volume	= 1000–10 000
Medium to high volume	= 10 000–100 000
High volume	= 100 000 +
All quantities.	

Due to page size constraints and the number of processes involved, each manufacturing process has been assigned an identification code rather than using process names, as shown at the bottom of Figure 2.2. There may be just one or a dozen processes at each node in the selection matrix representing the possible candidates for final selection.

As seen in Figure 1.11, there are many cost drivers in manufacturing process selection, not least component size, geometry, tolerances, surface finish, capital equipment and labor costs. The justification for basing the matrix on material and production quantity is that it combines technological and economic issues of prime importance. Many manufacturing processes are only viable for low-volume production due to the time and labor involved. On the other hand, some processes require expensive equipment and are, therefore, unsuitable for low production volumes. By considering production quantities in the early stages, the process that will prove to be the most economical later in the development process can be identified and selected. The boundaries of economic production, however, can be vague when so many factors are relevant, therefore the matrix concentrates rather more on the use of materials. By limiting itself in this way, the matrix cannot be regarded as comprehensive and should not be taken as such. It represents the main common industrial practice, but there will always be exceptions at this level of detail. It is not intended to represent a process selection methodology in itself. It is essentially a first-level filter. The matrix is aimed at focusing attention on those PRIMAs that are most appropriate based on the important consideration of material and production quantity. It is the PRIMAs that do the task of guiding final manufacturing process selection.

Note that conventional and Non-Traditional Machining (NTM) processes are often considered as secondary rather than primary manufacturing processes, although they can be applicable to both situations. The user should be aware of this when using the PRIMA selection matrix. Also, the conventional machining processes are grouped under just two headings in the matrix, manual and automatic machining. Reference should be made to the individual processes for more detail.

2.3.2 Assembly system selection

Assemblies involve two or more combined components of varying degrees of build complexity and spatial configuration. The assembly technologies used range from simple manual operations through to dedicated and fully mechanized systems. The final system or combination of systems selected has the task of reproducing the product at the volume dictated by the customer, in a cost-effective way for the producer, being technically appropriate for the components manipulated and composed, and ultimately satisfying the functional requirements dictated by the specification.

The assembly phase represents a significant proportion of the total production cost of a product and can outweigh manufacturing costs in some industries (2.3). Through the identification of the most effective manufacturing and assembly technologies early in the development process, downstream activity, inefficiency and costs can be reduced. Significantly, assembly is a major source of late engineering change, rework and production variability in product development (2.4). The cost of recovering from these problems during assembly is high and is estimated to be in the range of 5–10 per cent of the final cost (2.5). In part, this is due to the fact that assembly is governed by much less controllable and less tangible issues than manufacturing, such as assembly actions and fixture design (2.6).

In practice, assembly selection is a very difficult task. It does not mean, however, that we cannot make a sound decision about the most appropriate assembly technology to use for a given set of conditions or requirements. A number of researchers have proposed strategies for assembly system selection. The reader interested in this topic can find more information in References (2.7–2.9).

Prior to the selection of an assembly technology, a number of activities should be undertaken and factors considered, some of which also help drive the final quality of the assembly:

- *Business level* – Identification and availability of assembly technologies/expertise in-house, integration into business practices/strategy, geographical location and future competitive issues, such as investment in equipment.
- *Product level* – Anticipated lead times, product life, investment return time-scale, product families/variants and product volumes required.
- *Supplier level* – Component quality (process capability, gross defects) and timely supply of bought-in and in-house manufactured parts.

The final point is of particular importance. A substantial proportion of a finished product, typically, two-thirds, consists of components or sub-assemblies produced by suppliers (2.10). The original equipment manufacturer is fast becoming purely an assembler of these bought-in parts, and therefore it is important to realize the key role suppliers have in developing products that are also 'assembly friendly'. Consideration must be given to the tolerances and process variability associated with component parts from a very early stage, especially when using automated assembly technologies, because production variability is detrimental to an assembly process.

From the above, a number of drivers for assembly technology selection can be highlighted:

- Availability of labor
- Operating costs
- Production quantity
- Capital cost of assembly equipment
- Production rate required
- Number of components in the assembly
- Number of product variants
- Handling characteristics (safety, environmental hazards, supply logistics)
- Complexity of components and assembly operations
- Size and weight of components to be assembled.

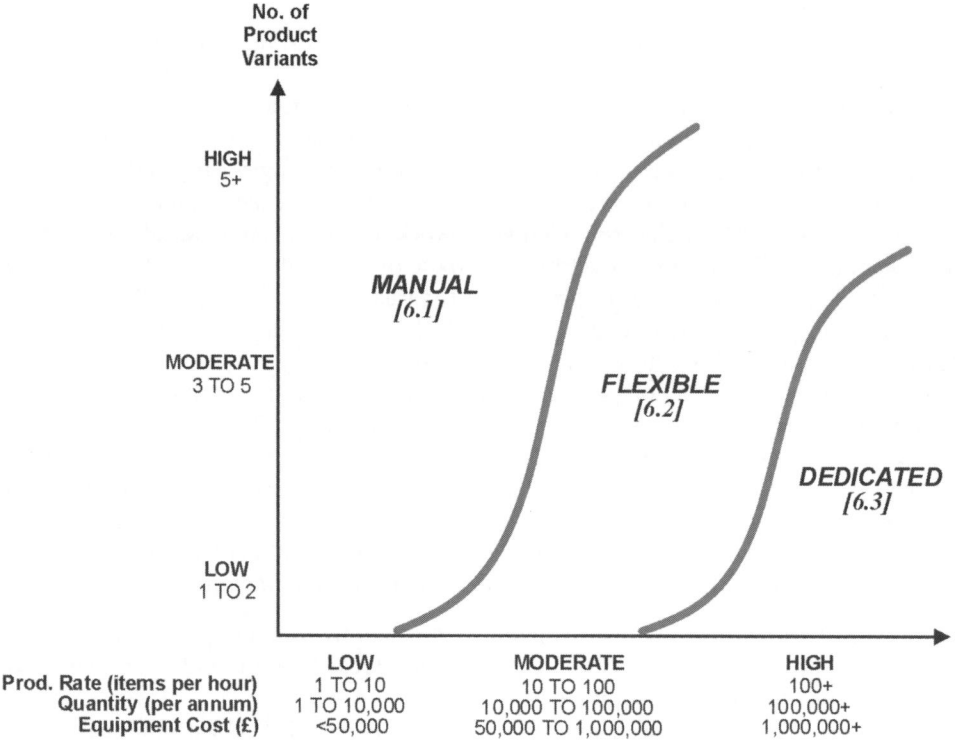

Fig. 2.3 Assembly system selection chart.

Figure 2.3 maps several of the important selection drivers with assembly technology. It is a general guide for the selection of the most appropriate assembly system based on:

- Number of product variants (flexibility), and
- Production rate, or
- Production quantity *per annum*, or
- Capital cost of the assembly equipment (although this is more of an outcome than a requirement).

Three basic assembly systems can be identified and are classified in Figure 1.14 and with their respective PRIMA number below:

- *Manual* (with or without mechanized assistance) [6.1]
- *Flexible* (programmable, robots) [6.2]
- *Dedicated* (special purpose) [6.3].

Upon candidate selection, further reference is made to the individual PRIMAs for each assembly system type in order to fully understand the technical and economic implications of the final decision and explore system variants available. This is particularly advantageous when Figure 2.3 shows that a set of requirements is on the boundary of two assembly system types.

2.3.3 Joining process selection

There is extensive evidence to suggest that many industrial products are designed with far too many parts. DFA case studies indicate that in many designs large proportions of excess components are only used for fastening (2.11). These non-value added components increase part-count and production costs without contributing to the product's functionality. In many cases, incorrect joining processes are used due to a lack of knowledge of such factors as availability, cost and functional performance of alternatives. As with primary and secondary manufacturing processes, selecting the most suitable joining process greatly influences the manufacturability of a design, but the selection of the joining technology to be used can also greatly influence the assemblability of a design. The method chosen can also have a significant influence on the product architecture and assembly sequence and it is well known that complicated joining processes lead to incorrect, incomplete and faulty assemblies (2.12).

Selecting the most appropriate joining technique requires consideration of many factors relating to joint design, material properties and service conditions. During the selection procedure the designer is required to scrutinize large quantities of data relating to many different technologies. Several selection methods exist for the selection of the process variants within individual joining technologies. However, selecting the most appropriate technology itself remains a design-orientated task that often does not get the attention it deserves. It can be concluded that a selection methodology that incorporates joining processes and technologies that can be applied at an early stage in the design process is a useful tool to support designing and particularly DFA. Considering joining processes prior to the development of detailed geometry enables components to be tailored to the selected process rather than limiting the number of suitable processes. Addressing such issues during the early stages of product development actively encourages designers to employ good DFA practice and reduces the need for costly redesign work.

As mentioned above, a number of other selection methods exist for different joining technologies, and the reader interested in further information is referred to:

- Adhesive bonding (2.13)(2.14)
- Welding, soldering and brazing (1.6)(1.7)(2.15)
- All joining technologies (1.10)(2.16).

Currently, available selection techniques tend to focus on particular joining technologies or do not offer the designer a wide range of suitable joining processes or in enough detail to support the selection process. The aim of the joining process selection methodology presented here is to provide a means of identifying feasible methods of joining regardless of their fundamental technology. The methodology is not intended to select a specific joining method, for example, torch brazing or tubular rivet, but to highlight candidate processes that are capable of joining under the given conditions. The final selection can be made after considering process specific data and detailed data against design requirements from the PRIMAs.

Joining process classification

Due to the large number of different joining processes and variants, only the most commonly used and well-established processes in industry are included. Investigations highlighted 73 major joining techniques, as shown in Figure 1.15. In order to classify them, a common factor is used, based on technology and process. Technology class refers to the collective group that a process belongs to, for example, welding or adhesive bonding. The process class refers to the specific joining technique, for example, Metal Inert-gas Welding (MIG) or anaerobic adhesive. Each process is derived from a particular fundamental technology providing a means for classification. From this, the joining processes have been divided into five main categories: welding, brazing, soldering, mechanical fastening and adhesive bonding.

Technical classes can be separated into sub-categories based on distinct differences in underlying technology. Although the basic premise of all welding processes is the same, specific techniques differ considerably due to the particular processes involved in generating heat and/or enabling the fusion process. This can be used as a means of classifying sub-sets. Both brazing and soldering have a number of different processes, hence they have been split into two sub-sets. Mechanical fasteners can be divided in two ways, by group technologies and degree of permanence. The latter has been chosen as it relates to the functionality of the fastener in service and therefore product requirements. Due to the large number of specific adhesives, which in many cases are exclusive to the producer, adhesive bonding has been viewed from a generic level, therefore, only the adhesive group can be selected.

Joining process selection criteria

In order to select the most appropriate joining process, it is necessary to consider all processes available within the methodology. As technology specific selection criteria tend to be non-transportable between domains, evaluating the merits of joining processes that are based on fundamentally dissimilar technologies requires a different approach. Differentiating between technology classes and process classes requires the comparison of specifically selected parameters. In order to evaluate a joint, consideration must be given to its functional, technical, spatial and economic requirements. A review of important joining requirements has identified a number of possible selection criteria, as shown in Figure 2.4 and discussed below.

- *Functional* – Functional requirements define the working characteristics of the joint. The functional considerations for a joint are degree of permanence, load type and strength. Degree of permanence identifies whether a joint needs to be dismantled or not. In most cases the permanence of a joining process is independent of its technology class. Degree of permanence provides a suitable high-level selection criterion that is not reliant on detailed geometry. Load type and strength are often mutually dependent and can be influenced by the geometric characteristics of the joint interface. As joint design is dissimilar for different technology classes, it is difficult to use load type or strength as a universal selection criteria. However, these considerations must be taken into account when evaluating suitable joining processes for final selection when appropriate.
- *Technical* – Specific needs of components to be joined are categorized by the joint's technical requirements. The technical considerations for a joint are material type, joint design and operating temperature. Material type is selected based on parameters defined by the

product's operating environment such as corrosion resistance. The material type is relevant to all joining technologies because they need to be compatible. Joint design is often defined by the geometry. However, if joining is considered prior to detailed geometry, the selected process can influence the design. Due to the fundamental differences in joint configurations, it is not suitable as a selection criterion for non-technology specific selection. Operating temperature influences the performance of most joining processes, although it should be considered during material selection. While an important aspect, its effect varies for different joining technologies. Therefore, consideration of operating temperature is more appropriate during final selection.

- *Spatial* – Geometric characteristics of the joint are accounted for by the spatial requirements. The spatial requirements identified are size, weight, geometry and material thickness. The size and weight of components to be joined is considered and determined when their material is selected. As the selection methodology is intended for use prior to the development of detailed geometry, using geometry as a selection criterion would be contradictory. Material thickness has already proven to be a successful criterion in other selection methodologies, and the suitability of joining processes is easily classified for different thicknesses of material.
- *Economic* – The economics of joining processes aligns the design with the business needs of the product. Economic considerations can be split into two sections: tooling and

CATEGORY		CRITERIA	DESCRIPTION
Technological	Functional	• Degree of permanence • Loading type (static, cyclic, impact) • Strength	Functional requirements define the working characteristics of the joint.
	Technical	• Joint configuration/design • Operating temperature • Material type • Material compatibility • Accuracy	Specific needs of components to be joined are categorised by a joint's technical requirements.
	Spatial	• Material thickness • Size, weight • Geometry	Geometric characteristics of the joint are accounted for by the spatial requirements.
	Misc.	• Complexity • Flexibility (assembly/orientation) • Safety • Joint accessibility • Quality	Other important issues not considered by the above groups.
Economic		• Production quantity • Production rate • Availability of equipment • Ease of automation • Skill required • Tooling requirements • Cost	The economics of joining processes aligns the design with the business needs of the product.

Fig. 2.4 Classification of joint requirements.

product. Tooling refers to the ease of automation, availability of equipment, skill required, tooling requirements and cost. Product economics relate to production rates and quantity. These business considerations are driven by the product economics as they determine the need for tooling and its complexity, levels of automation and labor requirements. Production rate and quantity are very closely linked. They can both be used to determine the assembly speed and the need for and feasibility of automation. However, as the selection methodology is to be used in the early stages of product development it is more likely that quantity will be known from customer requirements or market demand.

In order for the selection methodology to be effective in the early stages of design appraisal, the chosen parameters must apply to all joining processes. Also, it is essential that the parameters relate to knowledge that is readily available and appropriate to the level of selection. Having reviewed the requirement against the joining processes, four selection parameters have been chosen for initial stages of the methodology:

- *Material type* – Accounts for the compatibility of the parent material with the joining process. A large proportion of the materials used in engineering manufacture have been included in the selection methodology, from ferrous alloys to precious metals. In situations involving multiple material types the selection methodology must be applied for each.
- *Material thickness* – Divided into three ranges: thin ≤ 3 mm, medium from 3 mm to 19 mm and thick ≥ 19 mm. When selecting the material type and thickness, the designer considers many other factors that can be attributed to the joint requirements, such as corrosion resistance, operating temperature and strength. Consequently, the requirements should be known and can be compared to joining process design data for making the final choice at a later stage.
- *Degree of permanence* – This is a significant factor in determining appropriate joining processes, as it relates to the in-service behavior of the joint and considers the need for a joint to be dismantled. This selection criterion is divided into three types:

  1 Permanent joint – Can only be separated by causing irreparable damage to the base material, functional element or characteristic of the components joined, for example, surface integrity. A permanent joint is intended for a situation where it is unlikely that a joint will be dismantled under any servicing situation.
  2 Semi-permanent joint – Can be dismantled on a limited number of occasions, but may result in loss or damage to the fastening system and/or base material. Separation may require an additional process, for example, re-heating a soldered joint or plastic deformation. A semi-permanent joint can be used when disassembly is not performed as part of regular servicing, but for some other need.
  3 Non-permanent joint – Can be separated without special measures or damage to the fastening system and/or base material. A non-permanent joint is suited to situations where regular dismantling is required, for example, at scheduled maintenance intervals.

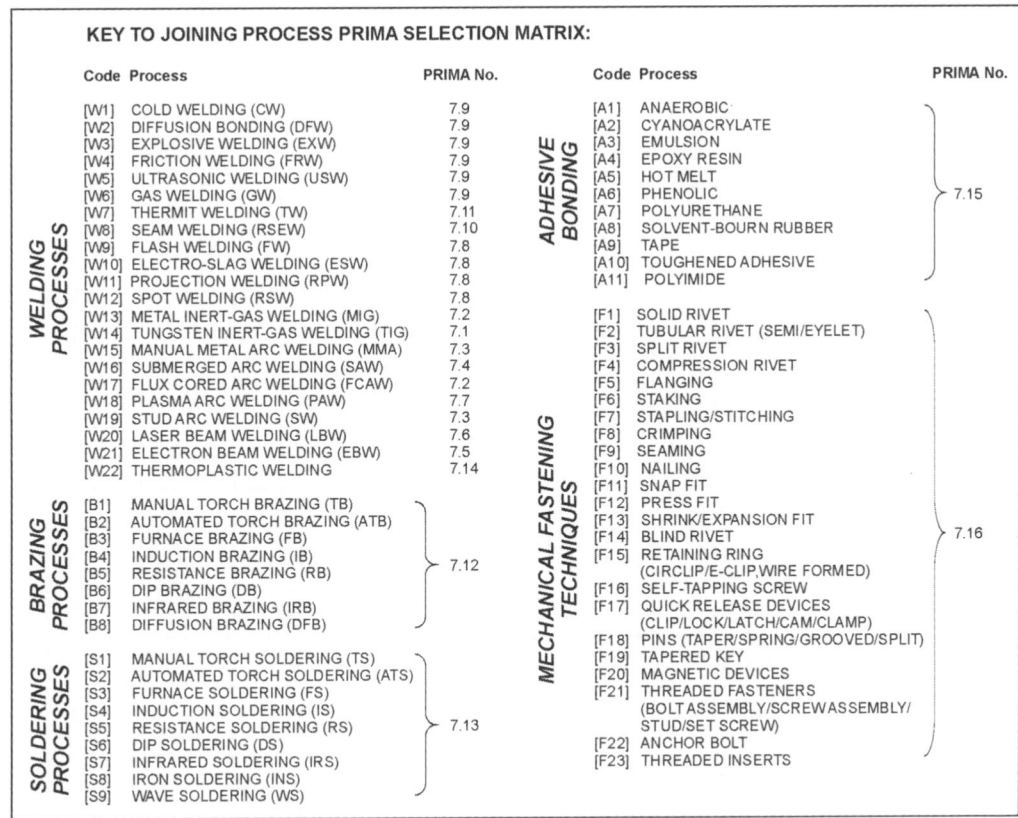

KEY TO JOINING PROCESS PRIMA SELECTION MATRIX:

Code	Process	PRIMA No.
WELDING PROCESSES		
[W1]	COLD WELDING (CW)	7.9
[W2]	DIFFUSION BONDING (DFW)	7.9
[W3]	EXPLOSIVE WELDING (EXW)	7.9
[W4]	FRICTION WELDING (FRW)	7.9
[W5]	ULTRASONIC WELDING (USW)	7.9
[W6]	GAS WELDING (GW)	7.9
[W7]	THERMIT WELDING (TW)	7.11
[W8]	SEAM WELDING (RSEW)	7.10
[W9]	FLASH WELDING (FW)	7.8
[W10]	ELECTRO-SLAG WELDING (ESW)	7.8
[W11]	PROJECTION WELDING (RPW)	7.8
[W12]	SPOT WELDING (RSW)	7.8
[W13]	METAL INERT-GAS WELDING (MIG)	7.2
[W14]	TUNGSTEN INERT-GAS WELDING (TIG)	7.1
[W15]	MANUAL METAL ARC WELDING (MMA)	7.3
[W16]	SUBMERGED ARC WELDING (SAW)	7.4
[W17]	FLUX CORED ARC WELDING (FCAW)	7.2
[W18]	PLASMA ARC WELDING (PAW)	7.7
[W19]	STUD ARC WELDING (SW)	7.3
[W20]	LASER BEAM WELDING (LBW)	7.6
[W21]	ELECTRON BEAM WELDING (EBW)	7.5
[W22]	THERMOPLASTIC WELDING	7.14

Code	Process	PRIMA No.
BRAZING PROCESSES		
[B1]	MANUAL TORCH BRAZING (TB)	
[B2]	AUTOMATED TORCH BRAZING (ATB)	
[B3]	FURNACE BRAZING (FB)	
[B4]	INDUCTION BRAZING (IB)	
[B5]	RESISTANCE BRAZING (RB)	7.12
[B6]	DIP BRAZING (DB)	
[B7]	INFRARED BRAZING (IRB)	
[B8]	DIFFUSION BRAZING (DFB)	

Code	Process	PRIMA No.
SOLDERING PROCESSES		
[S1]	MANUAL TORCH SOLDERING (TS)	
[S2]	AUTOMATED TORCH SOLDERING (ATS)	
[S3]	FURNACE SOLDERING (FS)	
[S4]	INDUCTION SOLDERING (IS)	
[S5]	RESISTANCE SOLDERING (RS)	7.13
[S6]	DIP SOLDERING (DS)	
[S7]	INFRARED SOLDERING (IRS)	
[S8]	IRON SOLDERING (INS)	
[S9]	WAVE SOLDERING (WS)	

Code	Process	PRIMA No.
ADHESIVE BONDING		
[A1]	ANAEROBIC	
[A2]	CYANOACRYLATE	
[A3]	EMULSION	
[A4]	EPOXY RESIN	
[A5]	HOT MELT	
[A6]	PHENOLIC	7.15
[A7]	POLYURETHANE	
[A8]	SOLVENT-BOURN RUBBER	
[A9]	TAPE	
[A10]	TOUGHENED ADHESIVE	
[A11]	POLYIMIDE	

Code	Process	PRIMA No.
MECHANICAL FASTENING TECHNIQUES		
[F1]	SOLID RIVET	
[F2]	TUBULAR RIVET (SEMI/EYELET)	
[F3]	SPLIT RIVET	
[F4]	COMPRESSION RIVET	
[F5]	FLANGING	
[F6]	STAKING	
[F7]	STAPLING/STITCHING	
[F8]	CRIMPING	
[F9]	SEAMING	
[F10]	NAILING	
[F11]	SNAP FIT	
[F12]	PRESS FIT	
[F13]	SHRINK/EXPANSION FIT	7.16
[F14]	BLIND RIVET	
[F15]	RETAINING RING (CIRCLIP/E-CLIP,WIRE FORMED)	
[F16]	SELF-TAPPING SCREW	
[F17]	QUICK RELEASE DEVICES (CLIP/LOCK/LATCH/CAM/CLAMP)	
[F18]	PINS (TAPER/SPRING/GROOVED/SPLIT)	
[F19]	TAPERED KEY	
[F20]	MAGNETIC DEVICES	
[F21]	THREADED FASTENERS (BOLT ASSEMBLY/SCREW ASSEMBLY/ STUD/SET SCREW)	
[F22]	ANCHOR BOLT	
[F23]	THREADED INSERTS	

Fig. 2.5 Key to joining process PRIMA selection matrix.

• *Quantity* – Production quantity *per annum*, and consequently the number of joints to be produced, accounts for the economic feasibility of the joining process. The quantities specified for selection purposes are the same as for the manufacturing process selection strategy.

Joining process selection matrix

The joining process selection methodology is based on the same matrix approach used for manufacturing process selection. Again, due to page size constraints and the number of processes to be detailed, each process has been assigned an identification code rather than using process names. The key to the joining processes used in the matrix is shown in Figure 2.5 together with the relevant PRIMA number, where information can be found regarding that individual process or joining technology. Due to size constraints, the joining process selection matrix is divided into two parts; Figures 2.6(a) and (b) together show the complete matrix.

The matrix representation of the selection technique provides an intuitive way of navigating a large quantity of data. This makes the selection process simple and quick to use. Supporting the selection matrix with design advice through the use of the PRIMAs completes the

Note - The joining process PRIMA selection matrix cannot be considered as comprehensive and should not be taken as such. It represents the main common industrial practice, but there will always be exceptions at this level of detail. Also, the order in which the PRIMAs are listed in the nodes of the matrix has no significance in terms of preference. Dissimilar metals also accounts for joining metals with coatings.

MATERIAL & THICKNESS / QUANTITY & PERMANENCE		IRONS		STEEL (carbon)			STEEL (tool, alloy)			STAINLESS STEEL			COPPER & ALLOYS			ALUMINIUM & ALLOYS			MAGNESIUM & ALLOYS			ZINC & ALLOYS			TIN & ALLOYS		
		MED. 3 to 19mm	THICK ^{3}19mm	THIN £3mm	MED. 3 to 19mm	THICK ^{3}19mm	THIN £3mm	MED. 3 to 19mm	THICK ^{3}19mm	THIN £3mm	MED. 3 to 19mm	THICK ^{3}19mm	THIN £3mm	MED. 3 to 19mm	THICK ^{3}19mm	THIN £3mm	MED. 3 to 19mm	THICK ^{3}19mm	THIN £3mm	MED. 3 to 19mm	THICK ^{3}19mm	THIN £3mm	MED. 3 to 19mm	THICK ^{3}19mm	THIN £3mm	MED. 3 to 19mm	THICK ^{3}19mm
VERY LOW 1 TO 100	P	[W6][W13][W15][B1]	[W6][W13][W15][B1]	[W3][W6][W8][W9][W11][W13][W14][W15][B1]	[W3][W6][W9] [W11][W13][W14][W15] [B1]	[W6][W7][W9][W10][W13][W15]	[W2][W3][W6][W8][W9][W11][W13][W14][W15][B1]	[W2][W3][W9][W11][W13][W14][W15][B1]	[W3][W7][W9][W10][W13][W15]	[W2][W3][W6][W8][W9][W11][W13][W14][W15][B1]	[W2][W3][W9][W11][W13][W14][W15][W17][B1]	[W9][W10][W13][W15][W17]	[W1][W2][W3][W9][W13][W14][B1]	[W1][W2][W3][W9][W13][B1]	[W3][W7][W9] [W13]	[W1][W2][W3][W6][W8][W9][W11][W13][W14][B1]	[W1][W2][W3][W6][W9][W11][W13][W14][B1]	[W3][W9]	[W3][W8][W9][W13][W14][B1]	[W3][W9][W4][W13][W14]	[W13]				[W1]	[W1]	
	SP			[S1]						[S1]			[S1][S8]			[S1][S8]						[S1][S8]			[S1][S8]		
	NP	[F19][F20][F23]	[F19][F20][F23]	[F20]	[F19][F20]	[F19][F20]	[F20]	[F19][F20]	[F19][F20]	[F20]	[F19][F20]	[F19][F20]		[F19][F23]	[F19][F23]		[F23]	[F23]		[F23]	[F23]		[F23]	[F23]			
LOW 100 TO 1,000	P	[W13][W15][W16][W17][B1]	[W13][W15][W16][W17]	[W3][W8][W9][W11][W13][W14][W15][B1][B6]	[W3][W9][W11][W13][W14] [W15][W16][W17][B1]	[W9] [W13][W15][W16][W17]	[W2][W3][W8][W9][W11] [W13][W14][W15][B1][B6]	[W2][W3][W9][W11][W13][W14][W15] [W17][W16][B1]	[W3][W9][W13][W15][W16][W17]	[W2][W3][W8][W9][W11][W13][W14][W15][B1][B6]	[W2][W3][W9][W11][W13][W14][W15][W16][B1]	[W3][W9][W13][W15][W16][W17]	[W1][W2][W3][W9][W13] [W14][B1]	[W2][W3][W9][W13][B1]	[W3][W9][W13]	[W1][W2][W3][W8][W9][W11][W13][W14][B1][B6]	[W1][W2][W3][W9][W11][W13][W14][B1][B6]	[W3][W9]	[W3][W8][W9][W13][W14][B1][B6]	[W4][W3][W9][W13][W14]	[W13]				[W1]	[W1]	
	SP			[S1]						[S1]			[S1][S8]			[S1][S8]						[S1][S8]			[S1][S8]		
	NP	[F19][F20][F23]	[F19][F20][F23]	[F20]	[F19][F20]	[F19][F20]	[F20]	[F19][F20]	[F19][F20]	[F20]	[F19][F20]	[F19][F20]		[F19][F23]	[F19][F23]		[F23]	[F23]		[F23]	[F23]		[F23]	[F23]			
LOW TO MEDIUM 1,000 TO 10,000	P	[W16][W17][B2][B4]	[W16][W17][B4]	[W8][W11][W13][W14][W19][B2][B4][B6]	[W11][W13][W14][W16][W17][W19][B2] [B4]	[W13][W16][W17]	[W2][W8][W11][W13][W14][W19][B2][B4][B6]	[W2][W11][W13][W14][W16][W17][W19][B2][B4]	[W13][W16][W17]	[W2][W8][W11][W13][W14][W19][B2][B4][B6]	[W2][W11][W13][W14][W16][W17][W19][B2][B4]	[W13][W16][W17]	[W1][W2][W13][W14][W19][B2][B4]	[W2][W13][B2]		[W1][W2][W8][W11][W13][W14][W19][B2][B4][B6]	[W1][W2][W11][W13][W14][W19][B2][B6]		[W8][W13][W14][B2][B4][B6]	[W4][W13][W14]	[W13]	[W19]			[W1]	[W1]	
	SP			[S6]			[F11]			[F11][S6]			[S2][S6] [S8]			[S2][S6][S8]						[S2][S6][S8]			[S1][S6][S8]		
	NP																										
MEDIUM TO HIGH 10,000 TO 100,000	P	[W18][B2][B3][B4][B8]	[B2][B3][B4][B8]	[W8][W11][W13][W14][W18][W19][W20][B2][B3][B4][B5][B7][B8][F7] [F9]	[W4][W11][W13][W14][W16][W17][W18][W19][W20][B2][B3][B4][B8]	[W4][W13][W16][W17][B3][B8]	[W8][W11][W14][W18][W19][W20][B2][B3][B4][B5][B7][B8]	[W4][W11][W14][W16][W17][W18][W19][W20][B2][B3][B4][B8]	[W4][W16][W17][B3][B8]	[W8][W11][W14][W18][W19][W20][B2][B3][B4][B5][B7][B8][F7][F9]	[W4][W11][W14][W16][W17][W18][W19][W20][B2][B3][B4][B8]	[W4][W16][W17][B3][B8]	[W18] [W19][W20][B2][B3][B4][B5][B8][F7][F9]	[W4][W18][W20][B2][B3][B8]	[B3][B8]	[W4][W8][W11][W14][W18][W19][W20][B2][B3][B4][B5][B7][B8][F7][F9]	[W4][W11][W14][W19][B2][B3][B8]	[B8]	[W8][W14][W18][W20][B2][B3][B4][B8]	[W4][W14][W18][W20][B3][B8]		[W19]					
	SP			[S3]			[F11]			[F11][S3]			[S2][S3][S4][S5][S7] [S9]			[S2][S3][S4][S5][S7][S9]						[S2][S3][S4][S5][S7][S9]			[S2][S3][S4][S5][S7][S9]		
	NP						[F11]			[F11]																	
HIGH 100,000+	P	[B3][B8]	[B3][B8]	[W18][W20][W21] [B3][B8][F7][F9]	[W4][W18][W20] [W21][B3][B8]	[W4][W21][B3][B8]	[W18][W20][W21][B3][B8]	[W4][W18][W20][W21][B3] [B8]	[W4][W21][B3] [B8]	[W18][W20][W21][B3][B8][F7][F9]	[W4][W18][W20][W21][B3] [B8]	[W4][W21][B3][B8]	[W18][W20][W21] [B3][B8][F7][F9]	[W4][W18][W20][W21][B3][B8]	[W21][B3][B8]	[W4][W18][W20][W21][B3][B8][F7][F9]	[W4][W21][B3][B8]	[W21][B8]	[W18][W20][B3][B8]	[W18][W20][W21][B3][B8]	[W21]						
	SP			[S3]			[F11]			[F11][S3]			[S3] [S9]			[S3] [S9]						[S3][S9]			[S3][S9]		
	NP						[F11]			[F11]																	
ALL QUANTITIES	P	[F1][A1][A2][A4][A5][A8][A10]	[F1]	[W5][W12][A1][A2][A4][A5][A8][A10][F1][F5][F6][F8][F10]	[W12][A1][A2][A4][A5][A10][F1]	[F1]	[W5][W12][F1][A1][A2][A4][A5][A8][A10]	[F1][A1][A2][A4][A5][A10]	[F1]	[W5][W12][F1][F5][F6][F8][F10][A4][A10]	[W12][F1][A4] [A10]	[F1]	[W12][F1][F5][F6][F8][F10][A1][A2][A4][A5][A10]	[W12][F1][F10][A4][A10]	[F1]	[W5][W12][F1][F2][F5][F6] [F8][F10][A2][A4][A5][A7][A10]	[W5][W12][F1][F2][A2][A4][A5][A7][A10]	[F1][F2]	[W12][F1][F2][A1][A4][A5]	[W12][F1][F2][A4][A5]	[F1][F2]	[W12][F2][F10][F14]	[W12][F2]	[F2]	[W12][F2][F10][F14][A4]	[W12][F2]	[F2]
	SP	[F12][F13]	[F12][F13]	[A9][F14]	[F12][F13]	[F12][F13]	[F13][A9]	[F12][F13]	[F12][F13]	[F13][A9][F14]	[F12][F13]	[F12][F13]	[A9][F13][F14]	[F12][F13]	[F12][F13]	[F13][F14][A9]	[F12][F13]	[F12][F13]	[F13][A9]	[F12][F13]	[F12][F13]	[F14]			[F13][A9][F14]	[F13]	
	NP	[F15][F16][F18][F21]	[F15][F18][F21]	[F15][F17][F21]	[F15][F17][F18][F21]	[F15][F17][F21]	[F15][F17][F21]	[F15][F17][F21]	[F15][F17][F21]	[F15][F17][F21]	[F15][F17][F18][F21]	[F15][F17][F18][F21]	[F16][F21]	[F16][F18][F21]	[F18][F21]	[F15][F16][F17][F21]	F[15][F16][F17][F18][F21]	[F15][F17][F18][F21]	[F16][F21]	[F16][F18][F21]	[F18][F21]	[F15][F16][F21]	[F15][F16][F18][F21]	[F15][F18][F21]	[F16][F21]	[F16][F21]	[F21]

Fig. 2.6 (a) Joining process PRIMA selection matrix – part A.

Fig. 2.6 (b) Joining process PRIMA selection matrix — part B.

MATERIAL & THICKNESS		QUANTITY & PERMANENCE	VERY LOW 1 TO 100			LOW 100 TO 1,000			LOW TO MEDIUM 1,000 TO 10,000			MEDIUM TO HIGH 10,000 TO 100,000			HIGH 100,000+			ALL QUANTITIES		
			P	SP	NP	P	SP	NP	P	SP	NP	P	SP	NP	P	SP	NP	P	SP	NP
LEAD & ALLOYS	THIN 3mm 19mm		[W1]			[W1]			[W14][W14]									[W8][W14][P13]		[P14][P15]
	MED. 3 to 19mm		[W7]			[W7]												[P21]		[P16][P21]
	THICK 19mm		[W2]	[S1][S8]		[W22]	[S1][S8]		[S1][S8][S2]	[S1][S8][S2]		[W14]	[S8]					[P1]	[P15][P17]	[P16][P17][P21]
NICKEL & ALLOYS	THIN 3mm 19mm		[W1][W14]			[W1][W14]		[P20]											[W13]	[P13]
	MED. 3 to 19mm		[W13][W14]			[W13][W14]			[W13][W14]			[W13][W14]						[W13]	[P11]	[P11]
	THICK 19mm		[W13]			[W13]			[W13][S1][S8]	[S1][S8][S9]		[W13][S1][S8][S9]	[S1][S8]		[W11][W13]			[P11][P15]	[P11]	[P11]
TITANIUM & ALLOYS	THIN 3mm 19mm		[W13]			[W13]			[W13]									[W12]	[A4]	[P12]
	MED. 3 to 19mm		[W12]			[W13]			[W13]			[W13]						[W11]		[P11]
	THICK 19mm		[W22]			[W22]												[A4]		[A4]
THERMOPLASTICS	THIN 3mm 19mm		[W22]			[W22]												[W9][A16][A44][A38][A26]	[A4]	[P17][P18]
	MED. 3 to 19mm		[W22]															[A38][A39][A47][A1][A5]		[P19][P23]
	THICK 19mm													[P7]			[A10]		[P12][P23]
FR COMPOSITES	THIN 3mm 19mm																[A6][A47][A52]		[P11]
	MED. 3 to 19mm																[A6][A47][A53]		[P11]
	THICK 19mm					[S1]			[S18][S19]		[P22]	[S18][S19]					[A6][A47][A54][A55]		[P23]
CERAMICS	THIN 3mm 19mm		[P23]		[P22]												[A5]		[P11]
	MED. 3 to 19mm																[A5]		[P11]
	THICK 19mm				[P23]			[S18][S19]			[S18][S19]						[A5][A47][A51]	[P11][P12]	[P11]
THERMOSETS	THIN 3mm 19mm		[W2]			[W2]												[A4]	[P12][P13]	[P1]
	MED. 3 to 19mm		[W2][P6]			[W22][W29]			[S23][S24]			[S23][S24]						[A47][A48][A49][A50]	[P12]	[P16]
	THICK 19mm		[W22][W23]	[S1]		[W22][W23]			[S23][S24]			[S23][S24][S4]						[A52][A47]	[P12]	[P16]
REFRACTORY METALS	THIN 3mm 19mm		[W6][W23][P6][W24][W25]			[W23][P8][W24][P9][W25]			[S8][S23][S25]			[S8][S23]						[W4][W23][W24][W25]	[P12]	[P16][P18]
	MED. 3 to 19mm		[W2][W23]	[S1]		[W2][W23]			[S19]			[S18][S19]						[W4]	[P12][P13]	[P16][P17][P18]
	THICK 19mm		[W6]		[P30]	[W6]			[S19]			[W14][P8]						[W4]	[P12]	[P16][P17][P18]
PRECIOUS METALS	THIN 3mm		[W17][W22] [W28][W26] [W29][W27]	[S18][S19]		[W28][W25] [W18][W27] [S1][P8]	[S18]		[W1] [W14][S18] [W28][S19]	[S25][S26]	[P22]	[W1][W14] [W15][S18] [W28][S19]	[S9][S25] [S26][S27]		[P11]		[P10] [P16][P5] [P13][P14]	[P21][W12] [W13][P15] [P16][P20]	[P12][P13]	[P16][P20][P21] [P18][P19][P14] [P10][P17][P16]
DISSIMILAR MATERIALS METALS	THIN 3mm 19mm		[W17][W28] [W29][W30] [S20][P23]	[S20][S21]	[P23] [P30][S20]	[W1][S1] [W28][S19] [W28][P8]	[S1]	[P23] [P30][S21]	[W1] [W28][S19] [W28]	[S1][S23]	[P23]	[W1][W28] [S19][W28]	[S1][S23] [S24][S25]		[P1]		[P1]	[P11][P12] [P13][P14] [W28][W29]	[P12][P13][P14]	[P16][P17][P18] [P19][P20][P21] [P10][P14]

Note – The joining process PRIMA selection matrix cannot be considered as comprehensive and should not be taken as such. It represents the main common industrial practice, but there will always be exceptions at this level of detail. Also, the order in which the PRIMAs are listed in the nodes of the matrix has no significance in terms of preference. Dissimilar metals also accounts for joining metals with coatings.

methodology, allowing the user to both identify and gain an understanding of suitable joining processes. Combining material thickness with material type is a logical way of allowing the user to describe the material. It is also convenient to state quantity and then permanence in the rows of the matrix.

▪ 2.4 PRIMA categories

A complete list of all PRIMAs presented in this section of the book is shown in Figure 2.7 for reference.

1	Casting Processes	4	Machining Processes
	1.1 Sand Casting		4.1 Turning and Boring
	1.2 Shell Moulding		4.2 Milling
	1.3 Gravity Die Casting		4.3 Planing and Shaping
	1.4 Pressure Die Casting		4.4 Drilling
	1.5 Centrifugal Casting		4.5 Broaching
	1.6 Investment Casting		4.6 Reaming
	1.7 Ceramic Mould Casting		4.7 Grinding
	1.8 Plaster Mould Casting		4.8 Honing
	1.9 Squeeze Casting		4.9 Lapping
2	Plastic & Composite Processing	5	Nontraditional Machining (NTM) Processes
	2.1 Injection Moulding		5.1 Electrical Discharge Machining (EDM)
	2.2 Reaction Injection Moulding		5.2 Electrochemical Machining (ECM)
	2.3 Compression Moulding		5.3 Electron Beam Machining (EBM)
	2.4 Transfer Moulding		5.4 Laser Beam Machining (LBM)
	2.5 Vacuum Forming		5.5 Chemical Machining (CM)
	2.6 Blow Moulding		5.6 Ultrasonic Machining (USM)
	2.7 Rotational Moulding		5.7 Abrasive Jet Machining (AJM)
	2.8 Contact Moulding		
	2.9 Continuous Extrusion (Plastics)	6	Assembly Systems
			6.1 Manual Assembly
3	Forming Processes		6.2 Flexible Assembly
	3.1 Forging		6.3 Dedicated Assembly
	3.2 Rolling		
	3.3 Drawing	7	Joining Processes
	3.4 Cold Forming		7.1 Tungsten Inert-gas Welding (TIG)
	3.5 Cold Heading		7.2 Metal Inert-gas Welding (MIG)
	3.6 Swaging		7.3 Manual Metal Arc Welding (MMA)
	3.7 Superplastic Forming		7.4 Submerged Arc Welding (SAW)
	3.8 Sheet-metal Shearing		7.5 Electron Beam Welding (EBW)
	3.9 Sheet-metal Forming		7.6 Laser Beam Welding (LBW)
	3.10 Spinning		7.7 Plasma Arc Welding (PAW)
	3.11 Powder Metallurgy		7.8 Resistance Welding
	3.12 Continuous Extrusion (Metals)		7.9 Solid State Welding
			7.10 Thermit Welding (TW)
			7.11 Gas Welding (GW)
			7.12 Brazing
			7.13 Soldering
			7.14 Thermoplastic Welding
			7.15 Adhesive Bonding
			7.16 Mechanical Fastening

Fig. 2.7 PRIMA categories and individual processes.

1 Casting processes

1.1 Sand casting

Process description

- Moist bonding sand is packed around a pattern. The pattern is removed to create the mold, and molten metal poured into the cavity. Risers supply necessary molten material during solidification. The mold is then broken to remove the part (see 1.1F).

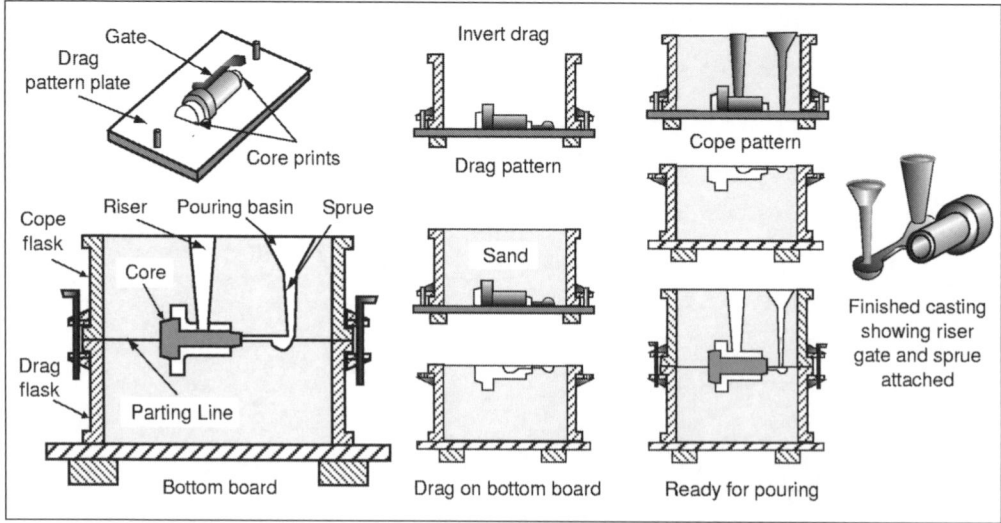

1.1F Sand casting process.

Materials

- Most metals, particularly ferrous and aluminum alloys. Some difficulty encountered in casting lead, tin and zinc alloys, also refractory alloys, beryllium, titanium and zirconia alloys.

Process variations

- Green sand casting: the most common and the cheapest. Associated problems are that the mold has low strength and high moisture content.
- Dry sand: core boxes are used instead of patterns, and an oven is used to cure the mold. Expensive and time consuming.
- Skin-dried sand: the mold is dried to a certain depth. Used in the casting of steels.
- Patterns: one-piece solid patterns are the cheapest to make; split patterns for moderate quantities; match plate patterns for high volume production.
- Wooden patterns: for low-volume production only.
- Metal patterns: for medium to high-volume production. Hard plastics are also being used increasingly.
- Cosworth casting: low pressure filling of mold used for better integrity, accuracy and porosity of casting. Longer production times and higher tooling costs, however.

Economic considerations

- Production rates of 1–50/h, but dependent on size.
- Lead time typically days, but depending on complexity and size of casting.
- Material utilization low to moderate. Twenty to fifty per cent of material lost in runners and risers.
- Both mold material and runners and risers may be recycled.
- Patterns easy to make and set, and reusable.
- Pattern material dependent on the number of castings required.
- Easy to change design during production.
- Economical for low production runs of less than 100. Can be used for one-offs and high production volumes depending on degree of automation.
- Tooling costs low.
- Equipment costs low.
- Direct labor costs high. Can be labor intensive.
- Finishing costs can be high. Cleaning and fettling required to remove gates and risers before secondary processing. Parting lines may also need finishing by hand.

Typical applications

- Engine blocks
- Manifolds
- Machine tool bases
- Pump housings
- Cylinder heads

Design aspects

- High degree of shape complexity possible. Limited only by the pattern.
- Loose piece patterns can be used for holes and protrusions.
- All intersecting surfaces must be filleted: prevents shrinkage cracks and eliminates stress concentrations.
- Design of gating system for delivery of molten metal into mold cavity important.
- Placing of parting line important, i.e. avoid placement across critical dimensions.
- Bosses, undercuts and inserts possible, but at added cost.
- Steel inserts can be used as heat flow barriers.
- Cored holes greater than $\varnothing 6\,mm$.
- Machining allowances usually in the range 1.5–6 mm.
- Draft angle ranging 1–5°.
- Minimum section typically 3 mm for light alloys, 6 mm for ferrous alloys.
- Sizes ranging 25 g–400 t in weight.

Quality issues

- Molding sand must be carefully conditioned and controlled.
- Most casting defects can be traced to and rectified by sand content.
- Casting shrinkage and distortion during cooling governed by shape, especially when one dimension is much larger than the other two.
- Extensive flat surfaces prone to sand expansion defects.
- Inspection of castings important.

- High porosity and inclusion levels common in castings.
- Defects in castings may be filled with weld material.
- Castings generally have rough grainy surfaces.
- Material strength inherently poor.
- Castings have good bearing and damping properties.
- If production volumes warrant the cost of a die.
- Surface detail fair.
- Surface roughness a function of the materials used in making the mold and ranging 3.2–50 µm Ra.
- Not suitable for close specification of tolerances without secondary processing.
- Process capability charts showing the achievable dimensional tolerances using various materials provided (see 1.1 CC). Allowances of ±0.5–±2 mm should be added for dimensions across the parting line.

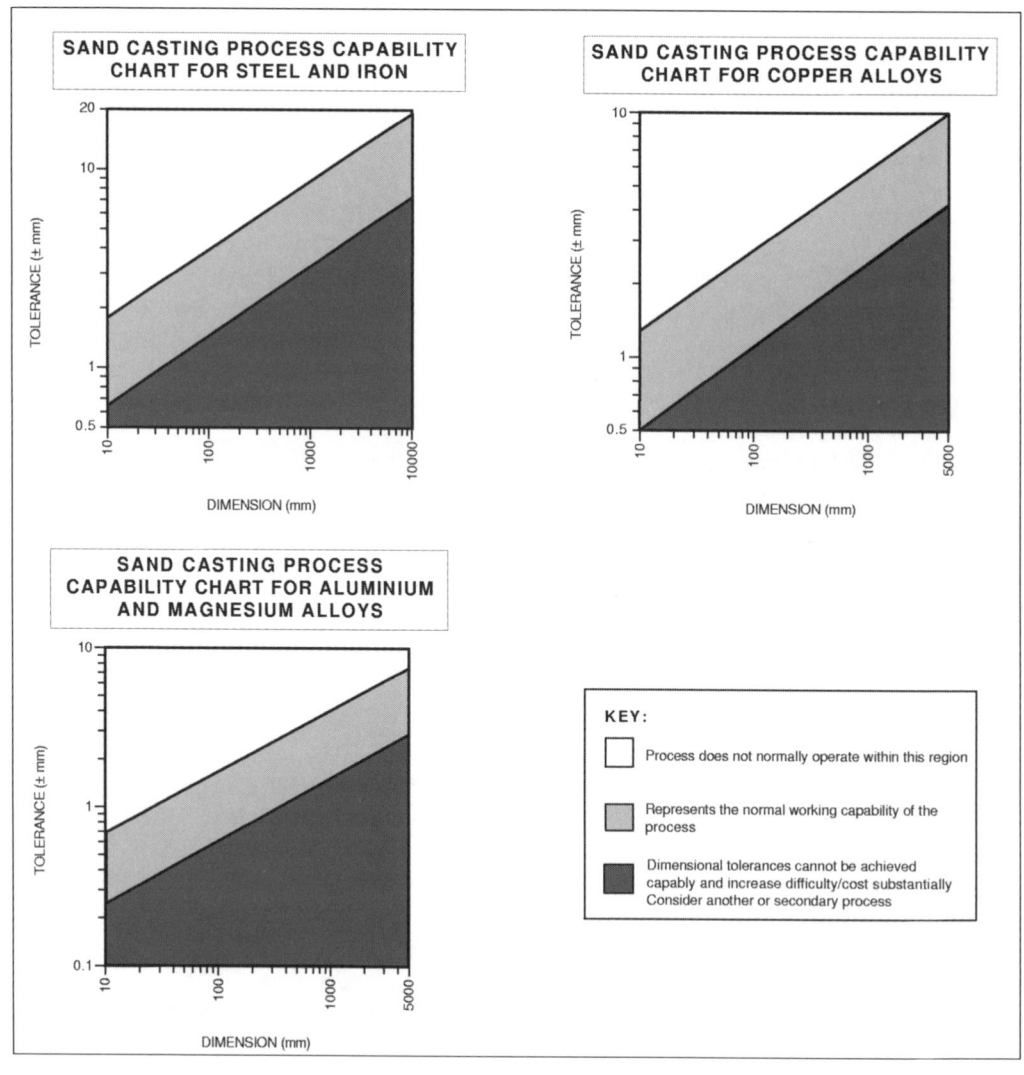

1.1CC Sand casting process capability charts.

1.2 Shell molding

Process description

- A heated metal pattern is placed over a box of thermosetting resin-coated sand. The box is inverted for a fixed time to cure the sand. The box is re-inverted, and the excess sand falls out. The shell is then removed from the pattern and joined with the other half (previously made). They are supported in a flask by an inert material ready for casting (see 1.2F).

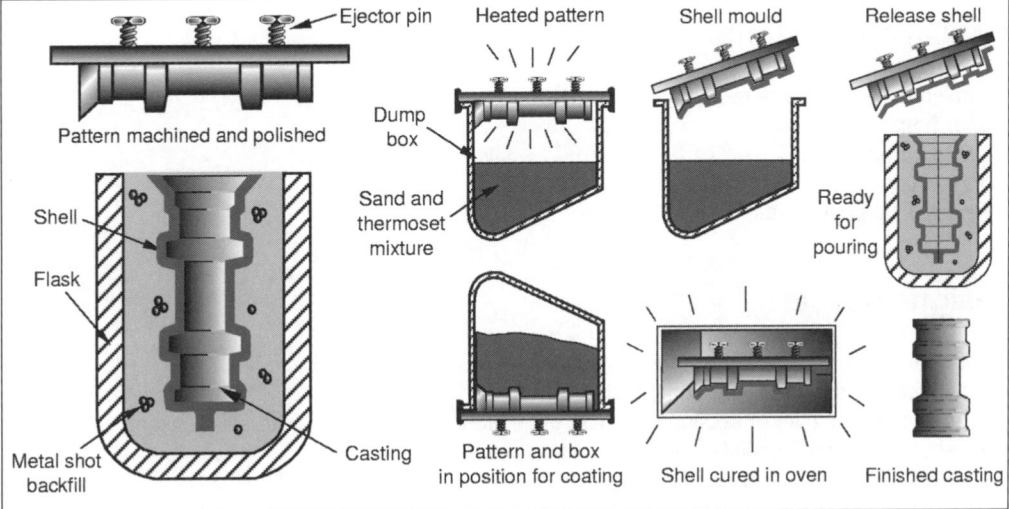

1.2F Shell molding process.

Materials

- Most metals, except: lead, zinc, magnesium and titanium alloys, also beryllium, refractory and zirconia alloys.

Process variations

- Molds produced from other casting processes may be joined with shell molds.
- Patterns are generally made of iron or steel giving good dimensional accuracy.
- Aluminum patterns may be used for low-volume production.
- Other pattern materials used are plaster, and graphite for reactive materials.

Economic considerations

- Production rates of 5–200/h, but dependent on size.
- Lead time several days to weeks depending on complexity and size.
- Material utilization high; little scrap generated.

- Potential for automation high.
- With use of gating systems several castings in a single mold possible.
- Resin binders cost more, but only 5 per cent as much sand used as compared to sand casting.
- Difficult to change design during production.
- More suited to moderate to high volume production, but production volumes of 100–500 may be economical.
- Considered the best of low cost casting methods for large quantities.
- Tooling costs low to moderate.
- Equipment costs moderate to high.
- Labor costs low to moderate.
- Low finishing costs. Often no finishing required.

Typical applications

- Small mechanical parts requiring high precision
- Gear housings
- Cylinder heads
- Connecting rods
- Transmission components

Design aspects

- Good for molding complex shapes, especially when using composite molds.
- Great variations in cross section possible.
- Sharper corners, thinner sections, smaller projections than possible with sand casting.
- Bosses and inserts possible.
- Undercuts difficult.
- Placing of parting line important, i.e. avoid placement across critical dimensions.
- Cored holes greater than $\varnothing 3\,mm$.
- Draft angle ranging 0.25–1°, depending on section depth.
- Maximum section $= 50\,mm$.
- Minimum section $= 1.5\,mm$.
- Sizes ranging 10 g–100 kg in weight. Better for small parts less than 20 kg.

Quality issues

- Blowing sand onto pattern makes depositing more uniform, especially good for intricate forms.
- Few castings scrapped due to blowholes or pockets. Gases are able to escape through thin shells or venting.
- Composite cores may include chills and cores to control solidification rate in critical areas.
- Moderate porosity and inclusions.
- Mechanical properties better than sand casting.
- Uniform grain structure.
- Surface detail good.
- Surface roughness ranging 0.8–12.5 μm Ra.
- Process capability charts showing the achievable dimensional tolerances using various materials provided (see 1.2CC). Allowances of $\pm0.25-\pm0.5\,mm$ should be added for dimensions across the parting line.

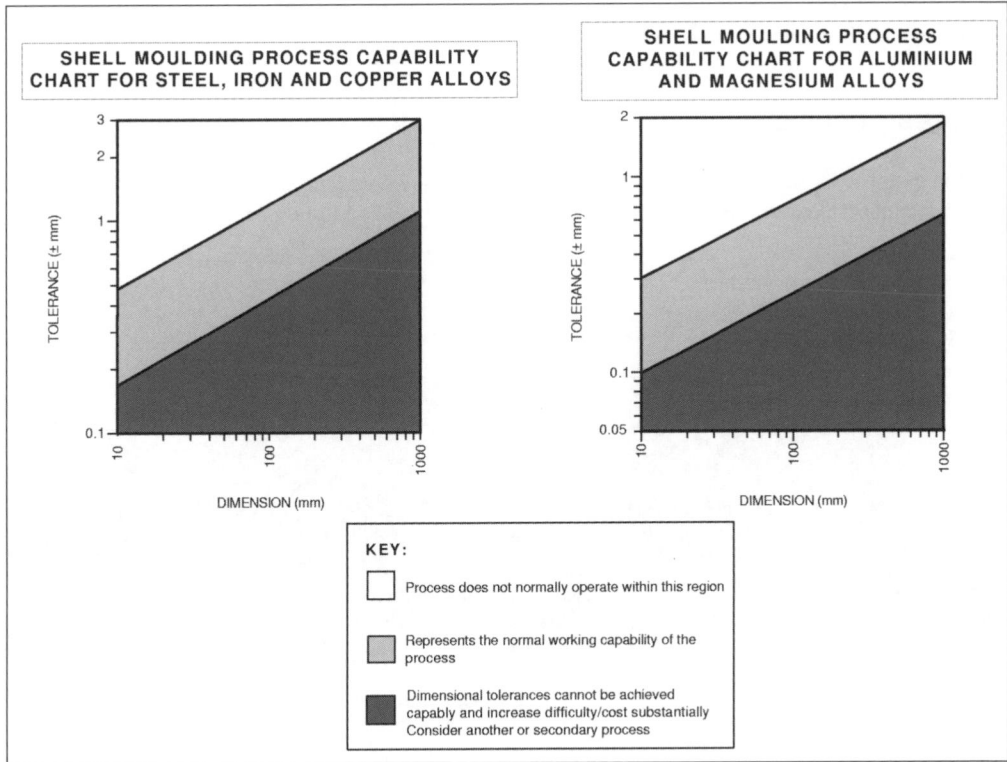

1.2CC Shell molding process capability chart.

1.3 Gravity die casting

Process description

- Molten metal is poured under gravity into a pre-heated die, where it solidifies. The die is then opened and the casting ejected. Also known as permanent mold casting (see 1.3F).

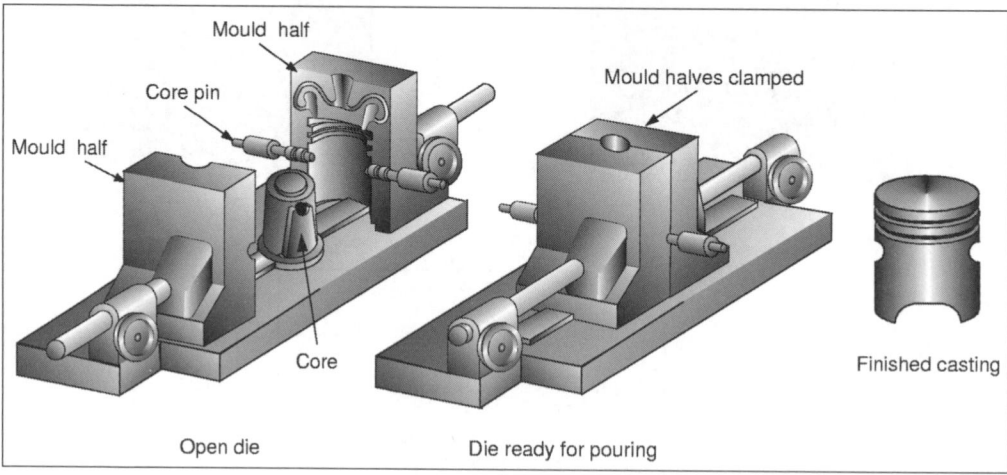

1.3F Gravity die casting process.

Materials

- Usually non-ferrous metals, for example: copper, aluminum, magnesium, but sometimes iron, lead, nickel, tin and zinc alloys. Carbon steel can be cast with graphite dies.

Process variations

- Dies typically cast iron, graphite or refractory material.
- Metal or sand cores can be used although surface finish can be poor.
- Low pressure die casting: uses low-pressure (1 bar) air to force the molten metal into the die cavity. Less popular than gravity die casting, and tends to be used purely for the production of car wheels. Gives lower production rates.
- Slush casting: for creating hollow parts without cores in low melting point metals such as lead, zinc and copper alloys.

Economic considerations

- Production rates of 5–50/h, but dependent on size.
- Lead times can be many weeks.
- Material utilization moderate to high (10–40 per cent lost in scrap, but can be recycled).
- If accuracy and surface finish not an issue, can use sand cores instead of metallic or graphite for greater economy.

- Production volumes of 500–1000 may be viable, but suited to higher volume production.
- Tooling costs moderate.
- Equipment costs moderate.
- Labor costs low to moderate.
- Finishing costs low to moderate. Gates need to be removed.

Typical applications

- Cylinder heads
- Engine connecting rods
- Pistons
- Gear and die blanks
- Kitchen utensils
- Gear blanks
- Gear housings
- Pipe fittings
- Wheels

Design aspects

- Shape complexity limited by that obtained in die halves.
- Undercuts are possible with large added cost.
- Inserts possible with small added cost.
- Machining allowances usually in the range 0.8–1.5 mm.
- Vertical parting lines commonly used.
- Placing of parting line important, i.e. avoid placement across critical dimensions.
- Cored holes greater than $\varnothing 5$ mm.
- Draft angle ranging 2–3°.
- Maximum section $= 50$ mm.
- Minimum section $= 2$ mm.
- Sizes ranging from 50 g to 300 kg in weight. Commonly used for castings less than 5 kg.

Quality issues

- Little porosity and inclusions: can be minimized by slow die filling to reduce turbulence.
- Redressing of the dies may be required after several thousand castings.
- Collapsible cores improve extraction difficulties on cooling.
- 'Chilling' effect of cold metallic dies on the surface of the solidifying metals needs to be controlled by pre-heating at correct temperature.
- Large castings sometimes require that the die is tilted as molten metal is being poured in to reduce turbulence.
- Mechanical properties fair to good.
- Surface detail is good.
- Surface roughness ranging 0.8–6.3 µm Ra.
- Process capability charts showing the achievable dimensional tolerances using various materials provided (see 1.3CC). Allowances of ± 0.25–± 0.75 mm should be added for dimensions across the parting line.

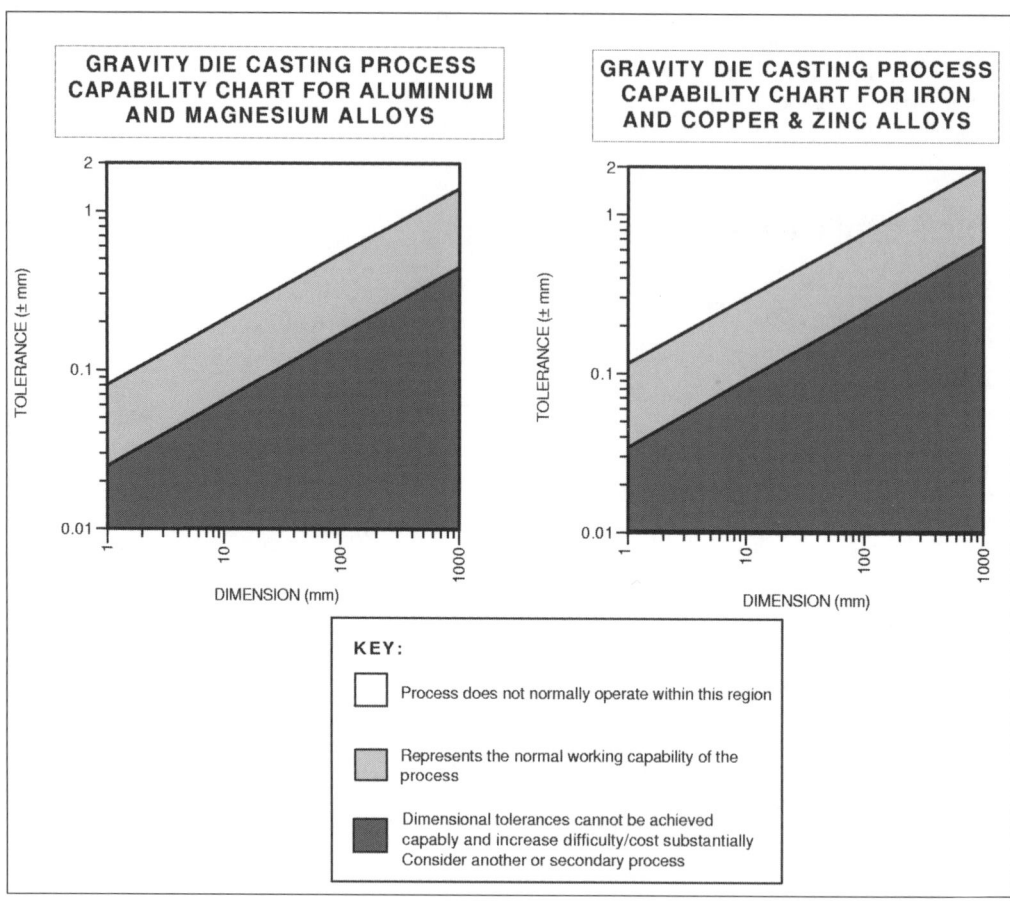

1.3CC Gravity die casting process capability chart.

1.4 Pressure die casting

Process description

- Molten metal is inserted into a metallic mold under very high pressures (100+ bar), where it solidifies. The die is then opened and the casting ejected (see 1.4F).

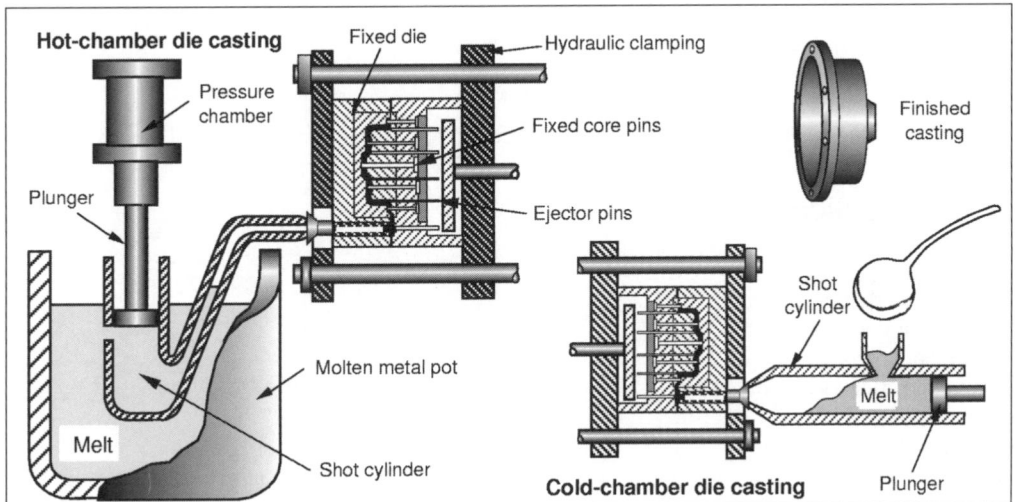

1.4F Pressure die process.

Materials

- Limited to non-ferrous metals, i.e. zinc, aluminum, magnesium, lead, tin and copper alloys.
- Zinc and aluminum alloys tend to be the most popular materials.
- High-temperature metals, such as copper alloys, reduce die life.
- Iron-based materials for casting are under development.

Process variations

- Dies and cores are made from hardened and tempered alloy steel.
- Cold-chamber die casting: shot cylinder filled with a ladle for each cycle. Used for high melting temperature metals.
- Hot-chamber die casting: shot cylinder immersed in molten metal and then forced using a separate ram. Used for low melting temperature metals due to erosive nature of molten metal. Can be either plunger or goose-neck type.
- Vacuum die casting: overcomes porosity for larger castings.
- Injection metal assembly: variant of hot-chamber die casting for the assembly of parts such as tubes and plates, cable terminations and to act as rivets by injecting zinc or lead alloys into a cavity in the assembly.

Economic considerations

- Very high production rates possible, up to 200/h.
- Lead time very long, months possibly.
- Material utilization high.
- Gates, sprues, etc., can be re-melted.
- High initial die costs due to high complexity and difficulty to manufacture.
- Full automation achievable. Robot machine loading and unloading common.
- Production quantities of up to 10 000 economically viable for copper alloys, but 100 000+ for aluminum, zinc and lead alloys.
- Tooling costs very high.
- Equipment costs very high.
- Direct labor costs low.
- Finishing costs low. Trimming operations are required to remove flash, gates and sprues.

Typical applications

- Transmission cases
- Machine and engine parts
- Pump components
- Electrical boxes
- Domestic appliance components
- Toy bodies
- Pump and impeller parts

Design aspects

- Shape complexity can be high. Limited by design of movable cores.
- Bosses, large threads, undercuts and inserts all possible with added cost.
- Molded-in bearing shells possible.
- Lettering possible.
- Wall thickness should be as uniform as possible; transitions should be gradual.
- Sharp corners should be avoided, but pressure die casting permits smaller radii, because metal flow is aided.
- Placing of parting line important, i.e. avoid placement across critical dimensions.
- Holes perpendicular to the parting line can be cast.
- Casting holes for subsequent tapping generally more economical than drilling.
- Cored holes greater than $\varnothing 0.8$ mm.
- Machining allowance normally in the range 0.25–0.8 mm.
- Draft angle ranging 0.25–3°, depending on section depth.
- Maximum section = 13 mm.
- Minimum section ranging 0.4 mm for zinc alloys, 1.5 mm for copper alloys.
- Sizes ranging 10 g–50 kg. Castings up to 100 kg made in zinc. Copper, tin and lead castings normally less than 5 kg.

Quality issues

- Low porosity in small castings typically, but can be a problem in castings with thick or long sections.
- Particularly suited where casting requires high mechanical properties or absence of creep.

- The high melting temperature of some metals can cause significant processing difficulties and die wear.
- Ejector pins may leave small marks and should be positioned at points of strength on the casting.
- Process variables need to be controlled. Variation in temperature, pressure and cycle time especially important for consistency.
- Difficulty is experienced in obtaining sound castings in the larger capacities due to gas entrapment.
- Close control of temperature, pressure and cooling times important in obtaining consistent quality castings.
- Mechanical properties are fair, but poorer than some other casting methods.
- Surface detail excellent.
- Surface roughness ranging 0.4–3.2 µm Ra.
- Process capability charts showing the achievable dimensional tolerances using various materials are provided (see 1.4CC). Allowances of ±0.05–±0.35 mm should be added for dimensions across the parting line.

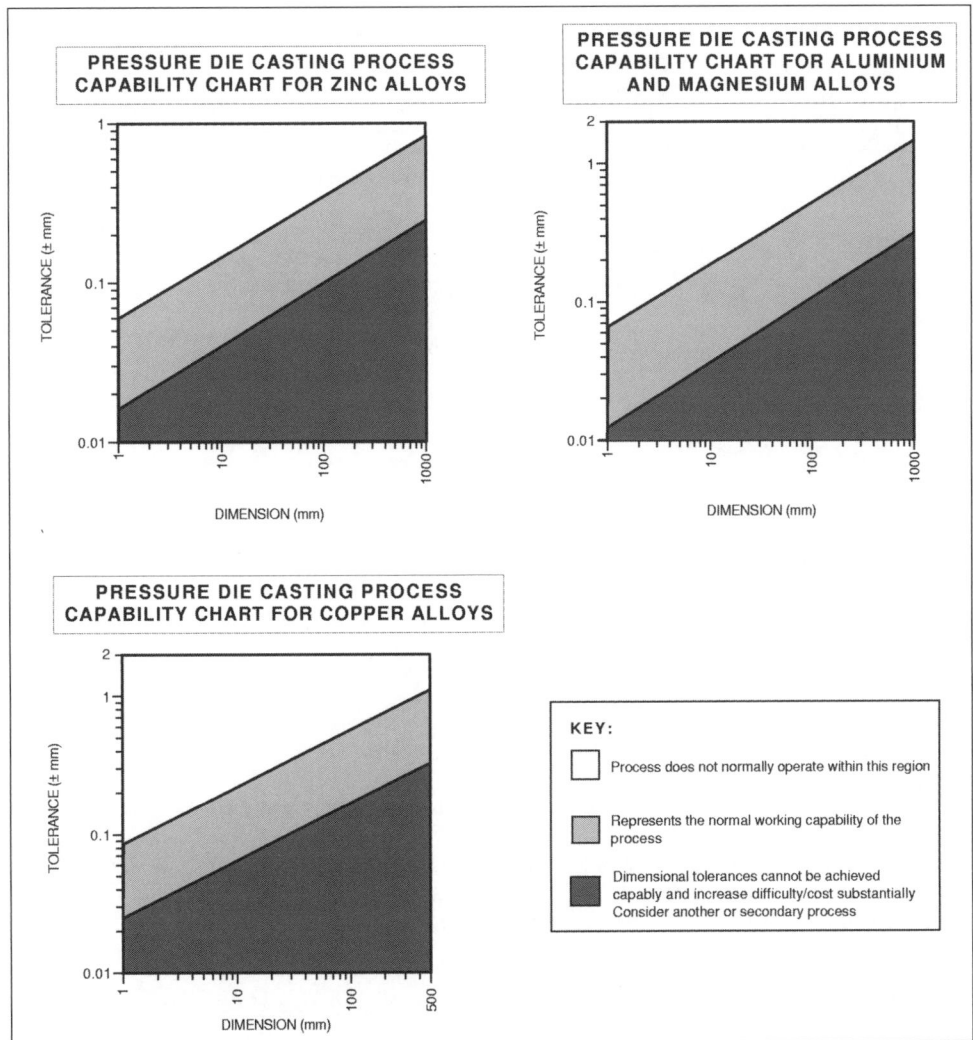

1.4CC Pressure die casting process capability chart.

1.5 Centrifugal casting

Process description

- Molten metal is poured into a high-speed rotating mold (300–3000 rpm depending on diameter) until solidification takes place. The axis of rotation is usually horizontal, but may be vertical for short work pieces (see 1.5F).

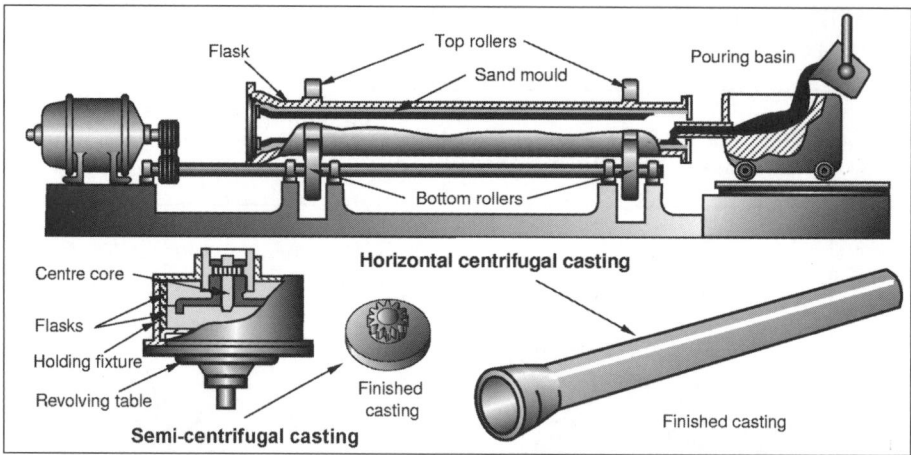

1.5F Centrifugal casting process.

Materials

- Most metals suitable for static casting are suitable for centrifugal casting: all steels, iron, copper, aluminum and nickel alloys.
- Also, glass, thermoplastics, composites and ceramics (metal molds sprayed with a refractory material) can be molded by this method.

Process variations

- Semi-permanent or expendable molds.
- Semi-centrifugal casting: used to cast parts with radial symmetry in a vertical axis of rotation at low speeds.
- Centrifuge casting: a number of molds are arranged radially around a central sprue. Molten metal is poured into the sprue and is forced into the mold cavities by centrifugal force due to high-speed rotation. Used for small gears mainly and parts of intricate detail.

Economic considerations

- Production rates of up to 50/h possible, but dependent on size.
- Lead time may be several weeks.

- Material utilization high (90–100 per cent). No runners or risers.
- Economic when the mechanical properties of thick-walled tubes are important and high alloy grades of steel are required.
- In large quantities, production of other than circular external shapes becomes more economical.
- Small diameter steel tubes made by this method not competitive with welded or rolled tubes.
- Selection of mold type (permanent or sand) determined by shape of casting, quality and number to be produced.
- Production volumes low, typically 100+. Can be used for one-offs.
- Tooling costs moderate.
- Equipment costs low to moderate.
- Direct labor costs low to moderate.
- Finishing costs low to moderate. Normally, machining of internal dimension necessary.

Typical applications

- Pipes
- Brake drums
- Pulley wheels
- Train wheels
- Flywheels
- Gun barrels
- Gear blanks
- Large bearing liners
- Engine-cylinder liners
- Pressure vessels
- Nozzles

Design aspects

- Shape complexity limited by nature of process, i.e. suited to parts with rotational symmetry.
- Contoured surfaces possible.
- Circular bore remains in the finished part.
- Dual metal tubes that combine the properties of two metals in one application possible.
- Inserts and bosses possible, but undercuts are not.
- Placing of parting line important, i.e. avoid placement across critical dimensions.
- Cored holes greater than $\varnothing 25\,mm$.
- Machining allowances ranging 0.75–6 mm.
- Draft angle approximately 1°.
- Maximum section thickness approximately 125 mm.
- Minimum section ranging 2.5–8 mm, depending on material cast.
- Maximum length = 15 m.
- Sizes ranging $\varnothing 25\,mm$–$\varnothing 2\,m$.
- Sizes up to 5 t in weight have been cast.

Quality issues

- Properties of castings vary by distance from the axis of rotation.
- Due to density differences in the molten material, dross, impurities and pieces of the refractory lining tend to collect on the inner surface of the casting. This is usually machined away.

- Tubular castings have higher structural strengths and more distinct cast impressions than gravity die cast or sand cast parts.
- Castings are free of shrinkage due to one-directional cooling.
- The mechanical properties of dense castings are comparable with that of forgings. Fine grain castings and low porosity is an advantage.
- Good mechanical properties and fine grain structure.
- Surface detail is fair to good.
- Surface roughness ranging 1.6–12.5 μm Ra.
- A process capability chart showing the achievable dimensional tolerances is provided (see 1.5CC). Allowances of approximately ±0.25–±0.75 mm should be added for dimensions across the parting line. Note, the chart applies to outside dimensions only. Internal dimensions are approximately 50 per cent greater.

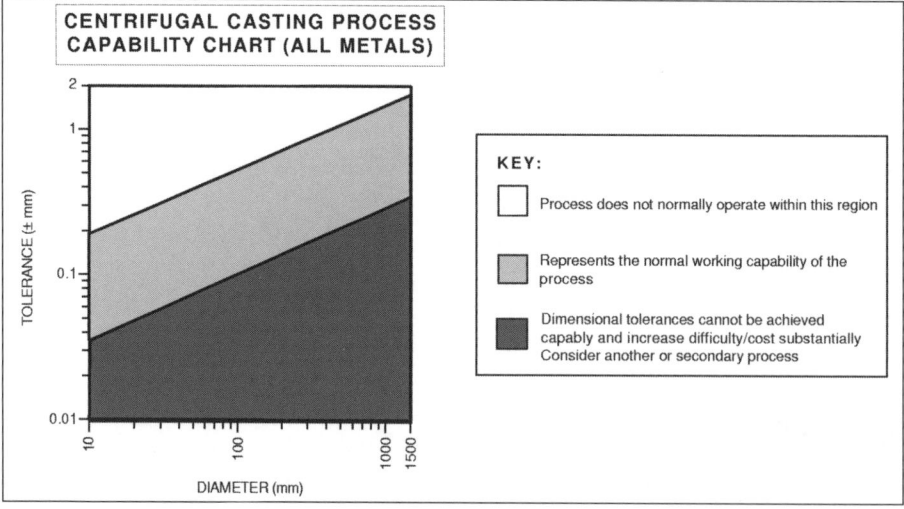

1.5CC Centrifugal casting process capability chart.

1.6 Investment casting

Process description

- A mold is used to generate a wax pattern of the shape required. A refractory material zircon, then a ceramic slurry and finally a binder is used to coat the pattern which is slow fired in an oven to cure. The wax is melted out and the metal cast in the ceramic mold. The mold is then destroyed to remove the casting. Process often known as the 'Lost Wax' process (see 1.6F).

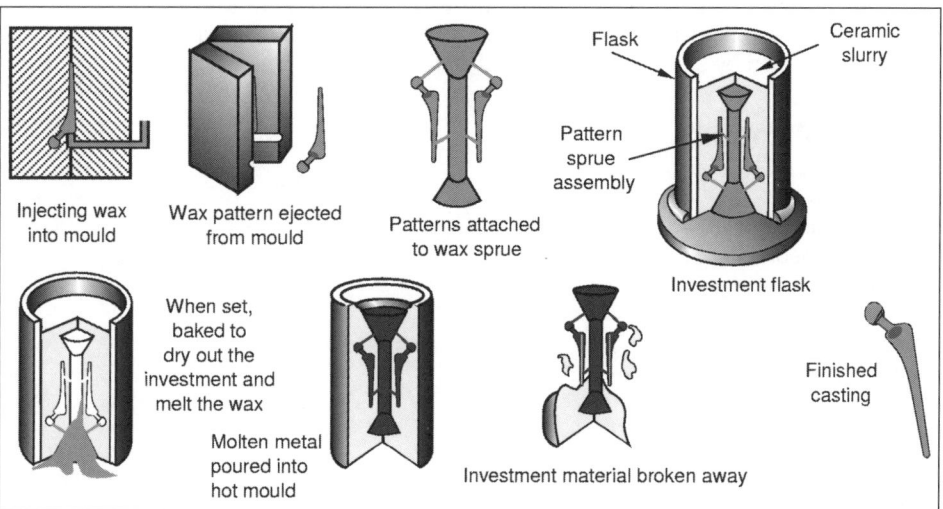

1.6F Investment casting process.

Materials

- All metals, including precious, refractory and reactive alloys (cast in vacuum).

Process variations

- Blends of resin, filler and wax used.
- Use of thermoplastic resin instead of wax.
- Ceramic and water soluble cores can be used.

Economic considerations

- Production rates of up to 1000/h, depending on size.
- Lead times usually several weeks, but can be shorter.
- Slow process due to many steps in production. Cure time can be as high as 48 h.
- Wax or plastic patterns can be injection molded for high production volumes.
- Best suited to metals having high melting temperatures, and/or which difficult to machine or which have high cost.

- Material utilization high.
- Some automation possible.
- Pattern costs can be high for low quantities.
- Ceramic and wax cores allow complex internal configurations to be produced, but increase the cost significantly.
- A 'tree' of wax patterns enables many small castings to be handled together.
- Most suitable for small batches (10–1000) using manual labor, but also high-volume production with automation.
- Sometimes used for one-offs, especially in decorative work.
- Tooling costs low to moderate, but dependent on complexity.
- Equipment costs low to moderate (high when processing reactive materials).
- Labor costs very high. Can be labor intensive as many operations involved.
- Finishing costs low to moderate. Gates and feeders are removed by machining or grinding. As cast part typically cleaned by shot, bead or sand blasting.

Typical applications

- Turbine blades
- Machine tool parts
- Aerospace components
- Valve and pump casings
- Pipe fittings
- Automotive engine components
- Decorative work, e.g. figurines
- Optical instrument parts
- Small arms parts
- Gear blanks
- Levers
- Jewelry

Design aspects

- Very complex castings with unusual internal configurations possible.
- Wax pattern must be easily removable from its mold.
- Complex shapes assembled from several simpler shapes.
- Practical way of producing threads in hard to machine materials, or where thread design unusual.
- Uniform sections preferred. Abrupt changes should be gradually blended in or designed out.
- Avoid sharp corners.
- Fillets should be as generous as possible.
- Bosses and undercuts possible with added cost.
- Inserts not possible, but integral rivets are.
- Lettering possible, either in relief or inset.
- Molded-in holes, both blind and through possible, but difficult.
- Length to diameter ratio for blind holes typically 4:1.
- Minimum hole = $\varnothing 0.5$ mm.
- Machining allowance usually between 0.25 mm and 0.75 mm, depending on size.
- Draft angle usually zero, but 0.5–1° desirable on long extended surfaces, or if mold cavity is deep.
- Minimum section ranging from 1 mm for aluminum alloys and steels, 2 mm for copper alloys, but can be as low as 0.6 mm for some applications.
- Maximum section = 75 mm.

- Maximum dimension = 1 m.
- Sizes ranging 0.5 g–100 kg in weight, but best for parts less than 5 kg.

Quality issues

- Moderate porosity.
- High strength castings can be produced.
- Grain growth more pronounced in longer sections which may limit the toughness and fatigue life of the part.
- Quality of casting to a large degree dependent upon the characteristics of wax.
- Very good to excellent surface detail possible.
- Surface roughness ranging 0.4–6.3 μm Ra.
- Flatness tolerances typically ±0.13 mm per 25 mm, but dependent on surface area.
- Minimum angular tolerance = ±0.5°.
- A process capability chart showing the achievable dimensional tolerances is provided (see 1.6CC).
- No parting line on casting.

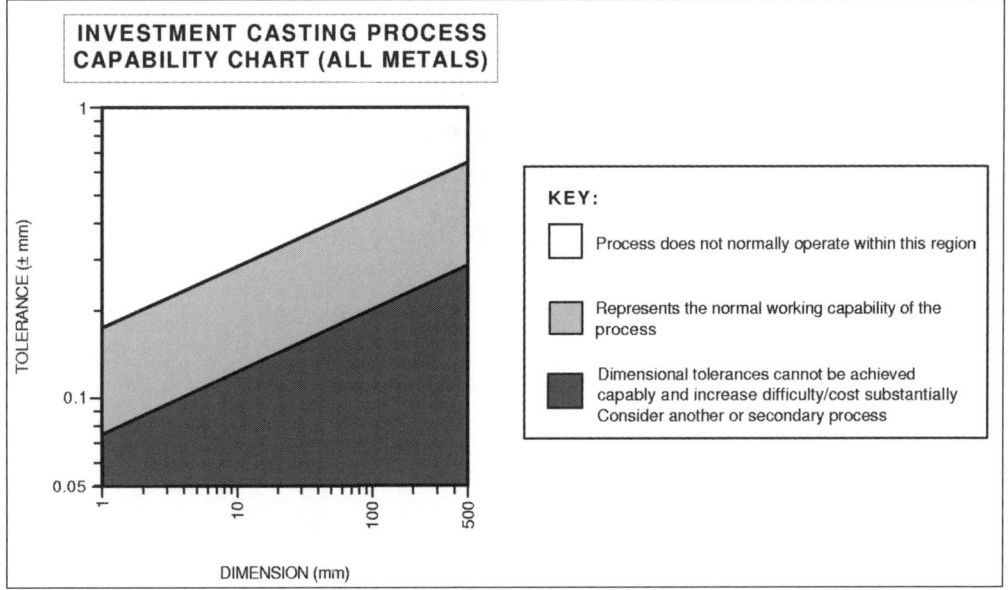

1.6CC Investment casting process capability chart.

1.7 Ceramic mold casting

Process description

- A precision pattern generates the mold which is coated with a ceramic slurry. The mold is dried and baked. The molten metal is then poured into the mold and allowed to solidify. The mold is broken to remove the part (see 1.7F).

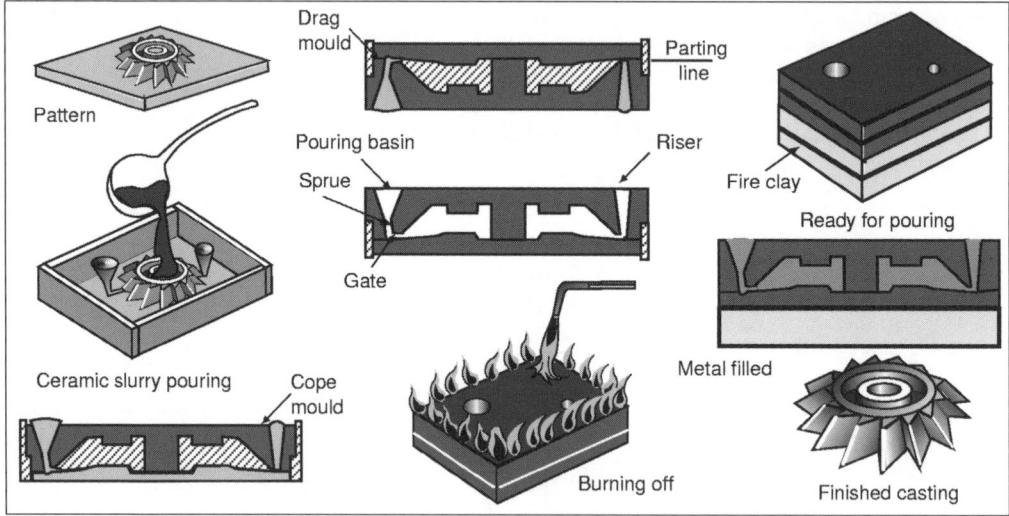

1.7F Ceramic mold casting process.

Materials

- All metals, but to a lesser degree, aluminum, magnesium, zinc, tin and copper alloys.

Process variations

- Variations on the composition of the ceramic slurry and curing mechanism.
- Plaster, wood, metal or rubber are used for patterns.

Economic considerations

- Production rates of up to 10/h typical.
- Lead times can be several days.
- Material utilization high.
- Low scrap losses.
- Best suited to metals having high melting temperatures and/or that are difficult to machine.
- Can be combined with investment casting to produce parts with increased complexity with reduced cost.

- Suitable for small batches and medium-volume production.
- Can be used for one-offs.
- Tooling costs moderate.
- Equipment costs moderate to high.
- Direct labor costs moderate to high.
- Finishing costs low. Usually no machining is required.

Typical applications

- All types of dies and molds for other casting and forming processes
- Cutting tool blanks
- Components for food handling machines
- Pump impellers
- Aerospace and atomic reactor components

Design aspects

- High complexity possible – almost any shape possible.
- Use of cores increases complexity obtainable.
- Inserts, bosses and undercuts possible.
- Placing of parting line important, i.e. avoid placement across critical dimensions.
- Cored holes greater than $\varnothing$0.5 mm.
- Where machining required, allowances of up to 0.6 mm should be observed.
- Draft angle usually zero, but 0.1–1° preferred.
- Minimum section ranging 0.6–1.2 mm, depending on material used.
- Sizes ranging 100 g–3 t in weight, but less than 50 kg better.

Quality issues

- Low porosity.
- Mechanical properties are good.
- Good surface detail possible.
- Surface roughness ranging 0.8–6.3 µm Ra.
- A process capability chart showing the achievable dimensional tolerances is provided (see 1.7CC). An allowance of ±0.25 mm should be added for dimensions across the parting line.
- Parting lines sometimes pronounced on finished casting.

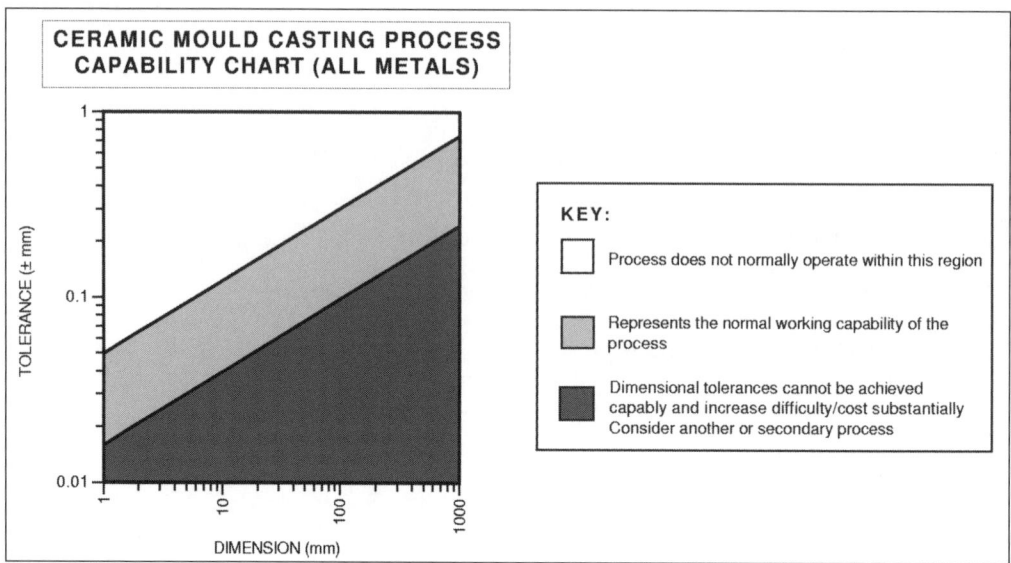

1.7CC Ceramic mold casting process capability chart.

1.8   Plaster mold casting

Process description

- A precision metal pattern (usually brass) generates the two part mold which is made of a gypsum slurry material. The mold is removed from the pattern and baked to remove the moisture. The molten metal is poured into the mold and allowed to cool. The mold is broken to remove the part (see 1.8F).

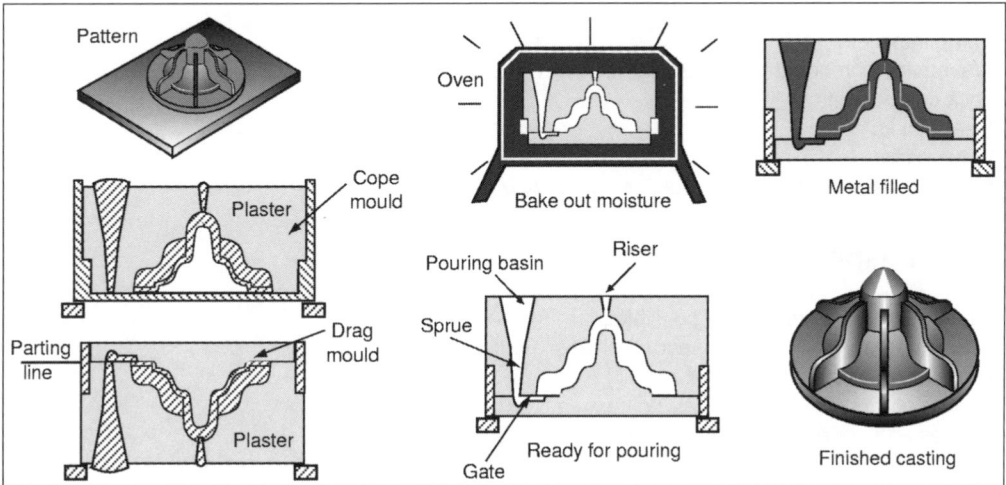

1.8F Plaster mold casting process.

Materials

- Limited to low melting temperature metals, i.e. aluminum, copper, zinc and magnesium alloys due to degradation of the plaster mold at elevated temperatures.
- Tin and lead alloys are sometimes processed.

Process variations

- Patterns can be made from: metal, plaster, wood or thermosetting plastic. Wood has a limited life due to water absorption from the plaster slurry.
- Composition of plaster slurry varies. Additives are sometimes used to control mold expansion and fibers added to improve mold strength.

Economic considerations

- Production rates of up to 10/h typical.
- Lead times can be several days to weeks.
- Material utilization high.
- Low scrap losses. Waste recycled.

- Mold destroyed in removing casting.
- Easy to change design during production.
- Suitable for small batches of 100 and medium-volume production.
- Tooling costs low to moderate.
- Equipment costs moderate.
- Direct labor costs moderate to high. Some skilled operations necessary.
- Finishing costs low. Little finishing required except grinding for gate removal and sanding of parting line.

Typical applications

- Pump impellers
- Waveguide components (for use in microwave applications)
- Lock components
- Gear blanks
- Valve parts
- Molds for plastic and rubber processing, i.e. tyre molds

Design aspects

- Moderate to high complexity possible.
- Possible to make mold from several pieces.
- Deep holes not recommended.
- Sharp corners and features can be cast easily.
- Bosses and undercuts possible with little added cost.
- Placing of parting line important, i.e. avoid placement across critical dimensions.
- Cored holes greater than $\varnothing 13\,\text{mm}$.
- Where machining required, allowances of up to 0.8 mm should be observed.
- Draft angles ranging 0.5–2° preferred, but can be zero.
- Minimum section ranging 0.8–1.8 mm, depending on material used.
- Sizes ranging 25 g–50 kg in weight. However, castings up to 100 kg have been made.

Quality issues

- Little or no distortion on thin sections.
- Plaster mold has low permeability and can create gas evolution problems.
- Moderate to high porosity obtained.
- Mechanical properties fair.
- Surface detail good.
- Surface roughness ranging 0.8–3.2 µm Ra.
- A process capability chart showing the achievable dimensional tolerances is provided (see 1.8CC). An allowance of approximately ±0.25 mm should be added for dimensions across the parting line.

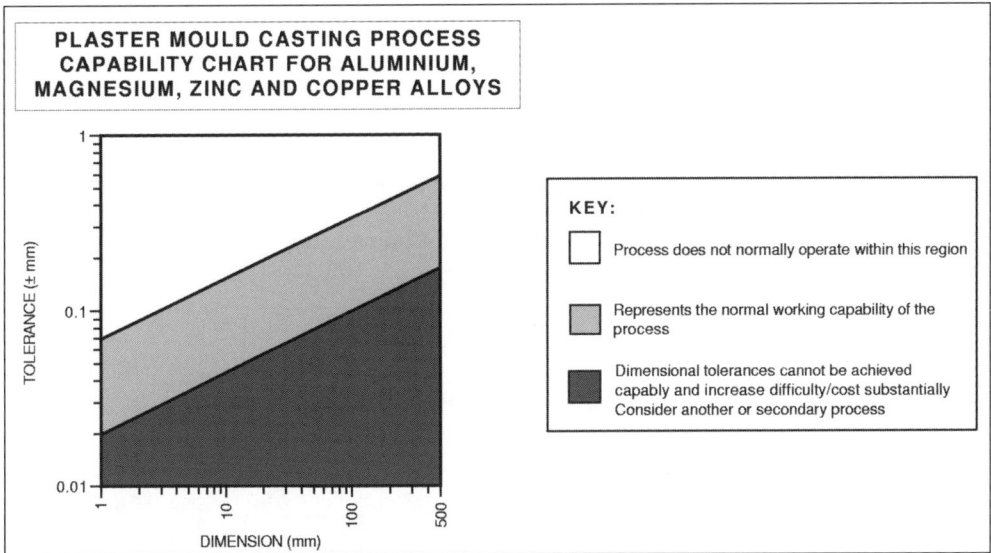

1.8CC Plaster mold casting process capability chart.

1.9 Squeeze casting

Process description

- Combination of casting and forging. Molten metal fills a preheated mold from the bottom and during solidification the top half of the mold applies a high pressure to compress the material into the final desired shape. Also known as liquid metal forging and load pressure casting (see 1.9F).

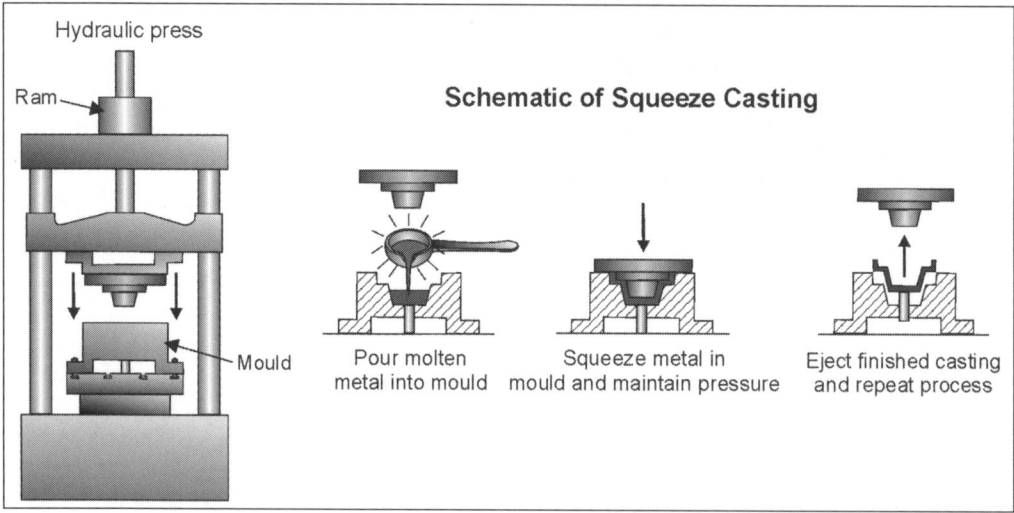

1.9F Squeeze casting process.

Materials

- Typically non-ferrous metals, but occasionally ferrous alloys.

Process variations

- Pouring can be performed automatically.

Economic considerations

- Production rates low due to mold filling for minimum turbulence.
- Long setup times.
- Lead time moderate to high.
- Material utilization excellent. Near-net shape achieved.
- High degree of automation possible.
- Economically viable for production volumes of 10 000+.
- Tooling costs high.
- Equipment costs high.

- Direct labor costs low to moderate.
- Finishing costs very low. Used to minimize or eliminate secondary processing.

Typical applications

- Aerospace components
- Suspension parts
- Steering elements
- Brake components

Design aspects

- Complex geometries possible.
- Retractable and disposable cores used to create complex internal features.
- Large variations in cross section possible.
- Undercuts, bosses, holes and inserts possible.
- Placing of parting line important, i.e. avoid placement across critical dimensions.
- Machining allowances usually in the range 0.6–1.2 mm.
- Draft angle ranging 0.1–3°, depending on section depth.
- Maximum section = 200 mm.
- Minimum section = 6 mm.
- Minimum dimension = $\varnothing$20 mm.
- Sizes ranging 25 g–4.5 kg in weight.

Quality issues

- Low porosity experienced.
- Low speed mold filling minimizes splashing.
- Accurate metering of molten metal required to flashing.
- Excellent mechanical properties can be obtained, similar to forging.
- Adequate process control important, i.e. metering of molten metal, pressures, solidification times, etc.
- Graphite releasing agent and ejector pins commonly used to aid removal of finished part.
- Surface detail good.
- Surface roughness ranging 1.6–12.5 µm Ra.
- Achievable dimensional tolerances approximately ±0.15 at 25 mm, ±0.3 at 150 mm. Allowances of ±0.25 mm should be added for dimensions across the parting line.

2 Plastic and composite processing

2.1 Injection molding

Process description

- Granules of polymer material are heated and then forced under pressure using a screw into the die cavity. On cooling, a rigid part or tree of parts is produced (see 2.1F).

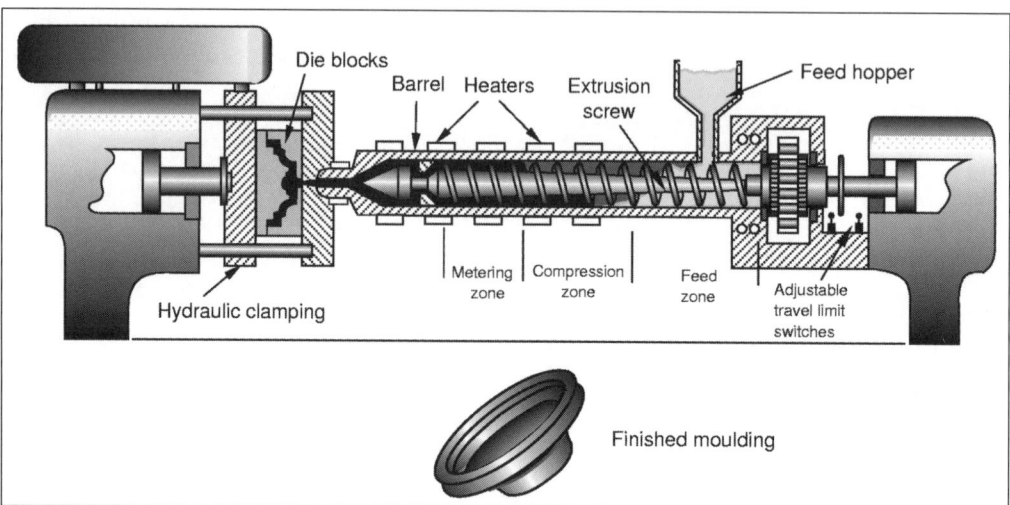

2.1F Injection molding process.

Materials

- Mostly thermoplastics, but thermosets, composites and elastomers can be processed.

Process variations

- Injection blow molding: allows small hollow parts with intricate neck detail to be produced.
- Co-injection: for products with rigid cores pre-placed in the die before injection or simultaneous injection of different materials into same die.

Economic considerations

- Production rates are high, 1–50/min, depending on size.
- Thermoset parts usually have a longer cycle time.
- Lead times can be several weeks due to manufacturing of complex dies.
- Material utilization is good. Scrap generated in sprues and risers.
- If material permits, gates and runners can be reused resulting in lower material losses.
- Flexibility limited by dedicated dies, die changeover and machine setup times.
- Economical for high production runs, typically 10 000+.
- Full automation achievable. Robot machine loading and unloading common.

- Tooling costs are very high. Dies are usually made from hardened tool steel.
- Equipment costs are very high.
- Direct labor costs are low to moderate.
- Finishing costs are low. Trimming is required to remove gates and runners.

Typical applications

- High precision, complex components
- Automotive and aerospace components
- Electrical parts
- Fittings
- Containers
- Cups
- Bottle tops
- Housings
- Tool handles

Design aspects

- Very complex shapes and intricate detail possible.
- Holes, inserts, threads, lettering, color, bosses and minor undercuts possible.
- Uniform section thickness should be maintained.
- Unsuitable for the production of narrow necked containers.
- Variation in thickness should not exceed 2:1.
- Marked section changes should be tapered sufficiently.
- Living hinges and snap features allow part consolidation.
- Placing of parting line important, i.e. avoid placement across critical dimensions.
- The clamping force required proportional to the projected area of the molded part.
- Radii should be as generous as possible. Minimum inside radii $= 1.5$ mm.
- Draft angle ranging from less than 0.25 to $4°$, depending on section depth.
- Maximum section, typically $= 13$ mm.
- Minimum section $= 0.4$ mm for thermoplastics, 0.9 mm for thermosets.
- Sizes ranging 10 g–25 kg in weight for thermoplastics, 6 kg maximum for thermosets.

Quality issues

- Thick sections can be problematic.
- Care must be taken in the design of the running and gating system, where multiple cavities used to ensure complete die fill.
- Control of material and mold temperature critical, also injection pressure and speed, condition of resin, dwell and cooling times.
- Adequate clamping force necessary to prevent the mold creating flash.
- Thermoplastic molded parts usually require no de-flashing: thermoset parts often require this operation.
- Excellent surface detail obtainable.
- Surface roughness a function of the die condition. Typically, 0.2–0.8 μm Ra is obtainable.
- Process capability charts showing the achievable dimensional tolerances using various materials are provided (see 2.1CC). Allowances of approximately ± 0.1 mm should be added for dimensions across the parting line. Note, that charts 1, 2 and 3 are to be used for components that have a major dimension, greater than 50 mm, and typically large production volumes. The chart titled 'Light Engineering' is used for components with a major dimension, less than 150 mm, and for small production volumes.

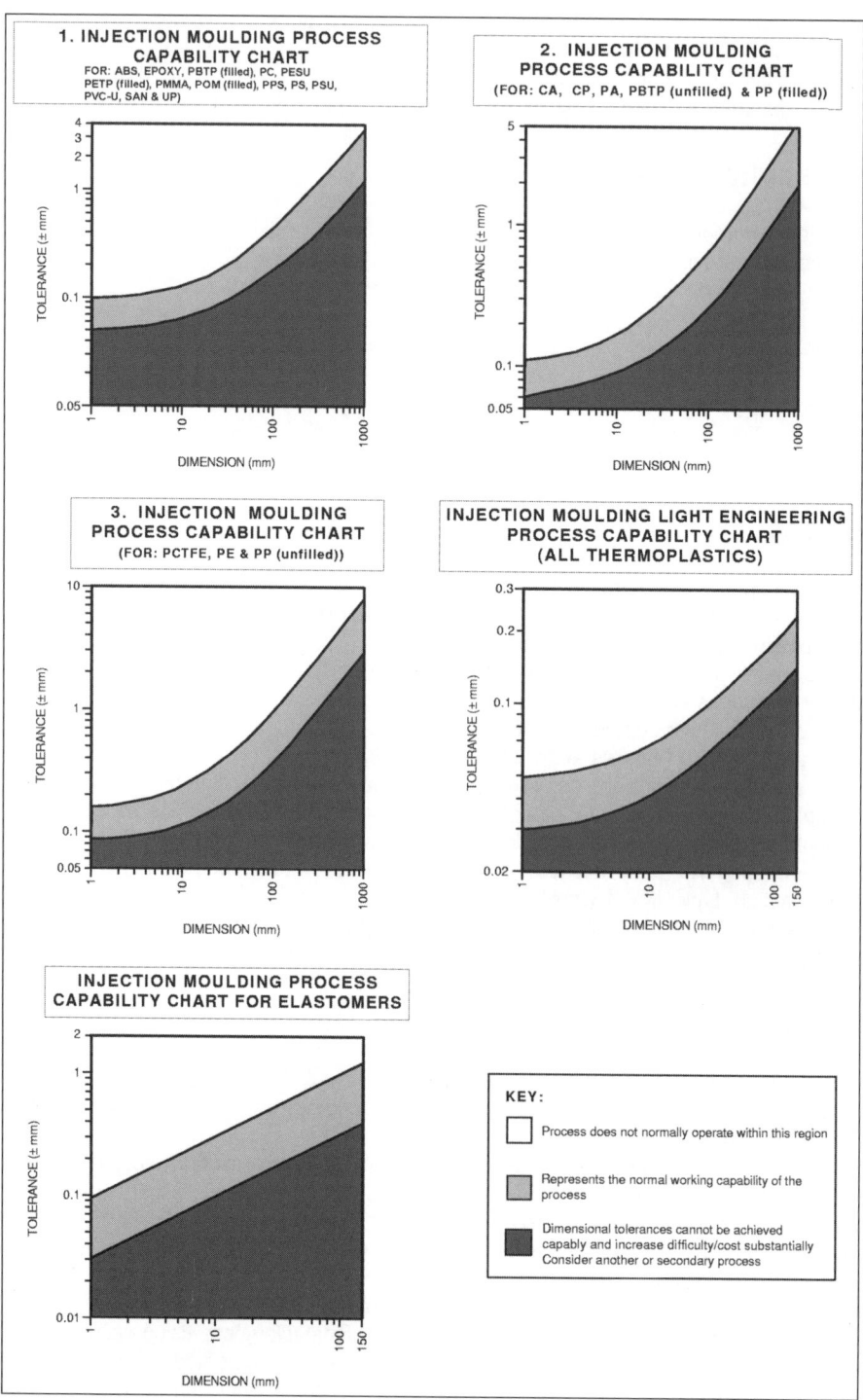

2.1CC Injection molding capability chart.

2.2 Reaction injection molding

Process description

- Two components of a thermosetting resin are injected into a mixing chamber and then injected into the mold at high speed where polymerization and subsequent solidification takes place (see 2.2F).

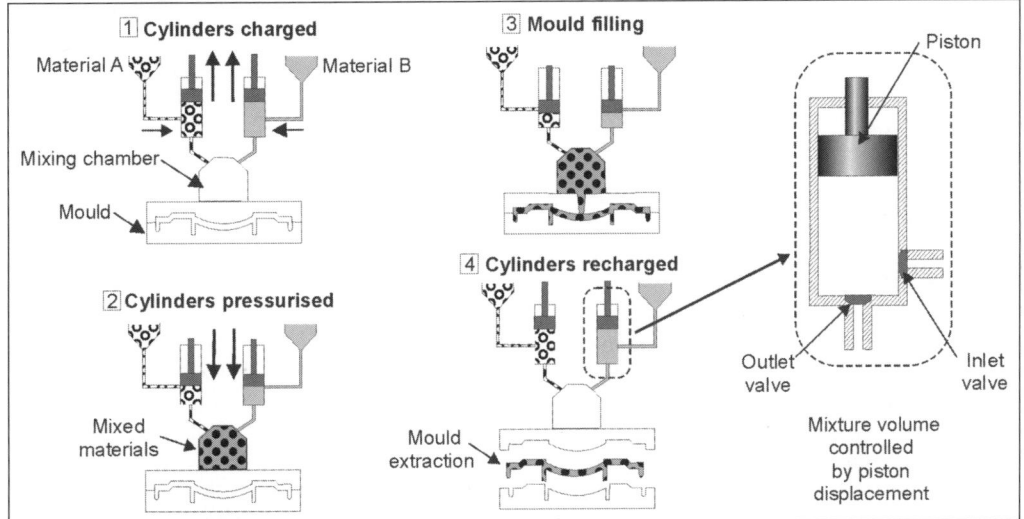

2.2F Reaction injection molding process.

Materials

- Mostly thermosets.
- Foamed materials possessing a solid skin can be created by setting up a pressure differential between mixing chamber and mold.
- Can add chopped fiber material (glass, carbon) for added stiffness to mixing to produce composites.

Process variations

- Mold material is usually aluminum. Can also use resin for low production runs or hardened tool steel for very high volumes.
- Further heating of resin components before mixing is dependent on material used.

Economic considerations

- Production rates from 1 to 10/h.
- Lead times can be several weeks.
- Material utilization good. Less than 1 per cent lost in scrap.
- Scrap cannot be recycled.

- Flexibility limited by dedicated dies, die changeover and machine setup times.
- Economical for low to medium production volumes (10–10 000).
- Can be used for one-offs, e.g. prototyping.
- Tooling costs low.
- Equipment costs high.
- Direct labor costs moderate to high.
- Finishing costs low. A little trimming required.

Typical applications

- Car bumpers
- Cups
- Containers
- Panels
- Housings
- Footwear
- Garden furniture

Design aspects

- Very complex shapes and intricate detail possible.
- Ribs, holes, bosses and inserts possible.
- Small re-entrant features possible.
- Radii should be as generous as possible.
- Uniform section thickness should be maintained.
- Marked section changes should be tapered sufficiently.
- Placing of parting line important, i.e. avoid placement across critical dimensions.
- Draft angles ranging 0.5–3°, depending on section depth.
- Maximum section = 10 mm.
- Minimum section = 1.5 mm; foamed material = 3 mm.
- Maximum dimension = 1.5 m.
- Sizes ranging 100 g–10 kg in weight.

Quality issues

- Thick sections can be problematic.
- Care must be taken in the design of the running and gating system, where multiple cavities are used to ensure complete die fill.
- Problems can be created by premature reaction before complete filling of mold.
- Excellent surface detail is obtainable.
- Surface roughness is variable, but mainly dependent on mold finish.
- Achievable dimensional tolerances are approximately ±0.05 at 25 mm, ±0.3 at 150 mm. Allowances of approximately ±0.2 mm should be added for dimensions across the parting line.

2.3 Compression molding

Process description

- A measured quantity of raw, unpolymerized plastic material is introduced into a heated mold which is subsequently closed under pressure, forcing the material into all areas of the cavity as it melts. Analogous to closed die forging of metals (see 2.3F).

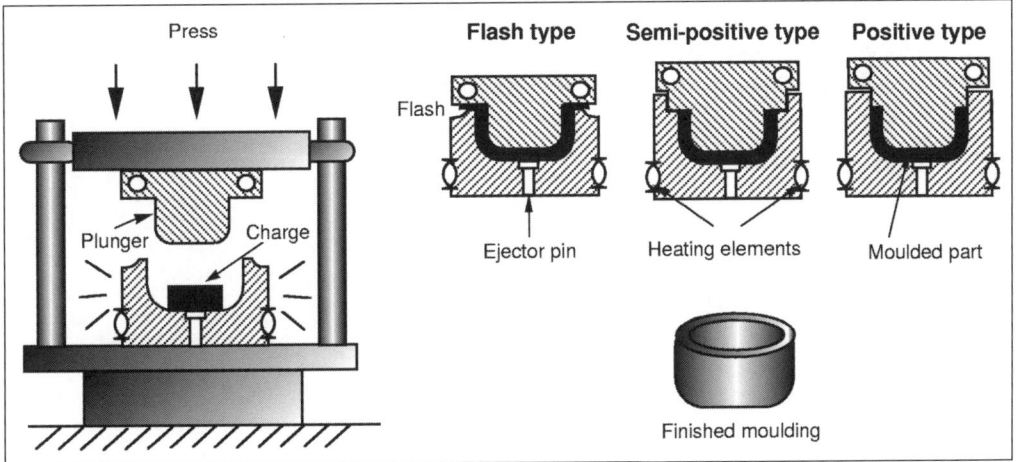

2.3F Compression molding process.

Materials

- Mainly thermosets, but also some composites, elastomers and a limited number of thermoplastics.
- Raw material supplied in either powder or liquid resin form.

Process variations

- Flash-type: for shallow parts, but more material lost.
- Semi-positive (partly positive, partly flash): used for closer tolerance work or when the design involves marked changes in section thickness.
- Positive: high density parts involving composite Sheet Molding Compounds (SMC), Bulk Molding Compounds (BMC) or impact-thermosetting materials.
- Cold-molding: powder or filler is mixed with a binder, compressed in a cold die and cured in an oven. Strictly for thermosets.

Economic considerations

- Production rates are from 20 to 140/h.
- Cycle time is restricted by material handling. Each cavity must be loaded individually.
- The greater the thickness of the part, the longer the curing time.

- Multiple cavity mold increases production rate.
- Mold maintenance is minimal.
- Certain amount of automation is possible.
- Time required for polymerization (curing) depends mainly on the largest cross section of the product and the type of molding compound.
- Lead times may be several weeks according to die complexity.
- Material utilization is high. No sprues or runners.
- Flexibility is low. Differences in shrinkage properties reduces the capability to change from one material to another.
- Production volumes are typically 1000+, but can be as low as 100 for large parts.
- Tooling costs are moderate to high.
- Equipment costs are moderate.
- Direct labor costs are low to moderate.
- Finishing costs are generally low. Flash removal required.

Typical applications

- Dishes
- Housings
- Automotive parts
- Panels
- Handles
- Container caps
- Electrical components and fittings

Design aspects

- Shape complexity limited to relatively simple forms. Molding in one plane only.
- Threads, ribs, inserts, lettering, holes and bosses possible.
- When molding materials with reinforcing fibers, directionality maintained enabling high strength to be achieved.
- Thin-walled parts with minimum warping and dimensional deviation may be molded.
- Placing of parting line important, i.e. avoid placement across critical dimensions.
- A draft angle of greater than $1°$ required.
- Maximum section, typically $= 13\,\text{mm}$.
- Minimum section $= 0.8\,\text{mm}$.
- Maximum dimension, typically $= 450\,\text{mm}$.
- Minimum area $= 3\,\text{mm}^2$.
- Maximum area $= 1.5\,\text{m}^2$.
- Sizes ranging from several grams to 16 kg in weight.

Quality issues

- Variation in raw material charge weight results in variation of part thickness and scrap.
- Air entrapment is possible.
- Internal stresses are minimal.
- Dimensions in the direction of the mold opening and the product density will tend to vary more than those perpendicular to the mold opening.
- Flash molds do not require that the quantity of material is controlled.

5 • Tumbling may be required as a finishing process to remove flash.
↑ • Surface detail is good.
6 • Surface roughness is a function of the die condition. Typically, 0.8 µm Ra is obtained.
— • Process capability charts showing the achievable dimensional tolerances using various materials
   are provided (see 2.3CC). Allowances of approximately ±0.1 mm should be added for dimensions
   across the parting line.

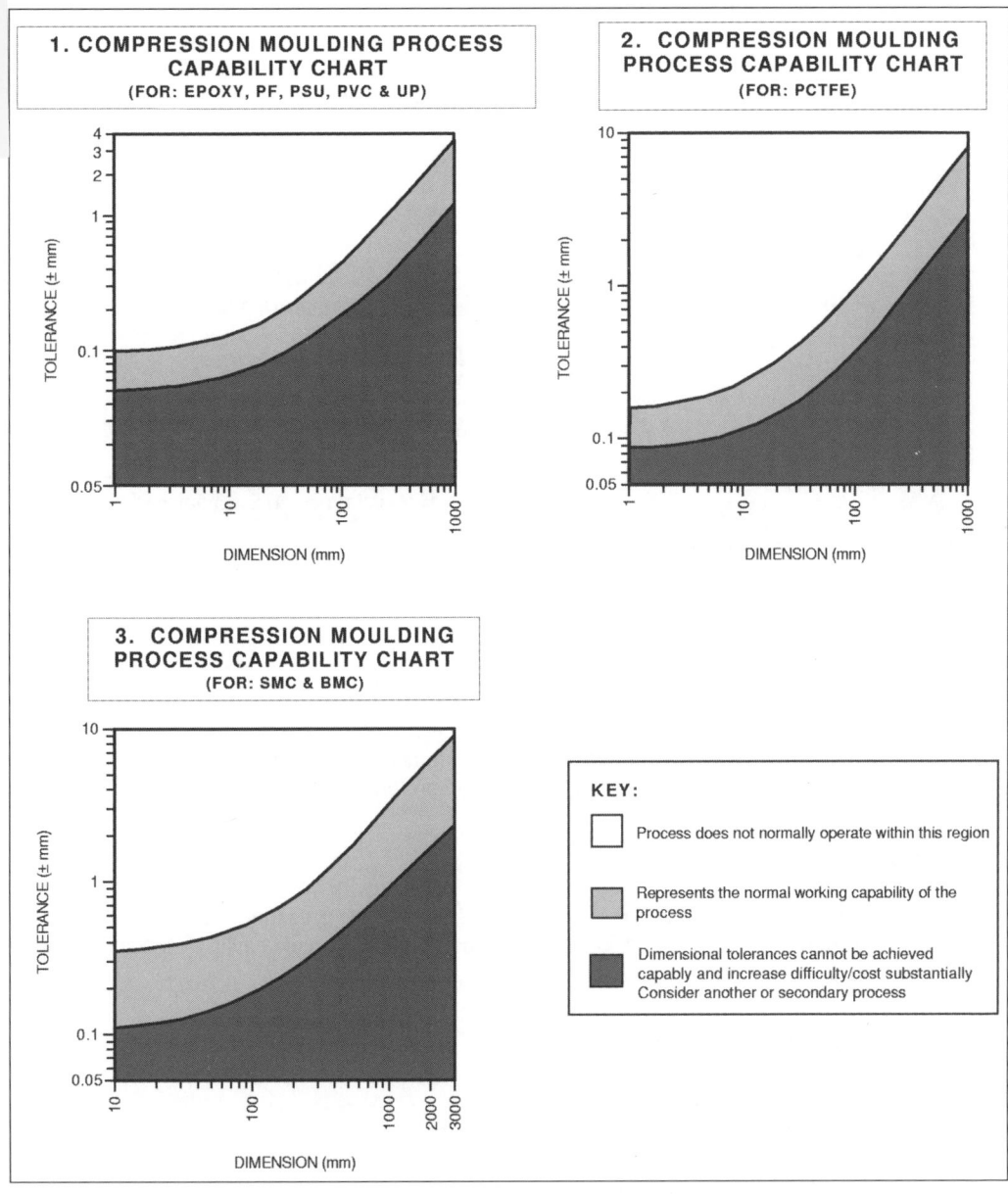

2.3CC Compression molding process capability chart.

2.4 Transfer molding

Process description

- A heated mold is closed under low pressure and then a liquid resin and catalyst is loaded into an adjacent mixing head and forced via a plunger into the cavity where curing takes place. Full name is resin transfer molding (see 2.4F).

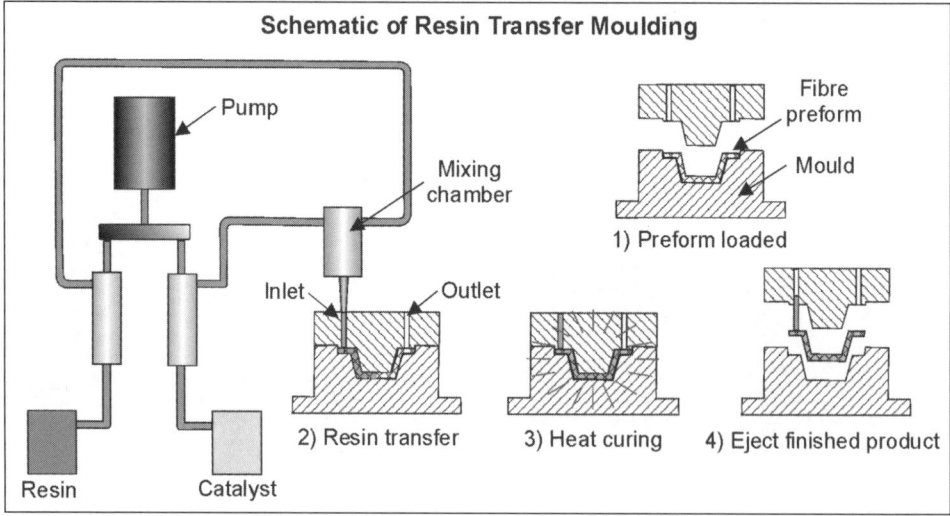

2.4F Transfer molding process.

Materials

- Limited to only several thermosetting plastics and elastomers, with or without fillers.
- Can use pre-pressed fiber-packs to fit the mold, called preforms. Fibers can be glass or carbon.

Process variations

- Powder material placed in a heated melting pot and forced under pressure into a heated mold.
- Vacuum assisted resin injection: additional vacuum can be used in mold cavity to assist resin filling of fiber preforms.

Economic considerations

- Production rates 20–300/h. Fast curing speed.
- Lead time typically days, depending on complexity of tool.
- Material utilization very good. Less than 3 per cent scrap typically.
- Scrap material cannot be recycled directly.
- High degree of automation possible.

- Economical for production runs of 1000–10 000.
- Tooling costs moderate to high.
- Equipment costs generally moderate.
- Direct labor costs low to moderate.
- Some skilled labor required, but easily reduced with automation.
- Finishing costs low, but no opportunity for in-mold trimming.

Typical applications

- Electrical cabinets
- Housings and panels
- Car body panels
- Wind turbine blades
- Seating
- Yacht hulls and decks
- Plant growing trays
- Garden ponds

Design aspects

- Complex geometries possible and hollow shapes.
- Cores possible for increased complexity.
- Can mold around inserts and delicate cores easily.
- Lettering, ribs, holes, inserts and threads possible.
- Undercuts possible, but at added cost.
- Thickness variation less than 2:1.
- Draft angles ranging 2–3° preferred, but can be as low as 0.5°.
- Minimum inside radius $= 6\,$mm.
- Minimum section ranging 0.8–1.5 mm, depending on material used.
- Maximum section $= 90\,$mm.
- Maximum dimension $= 450\,$mm.
- Minimum area $= 3\,$mm^2.
- Maximum size 16 kg in weight, but suited to smaller parts.

Quality issues

- Differential stress distribution may occur due to flow characteristics of mold resulting in minor distortion.
- High temperatures above resin melting temperatures must be maintained prior and during mold filling.
- Improperly placed fiber preforms can cause dry spots or pools of resin on surface of finished part.
- Fiber preforms can also move during injection mold filling without proper fixing arrangements within mold.
- Variation in resin/fiber concentration is difficult to control in sharp corners.
- It is not recommended for parts subjected to high loads in service.
- Surface detail is excellent.
- Surface roughness is a function of the die condition, with 0.8 µm Ra, readily obtainable.
- Achievable dimensional tolerances are ±0.05 at 25 mm, ±0.15 at 150 mm. Wall thickness tolerances are typically ±0.25 mm.

2.5 Vacuum forming

Process description

- A plastic sheet is softened by heating elements and pulled under vacuum on to the surface form of a cold mold and allowed to cool. The part is then removed (see 2.5F).

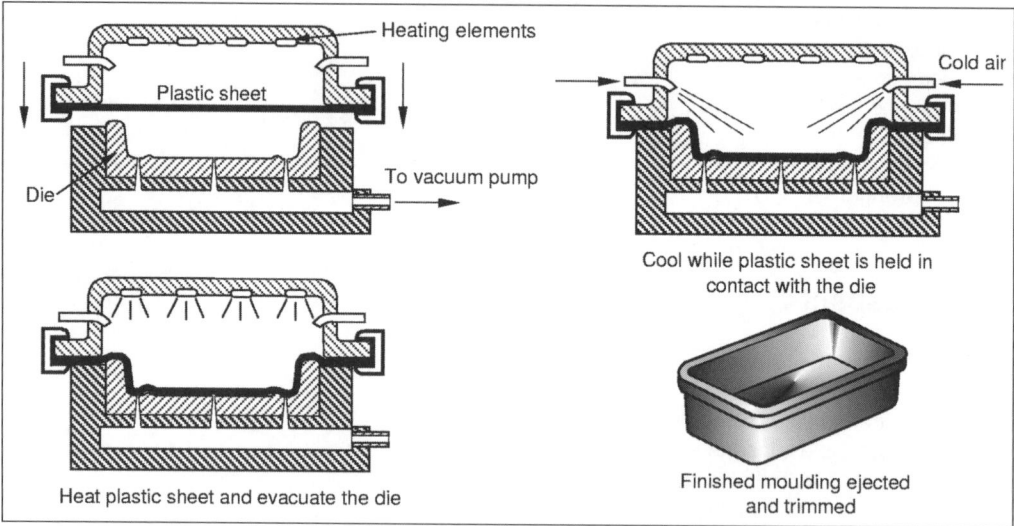

2.5F Vacuum forming process.

Materials

- Several thermoplastics that can be produced in sheet form.
- The material to be processed should exhibit high uniform elongation.
- Can also introduce some fiber reinforcing material to improve strength and rigidity.

Process variations

- Molds are usually made of cast aluminum or aluminum filled epoxy.
- Sheets can be heated by infrared heaters or in ovens.
- Can have top and bottom heating elements, or top heating element only.
- For thick sheets, a top enclosure and compressed air is used.
- Sheet is drawn over mold with additional force, other than provided by the vacuum, until cooled.
- Thermoforming: for thin-walled parts such as packaging.

Economic considerations

- Production rates from 60 to 360/h commonly. Cups can be produced at 3600/h.
- Lead times of a few days typically.

- Material utilization moderate to low. Unformed parts of the sheet are lost and cannot be directly recycled.
- Full automation achievable.
- Multiple molds may be used.
- Setup times and changeover times low.
- Sheet material much more expensive than raw pellet material.
- Production volumes economical in small batches of 10–1000.
- Tooling costs low to moderate, depending on complexity.
- Equipment costs low to moderate, but can be high if automated.
- Labor costs low to moderate.
- Finishing costs low. Some trimming of unformed material after molding.

Typical applications

- Open plastic containers and panels
- Pages of Braille text
- Vending cups
- Food packaging and containers
- Automotive parts
- Electrical cabinets and enclosures
- Bath tubs, sink units and shower panels
- Dinghy hulls
- Signs

Design aspects

- Shape complexity limited to moldings in one plane.
- Open forms of constant thickness.
- Undercuts possible with a split mold.
- Cannot produce parts with large surface areas.
- Bosses, ribs and lettering possible, but at large added cost.
- Parts with molded-in holes not possible.
- Corner radii should be large compared to thickness of material.
- Sharp corners should be avoided.
- No parting lines.
- Draft angles of 1° or greater recommended.
- Maximum section = 3 mm.
- Minimum section = 0.05–0.5 mm, depending on material used.
- Sizes ranging from 25 mm^2 to 2.5 × 7.5 m in area.

Quality issues

- Control of temperature, clamping force and vacuum pressure important if variability is to be minimized.
- Thermoplastic material must possess a high uniform elongation otherwise tearing at critical points in the mold may occur.
- Sheet material will have a plastic memory and so at high temperatures the formed part will revert back to original sheet profile. Operating temperature therefore important.
- Uniform temperature control of sheet important.

- If multiple molds used it is necessary that there is sufficient distance between cavities to avoid flow interference.
- Excessive thinning can occur, particularly at sharp corners.
- Surface detail fair.
- Surface finish good and related to the condition of mold surface.
- Achievable tolerances ranging ± 0.25–± 2 mm, and largely mold dependent. Wall thickness tolerances typically ± 20 per cent of the nominal.

2.6 Blow molding

Process description

- A hot hollow tube of plastic, called a parison, is extruded or injection molded downwards and then caught between two halves of a shaped mold which closes the top and bottom of the parison. Hot air is blown into the parison, expanding it until it uniformly contacts the inside contours of the cold mold. The part is allowed to cool and is then ejected (see 2.6F).

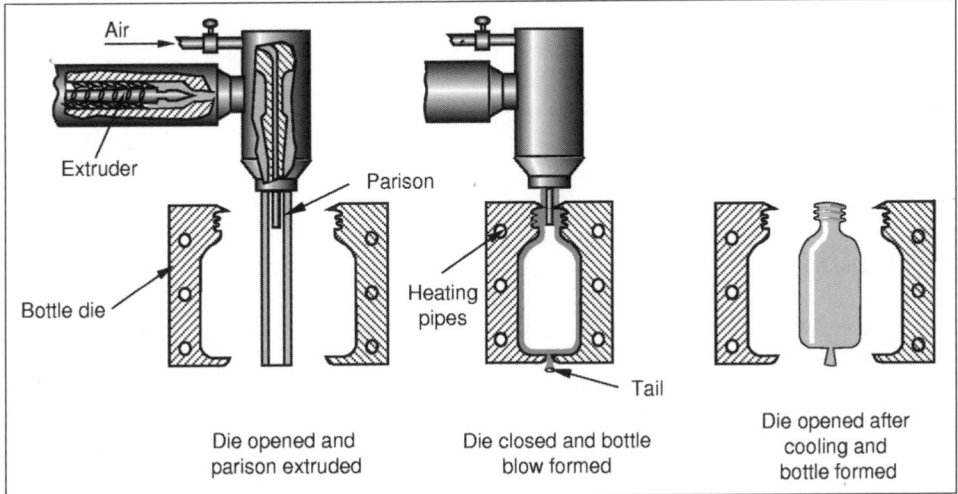

Die opened and parison extruded

Die closed and bottle blow formed

Die opened after cooling and bottle formed

2.6F Blow molding process.

Materials

- Most thermoplastics.

Process variations

- Extrusion blow molding: more applicable to asymmetrical parts, integrated handles possible.
- Injection blow molding: parison injection molded and then transferred to blow molding machine. For small parts with intricate neck detail.
- Multiple parisons: can create multi-layered parts. This requires close control since uneven parisons produce waste.
- Parisonless blowing: similar to dip-coating followed by expansion into the mold.
- Stretch blow molding: the simultaneous axial and radial expansion of a parison, yielding a biaxially orientated container.

Economic considerations

- Production rates between 100 and 2500/h, depending on size.
- Lead times a few days.

- Integration with extrusion process to produce parison provides continuous operation.
- There is generally little material waste, but can increase with some complex geometries using extrusion blow molding.
- Full automation readily achievable.
- Flexibility limited since molds are dedicated.
- Setup times and changeover times relatively short.
- Production volumes of 1000, but better suited to very high volumes.
- Tooling costs moderate to high.
- Equipment costs moderate to high, especially for full automation.
- Direct labor costs low. One operator can manage several machines.
- Finishing costs low. Some trimming required.

Typical applications

- Hollow plastic parts with relatively thin walls
- Bottles
- Bumpers
- Ducting

Design aspects

- Complexity limited to hollow, well rounded, thin walled parts with low degree of asymmetry.
- Asymmetrical moldings, e.g. off-set necks possible with movable blowing spigots.
- Undercuts, bosses, ribs, lettering, inserts and threads possible.
- Corner radii should be as generous as possible (>3 mm).
- Placing of parting line important, i.e. avoid placement across critical dimensions.
- Holes cannot be molded-in.
- Draft angles not required.
- Maximum section = 6 mm. Thick sections may need cooling aids (carbon dioxide or nitrogen gas).
- Minimum section = 0.25 mm.
- Sizes ranging 12 mm in length to volumes up to 3 m^3.

Quality issues

- Poor control of wall thickness, typically ±50 per cent of nominal.
- Creep and chemical stability of product important considerations.
- Residual stresses, e.g. non-uniform deformation, may relax in time causing distortion of the part.
- Good surface detail and finish possible.
- The higher the pressure the better the surface finish of the product.
- A process capability chart showing the achievable dimensional tolerances is provided (see 2.6CC). Allowances of approximately ±0.1 mm should be added for dimensions across the parting line.

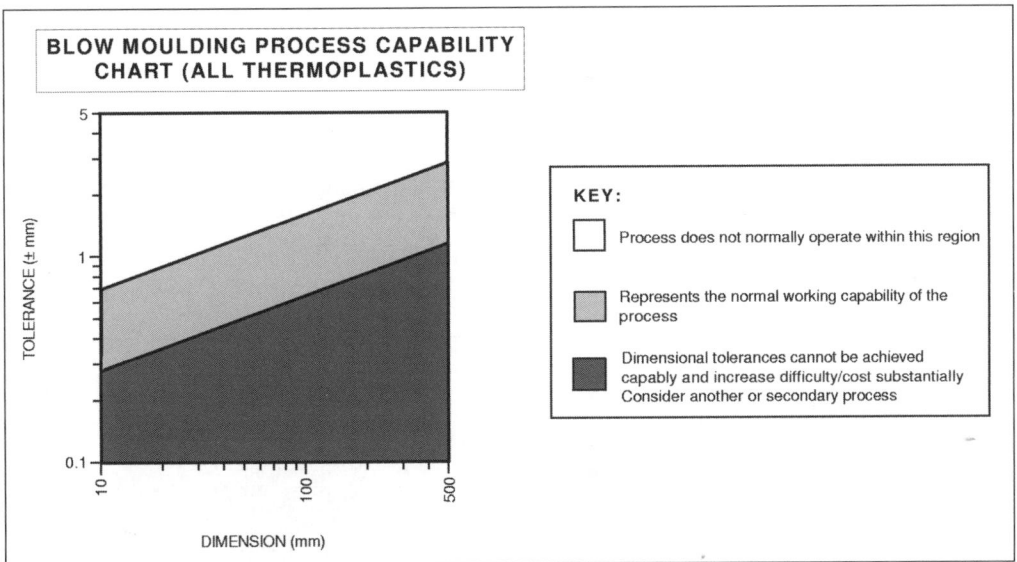

2.6CC Blow molding process capability chart.

2.7 Rotational molding

Process description

- Raw material is placed in the mold and simultaneously heated and rotated forcing the particles to deform and melt on the walls of a female mold without the application of external pressure or centrifugal forces. The part is cooled whilst rotating. The mold is designed to rotate about two perpendicular axes (see 2.7F).

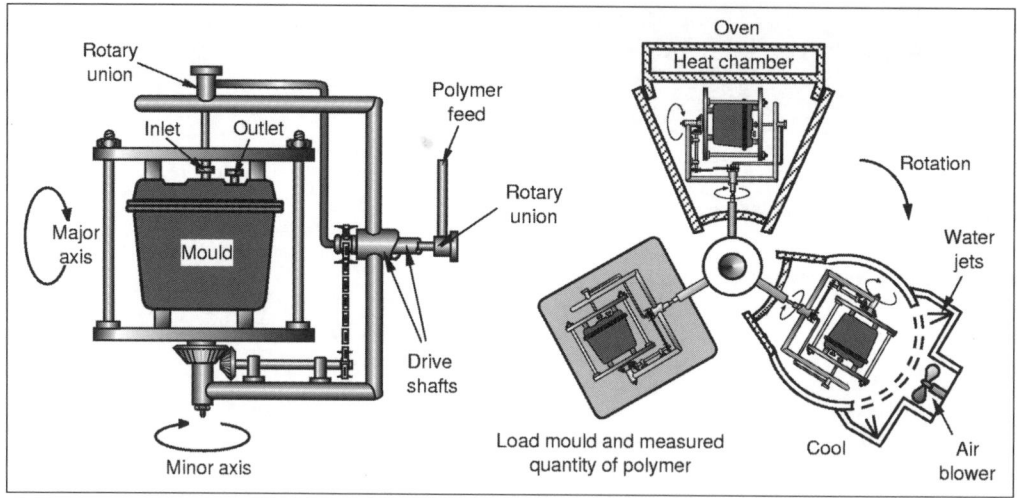

2.7F Rotational molding process.

Materials

- Several common thermoplastics, including difficult fluoropolymers.
- Raw material supplied as finely ground powder.

Process variations

- Slush molding: uses liquid polymers (plastisols) for small hollow parts.
- Air, water or mist can be used for mold cooling.

Economic considerations

- Production rates of 3–50/h, but dependent on size.
- To increase production rates, three-arm carousels often used with one mold each in the load-unload, heat and cool positions.
- Lead time several days.
- Material utilization very high. Little waste material.
- Production volumes in the range of 100–1000 typically.

- Tooling costs low.
- Equipment costs low to moderate.
- Labor costs moderate.
- Finishing costs low. Little finishing required.

Typical applications

- Water tanks
- Storage vessels
- Dust bins
- Buckets
- Housings
- Drums
- Prototypes

Design aspects

- Complexity limited to large, hollow parts of uniform wall thickness.
- Long, thin projections not possible.
- Large flat surfaces should be avoided due to distortion and difficulty to form. Use stiffening ribs.
- Internal walls need to be well spaced.
- Molded-in holes, bosses, finishes and lettering all possible at added cost and limited accuracy.
- With rotation speed variation, can build up thicker layers at key points in the mold.
- Integral handles possible.
- Large threads can be molded-in.
- Undercuts should be avoided.
- Sharp corners difficult to fill in the mold. Radii should be as generous as possible (greater than five times the wall thickness) and tend to become thicker than the wall thickness on molding.
- Metal or higher-melting point plastic inserts can be molded-in.
- Can clad the inside of the finished part using another polymer.
- Placing of parting line important, i.e. avoid placement across critical dimensions.
- Holes cannot be molded although open-ended articles possible.
- Thickness variation should be less than 2:1.
- Draft angles generally greater than 1°, typically 3°.
- Maximum section = 13 mm.
- Minimum section typically 2 mm, but can be as low as 0.5 mm for certain applications.
- Sizes up to 4 m^3.

Quality issues

- The part is practically free from residual stresses.
- Surface detail is fair.
- Outer surface finish of the part is a replica of the inside finish of the mold walls.
- Control of inside surface finish is not possible.
- Wall thickness is determined by the close control of the amount of raw material used.
- Dimensional variations can be large if sufficient setting time is not allowed before removal of the part.
- A process capability chart showing the achievable dimensional tolerances is provided (see 2.7CC). Allowances of approximately ±0.5 mm should be added for dimensions across the parting line.
- Wall thickness tolerances are generally between ±5 and ±20 per cent of the nominal.

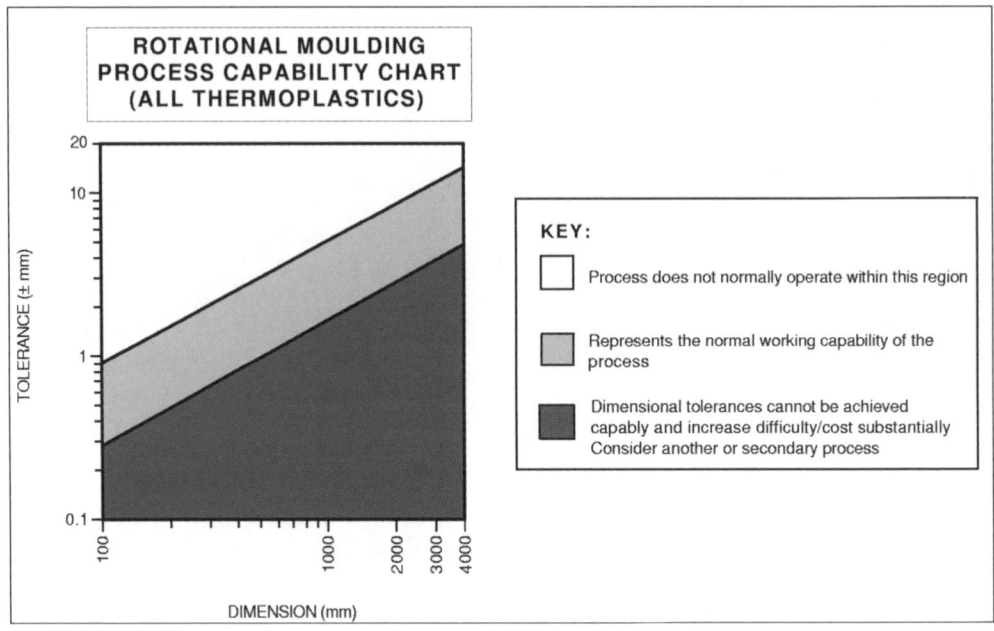

2.7CC Rotational molding process capability chart.

2.8 Contact molding

Process description

- Glass fiber reinforced material (30–45 per cent by volume) and a liquid thermosetting resin are simultaneously formed into a male or female mold and cured at room temperature or with the application of heat to accelerate the process (see 2.8F).

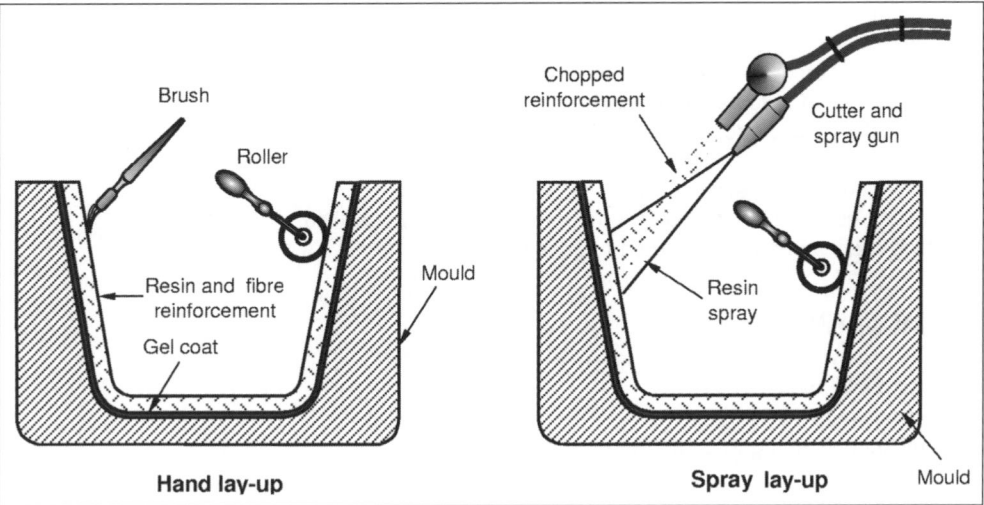

2.8F Contact molding process.

Materials

- Glass reinforced fiber in woven, continuous and chopped roving, mat and cloth forms.
- Can use pre-impregnated sheets of uncured resin and fiber, called SMC.
- Thermosetting liquid resin: commonly catalyzed polyester or epoxy.

Process variations

- Hand lay-up: manual laying of fiber reinforced material and application of resin to mold to build up the thickness. Hand or roller pressure removes any trapped air.
- Variations on hand lay-up are:
  - Vacuum bag molding: uses a rubber bag clamped over the mold. A vacuum is applied between the mold and the bag to squeeze the resin/reinforcement together removing any trapped air. Curing performed in an oven.
  - Pressure bag molding: as vacuum bag molding, but pressure is applied above the bag. Can be used for thicker section parts.
  - Hand lay-up using SMC: cured by heat and clamped if necessary to further reduce air pockets.

- Spray lay-up: use of an air spray gun incorporating a cutter that chops continuous rovings to a controlled length before being blown into the mold simultaneously with the resin.
- Molds can be made of wood, plaster, concrete, metal or glass fiber reinforced plastic.
- Cutting of composites can be performed using knives, disc cutters, lasers and water jets.

Economic considerations

- Production rates low. Long curing cycle typically.
- Production rates increased using SMC materials.
- Lead times usually short, depending on size and material used for the mold.
- Mold life approximately 1000 parts.
- Multiple molds incorporating heating elements should be used for higher production rates.
- Material utilization moderate. Scrap material cannot be recycled.
- Limited amount of automation possible.
- Economical for low production runs, 10–1000. Can be used for one-offs.
- Tooling costs low.
- Equipment costs generally low.
- Direct labor costs high. Can be very labor intensive, but not skilled.
- Finishing costs moderate. Some part trim is required.

Typical applications

- Hulls for boats and dinghies
- Large containers
- Swimming pools and garden pond moldings
- Bath tubs
- Small cabins and buildings
- Machine covers
- Car body panels
- Sports equipment
- Wind turbine blades
- Prototypes and mock-ups
- Architectural work

Design aspects

- High degree of shape complexity possible, limited only by ability to produce mold.
- Produces only one finished surface.
- Fibers should be placed in the expected direction of loading, if any. Random layering gives less strength.
- Avoid compressive stresses and buckling loads.
- Used for parts with a high surface area to thickness ratio.
- Molded-in inserts, ribs, holes, lettering and bosses are possible.
- Draft angles are not required.
- Undercuts are possible with flexible molds.
- Minimum inside radius $= 6\,$mm.
- Minimum section $= 1.5\,$mm.
- Maximum economic section $= 30\,$mm, but can be unlimited.
- Sizes ranging 0.01–500 m^2 in area.
- Maximum size depends on ability to produce the mold and the transport difficulties of finished part.

Quality issues

- Air entrapment and gas evolution can create a weak matrix and low strength parts.
- Non-reinforcing gel coat helps to create smoother mold surface and protects the molding from moisture.
- Resin and catalyst should be accurately metered and thoroughly mixed for correct cure times.
- Excessive thickness variation can be eliminated by sufficient clamping and adequate lay-up procedures.
- Toxicity and flammability of resin is an important safety issue, especially because of high degree of manually handling and application.
- Surface roughness and surface detail can be good on molded surface, but poor on opposite surface.
- Shrinkage increases with higher resin volume fraction.
- A process capability chart showing the achievable dimensional tolerances for hand/spray lay-up is provided (see 2.8CC). Wall thickness tolerances are typically ±0.5 mm.

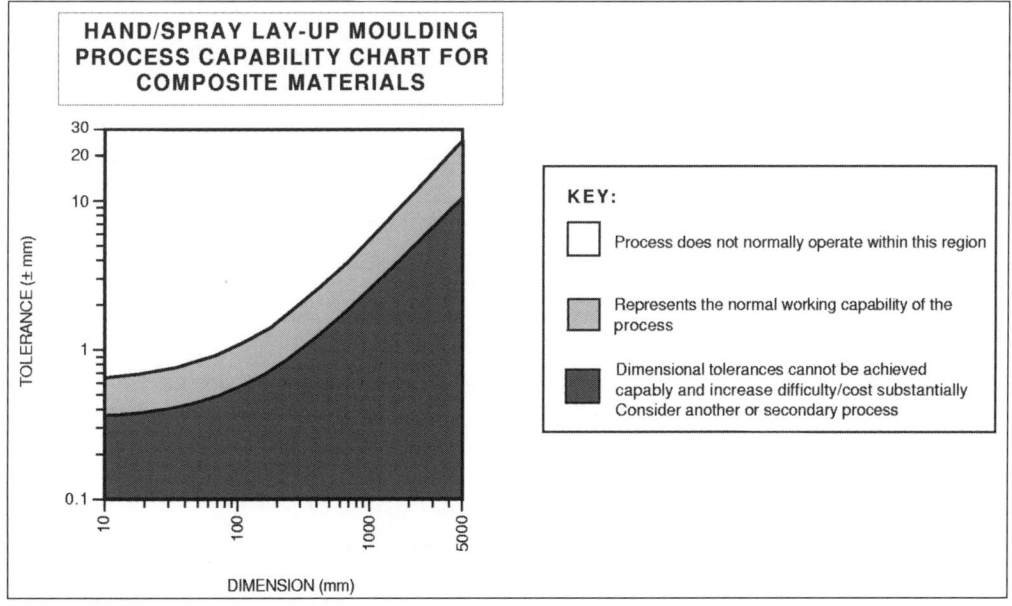

2.8CC Contact molding process capability chart.

2.9 Continuous extrusion (plastics)

Process description

- The raw material is fed from a hopper into a heated barrel and pushed along a screw-type feeder where it is compressed and melts. The melt is then forced through a die of the required profile where it cools on exiting the die (see 2.9F).

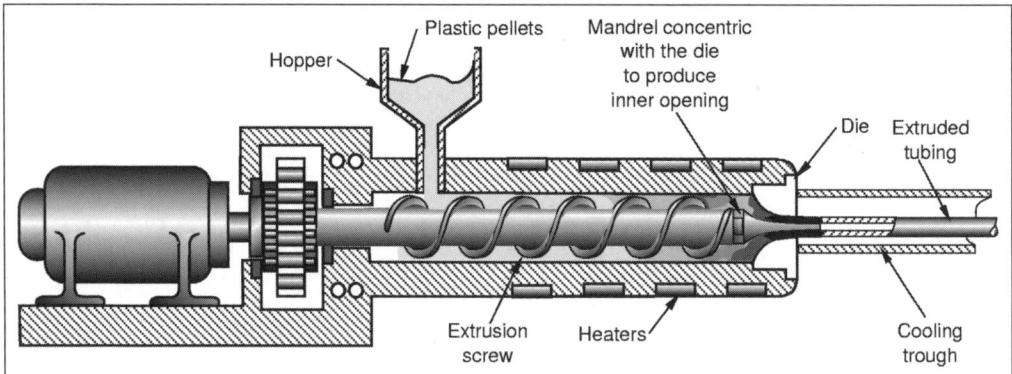

2.9F Continuous extrusion (plastics) process.

Materials

- Most plastics, especially thermoplastics, but also some thermosets and elastomers.
- Raw material in pellet, granular or powder form.

Process variations

- Most extruders are equipped with a single screw, but two-screw or more extruders are available. These are able to produce coaxial fibers or tubes and multi-component sheets.
- Metal wire, strips and sections can be combined with the extrusion process using an offset die to produce plastic coatings.
- Pultrusion: for fiber-reinforced rods, tubes and sections.

Economic considerations

- Production rates are high but are dependent on size. Continuous lengths up to 60 m/min for some tube sections and profiles, up to 5 m/min for sheet and rod sections.
- Extruders are often run below their maximum speed for trouble free production.
- It can have multiple holes in die for increased production rates.
- Extruder costs increase steeply at the higher range of output.
- Lead times are dependent on the complexity of the 2-dimensional die, but normally weeks.
- Material utilization is good. Waste is only produced when cutting continuous section to length.
- Process flexibility is moderate. Tooling is dedicated, but changeover and setup times are short.

- Production of 1000 kg of profile extrusion is economical, 5000 kg for sheet extrusions (equates to about 10 000 items).
- Tooling costs are generally moderate.
- Equipment costs are high.
- Some materials give off toxic or volatile gases during extrusion. Possible need for air extraction and washing plant which adds to equipment cost.
- Direct labor costs are low.
- Finishing costs are low. Cutting to length only real cost.

Typical applications

- Complex profiles. All types of thin walled, open or closed profiles
- Rods, bar, tubing and sheet
- Small diameter extruded bar which is cut into pellets and used for other plastic processing methods
- Fibers for carpets, tyre reinforcement, clothes and ropes
- Cling-film
- Plastic pipe for plumbing
- Plastic-coated wire, cable or strips for electrical applications
- Window frames
- Trim and sections for decorative work

Design aspects

- Dedicated to long products with uniform cross-sections.
- Cross-sections may be extremely intricate.
- Solid forms including re-entrant angles, closed or open sections.
- Section profile designed to increase assembly efficiency by integrating part consolidation features.
- Grooves, holes and inserts not parallel to the axis of extrusion must be produced by secondary operations.
- No draft angle required.
- Maximum section $= 150$ mm.
- Minimum section $= 0.4$ mm for profiles (0.02 mm for sheet).
- Sizes ranging 6 mm^2–1800 mm wide sheet, and $\varnothing$1–$\varnothing$150 mm for tubes and rods.

Quality issues

- The rate and uniformity of cooling are important for dimensional control because of shrinkage and distortion.
- Extrusion causes the alignment of molecules in solids.
- Die swell, where the extruded product increases in size as it leaves the die, may be compensated for by:

  - Increasing haul-off rate compared with extrusion rate
  - Decreasing extrusion rate
  - Increasing the length of the die land
  - Decreasing the melt temperature.

- There is a tendency for powdered materials to carry air into the extruder barrel: trapped gases have a detrimental effect on both the output and the quality of the extrusion.
- Surface roughness is good to excellent.
- Process capability charts showing the achievable dimensional tolerances for various materials are provided (see 2.9CC).

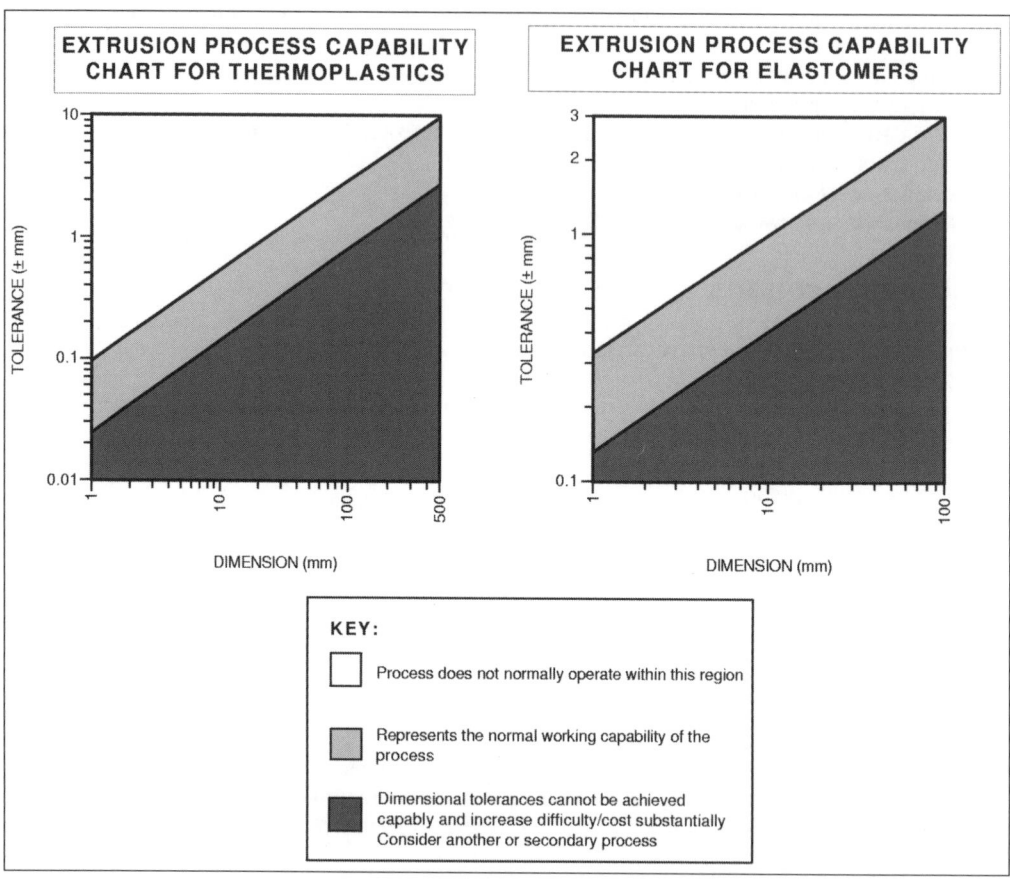

2.9CC Continuous extrusion (plastics) process capability chart.

3 Forming processes

3.1 Forging

Process description

- Hot metal is formed into the required shape by the application of pressure or impact forces causing plastic deformation using a press or hammer in a single or a series of dies (see 3.1F).

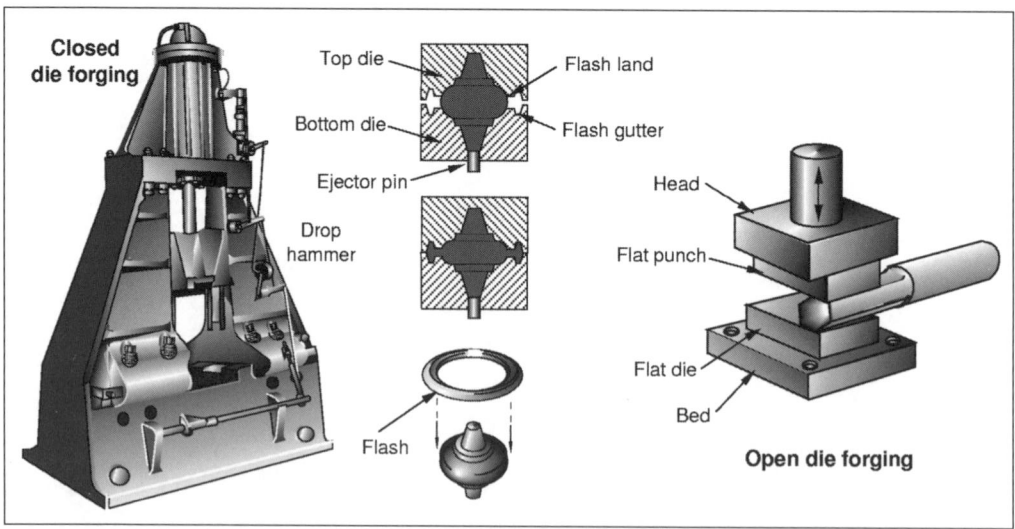

3.1F Forging process.

Materials

- Mainly carbon, low alloy and stainless steels, aluminum, copper and magnesium alloys. Titanium alloys, nickel alloys, high alloy steels and refractory metals can also be forged.
- Forgeability of materials important; must be ductile at forging temperature. Relative forgeability is as follows, with the easiest to forge first: aluminum alloys, magnesium alloys, copper alloys, carbon steels, low alloy steels, stainless steels, titanium alloys, high alloy steels, refractory metals and nickel alloys.

Process variations

- Presses can be mechanical, hydraulic or drop hammer type.
- Closed die forging: series of die impressions used to generate shape.
- Open die forging: hot material deformed between a flat or shaped punch and die. Sections can be flat, square, round or polygon. Shape and dimensions largely controlled by operator.
- Roll forging: reduction of section thickness of a doughnut-shaped preform to increase its diameter. Similar to ring rolling (see 3.2), but uses impact forces from hammers.
- Upset forging: heated metal stock gripped by dies and end pressed into desired shape, i.e. increasing the diameter by reducing height.

- Hand forging: hot material reduced, upset and shaped using hand tools and an anvil. Commonly associated with the blacksmith's trade, used for decorative and architectural work.
- Precision forging: near-net shape generation through the use of precision dies. Reduces waste and minimizes or eliminates machining.

Economic considerations

- Production rates from 1 to 300/h, depending on size.
- Production most economic in the production of symmetrical rough forged blanks using flat dies. Increased machining is justified by increased die life.
- Lead times typically weeks.
- Material utilization moderate (20–25 per cent scrap generated in flash typically).
- Economically viable quantities greater than 10 000, but can be as low as 100 for large parts.
- In the case of open die forging: lower material utilization, machining of the final shape necessary, slow production rate, low lead times, commonly used for one-offs and high usage of skilled labor.
- Tooling costs high.
- Equipment costs generally high.
- Direct labor costs moderate. Some skilled operations may be required.
- Finishing costs moderate. Removal of flash, cleaning and fettling important for subsequent operations.

Typical applications

- Engine components (connecting rods, crankshafts, camshafts)
- Transmission components (gears, shafts, hubs, axles)
- Aircraft components (landing gear, airframe parts)
- Tool bodies
- Levers
- Upset forging: for bolt heads, valve stems
- Open die forging: for die blocks, large shafts, pressure vessels

Design aspects

- Complexity is limited by material flow through dies.
- Deep holes with small diameters are better drilled.
- Drill spots caused by die impressions can be used to aid drill centralization for subsequent machining operations.
- Locating points for machining should be away from parting line due to die wear.
- Markings are possible at little expense on adequate areas that are not to be subsequently machined.
- Care should be taken with design of die geometry, since cracking, mismatch, internal rupture and irregular grain flow can occur.
- It is good practice to have approximately equal volumes of material both above and below the parting line.
- Inserts and undercuts are not possible.
- Placing of parting line is important, i.e. avoid placement across critical dimensions, keep along simple plane, line of symmetry or follow the part profile.
- Corner radii and fillets should be as large as possible to aid hot metal flow.
- Maximum length to diameter ratio that can be upset is 3:1.

- Avoid abrupt changes in section thickness. Causes stress concentrations on cooling.
- Minimum corner radii = 1.5 mm.
- Machining allowances range from 0.8 to 6 mm, depending on size.
- Drafts must be added to all surfaces perpendicular to the parting line.
- Draft angles ranging 0–8°, depending on internal or external features, and section depth, but typically 4°. Reduced by mechanical ejectors in dies.
- Minimum section = 3 mm.
- Sizes ranging 10 g–250 kg in weight, but better for parts less than 20 kg.

Quality issues

- Good strength, fatigue resistance and toughness in forged parts due to grain structure alignment with die impression and principal stresses expected in service.
- Low porosity, defects and voids encountered.
- Forgeability of material important and maintenance of optimum forging temperature during processing.
- Hot material in contact with the die too long will cause excessive wear, softening and breakage.
- Variation in blank mass causes thickness variation. Reduced by allowing for flash generation, but increases waste.
- Residual stresses can be significant. Can be improved with heat treatment.
- Die wear and mismatch may be significant.
- Surface roughness and detail may be adequate, but secondary processing usually employed to improve the surface properties.
- Surface roughness ranging 1.6–25 μm Ra.
- Process capability charts showing the achievable dimensional tolerances for closed die forging using various materials are provided. Note, the total tolerance on Charts 1–4 is allocated +2/3, −1/3. Allowances of +0.3–+2.8 mm should be added for dimensions across the parting line and mismatch tolerances ranging 0.3–2.4 mm, depending on part size (see 3.1CC).
- Tolerances for open die forging ranging ±2–±50 mm, depending on size of work and skill of the operator.

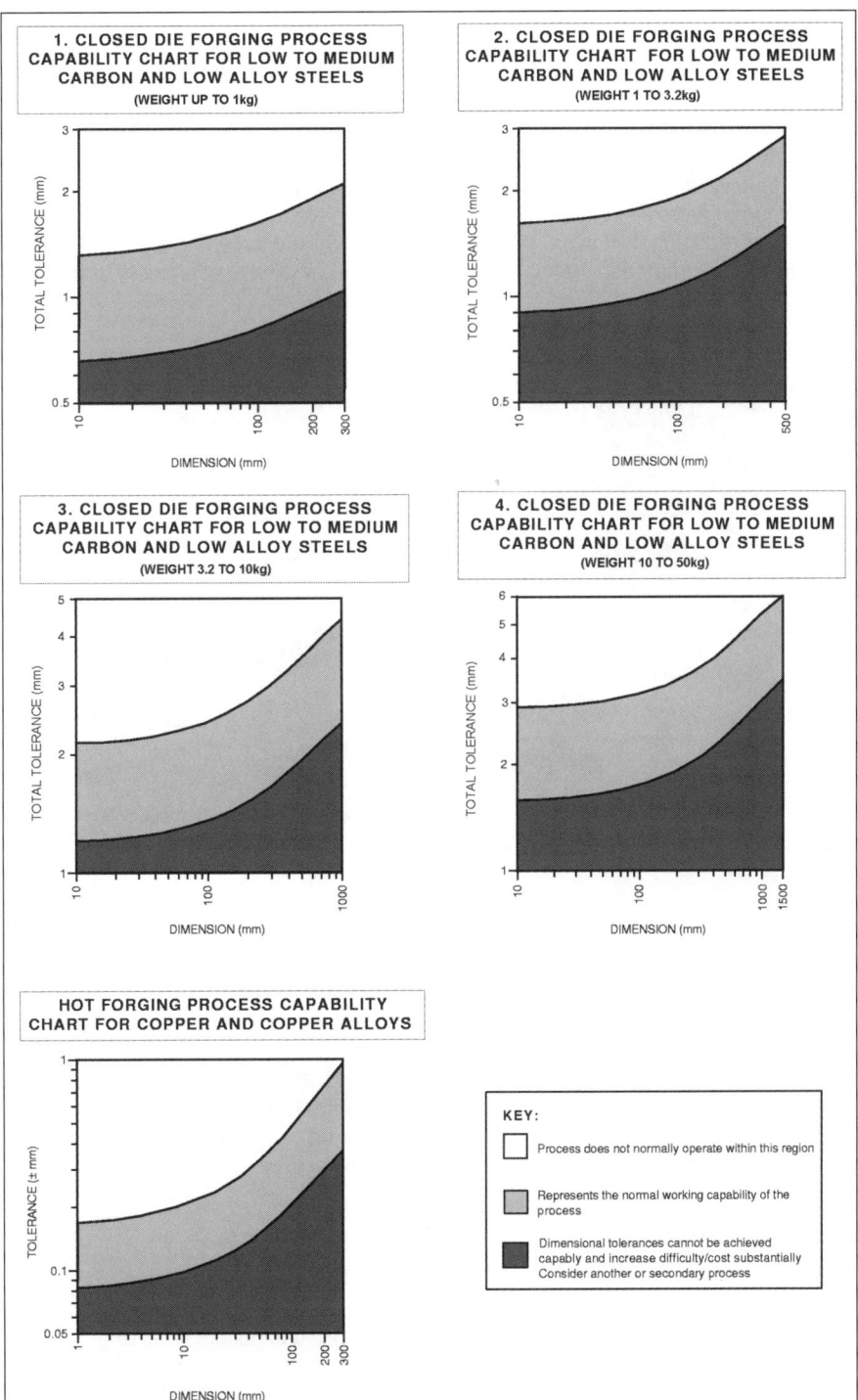

3.1CC Forging process capability chart.

3.2 Rolling

Process description

- Continuous forming of metal between a set of rotating rolls whose shape or height is adjusted incrementally to produce desired section through imposing high pressures for plastic deformation. It is the process of reducing thickness, increasing length without increasing the width markedly. Can be performed with the material at a high temperature (hot) or initially at ambient temperature (cold) (see 3.2F).

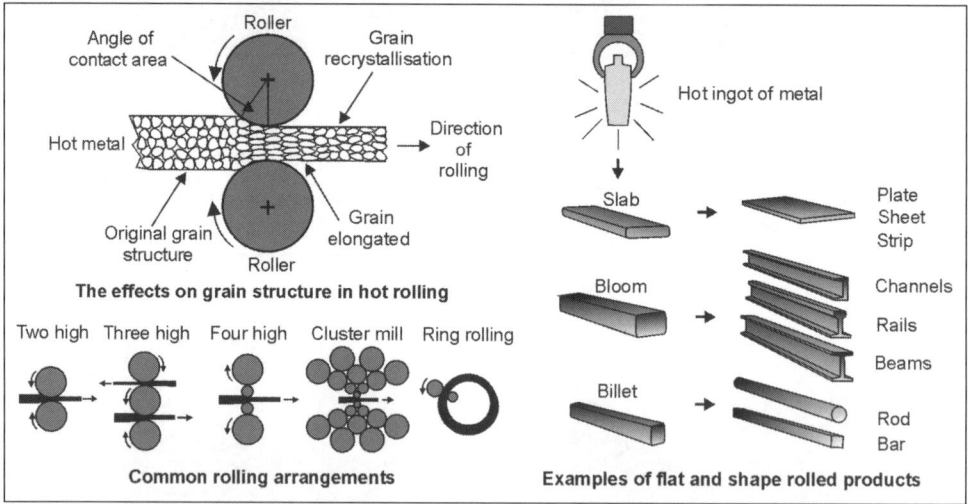

3.2F Rolling process.

Materials

- Most ductile metals such as low carbon, alloy and stainless steels, aluminum, copper and magnesium alloys.
- Metal ingots called blooms, slabs or billets, used to load the mill. Blooms are used to produce structural sections (beams, channels, rail sections), slabs are used to produce flat products such as sheets and plate, and billets are rolled into rods and bars using shaped rolls.
- Continuous casting also used for higher efficiency and lower cost.

Process variations

- Variety of roll combinations exist (called mills):
  - Two high: commonly used for hot rolling of plate and flat product, either reversing or non-reversing type.
  - Two high with vertical rolls: commonly used for hot rolling of structural sections. Vertical rolls maintain uniform deformation of section and prevent cracking.

- Three high: for reversing one length above the other simultaneously.
- Four high (tandem): backing rolls give more support to the rolls in contact with product for initial reduction of ingots.
- Cluster mills: very low roll deflection obtained due to many supporting rolls above the driven rolls that are in contact with product. For cold rolling thin sheets and foil to close dimensional tolerances.

- Leveling rolls: used to improve flatness of strip product after main rolling operations.
- Flat rolling: for long continuous lengths (long discontinuous lengths in reality) of flat product. The height between the rolls is adjusted lower on each reversing cycle, or the product is passed through a series of tandem rollers with decreasing roller gap and increasing speed, to reduce the product to its final thickness. Tandem roll system has higher production rates.
- Shape rolling: billet is passed through a series of shaped grooves on same roll or a set of rolls in order to gradually form the final shape. Typically used for structural sections.
- Transverse or cross rolling: wedge shaped forms in a pair of rolls create the final shape on short-cropped bars in one revolution. For parts with axial symmetry such as spanners.
- Ring rolling: an internal roller (idler) and external roller (driven) impart pressure on to the thickness of a doughnut-shaped metal preform. As the thickness decreases, the diameter increases. For creating seamless rings used for pressure vessels, jet engine parts and bearing races. Rectangular cross sections and contours are also possible. Can be readily automated.
- Pack rolling: operation where two or more layers of metal are rolled together.
- Thread rolling: wire or rod is passed between two flat plates, one moving and the other stationary, with a thread form engraved on surfaces. Used to produce threaded fasteners with excellent strength and surface integrity at high production rates and no waste.
- Roll forming: forming of long lengths of sheet metal into complex profiles using a series of rolls (see 3.9).
- Calandering: thermoplastic raw material is passed between a series of heated rollers in order to produce sheet product.

Economic considerations

- Production rates high. Continuous process with speeds ranging 20–500 m/min.
- Production rates for related processes: transverse rolling up to 100/h and thread rolling up to 30 000/h.
- Lead times typically months due to number of mills required and complexity of profile.
- Long set-up times for shaped rolls.
- Hot rolling requires less energy than cold rolling.
- Material utilization very good (rolling is a constant volume process). Less than 1 per cent scrap generated, commonly through line stoppages or when cutting to lengths. Can be recycled.
- High degree of automation possible.
- Plane rolls flexible in the range of flat products they can produce. Shaped rolls dedicated and therefore not flexible
- Economical for very high production runs. Minimum quantity 50 000 m of rolled product (equivalent to 100 000+).
- Tooling costs high.
- Equipment costs high.
- Direct labor costs low to moderate.
- Finishing costs very low.

Typical applications

- Rolling is an important process for producing the stock material for many other processes, e.g. machining, cold forming and sheet metal work. Around 90 per cent of all stock product used is produced by rolling for many industries:
    - Flat, square, rectangular and polygon sections
    - Structural sections, e.g. I-beams, H-beams, T-sections, channels, rails, angles and plate
    - Strip, foil and sheet
- Sheet for shipbuilding
- Structural fabrication
- Sheet metal for shearing and forming operations
- Tube forming
- Automotive trim

Design aspects

- Simple shapes using flat rolling, fairly complex 2-dimensional profiles using shape rolling and 3-dimensional shapes for transverse rolling.
- Re-entrant angles possible on profile.
- No draft angles required, except in transverse rolling.
- Hot rolling:
    Minimum section = 1.6 mm.
    Maximum section = 1 m.
- Cold rolling:
    Minimum section = 0.0025 mm.
    Maximum section = 200 mm.
- Maximum width = 5 m.

Quality issues

- Coarse grain structure and porosity of hot ingot or continuous casting is gradually improved and finer grain structure produced with little or no voids.
- Hot rolling takes place above recrystallization temperature, and therefore sections are free from residual stresses. No working hardening of material.
- Anisotropy in cold-rolled sections are due to directionality of grains during rolling and work hardening. Can be used to advantage, but does mean high compressive residual stresses that exist in surface are balanced by high tensile residual stresses in section bulk. Can lead to surface delamination.
- High sulfur contents in steels can cause cracking and flaring of rolled section ends. Possibility of jamming when introduced to a subsequent set of rolls. High scrap rates and downtime can be experienced if this occurs.
- Hot-rolled material is more difficult to handle than cold rolled. Cold-rolled strip product can be coiled for subsequent processing, hot rolled cannot.
- Rough surface finish of rolls is used in hot rolling to aid traction of metal through the rolls. Cold rolling rolls have a high surface finish.
- Lubrication can be used for ferrous alloys (graphite) and non-ferrous alloys (oil emulsion) to minimize friction during rolling.
- Cold rolling can be performed with low viscosity lubricants such as paraffin or oil emulsion.

- Hot rolling requires the preparation of stock material to remove surface oxides before processing.
- Maintenance of rolling temperature dictates quality. Too low and becomes difficult to deform. Too high and surface quality is reduced.
- Roll material must be highly wear resistant. Made to withstand 5 000 000 m of rolled section production. Can be re-coated and ground back to size.
- Surface defects may result from inclusions and impurities in the material (scale, rust, dirt, roll marks, and other causes related to prior treatment of ingots).
- Surface detail is poor in hot-rolled product (oxide layer called mill scale is always present). Oxide layer can be removed by pickling in acid.
- Surface detail is excellent for cold rolling.
- Surface roughness values ranging 6.3–50 μm Ra for hot rolling, 0.2–6.3 μm Ra for cold rolling.
- Process capability charts showing the achievable dimensional tolerances for cold rolling various materials are provided (see 3.2CC).
- Achievable tolerances ranging ±1–±2.5 per cent of the dimension for hot rolling. Dimensional variations are greater than cold rolling due to non-uniformities in material properties such as hardness, roll deflection and surface conditions.

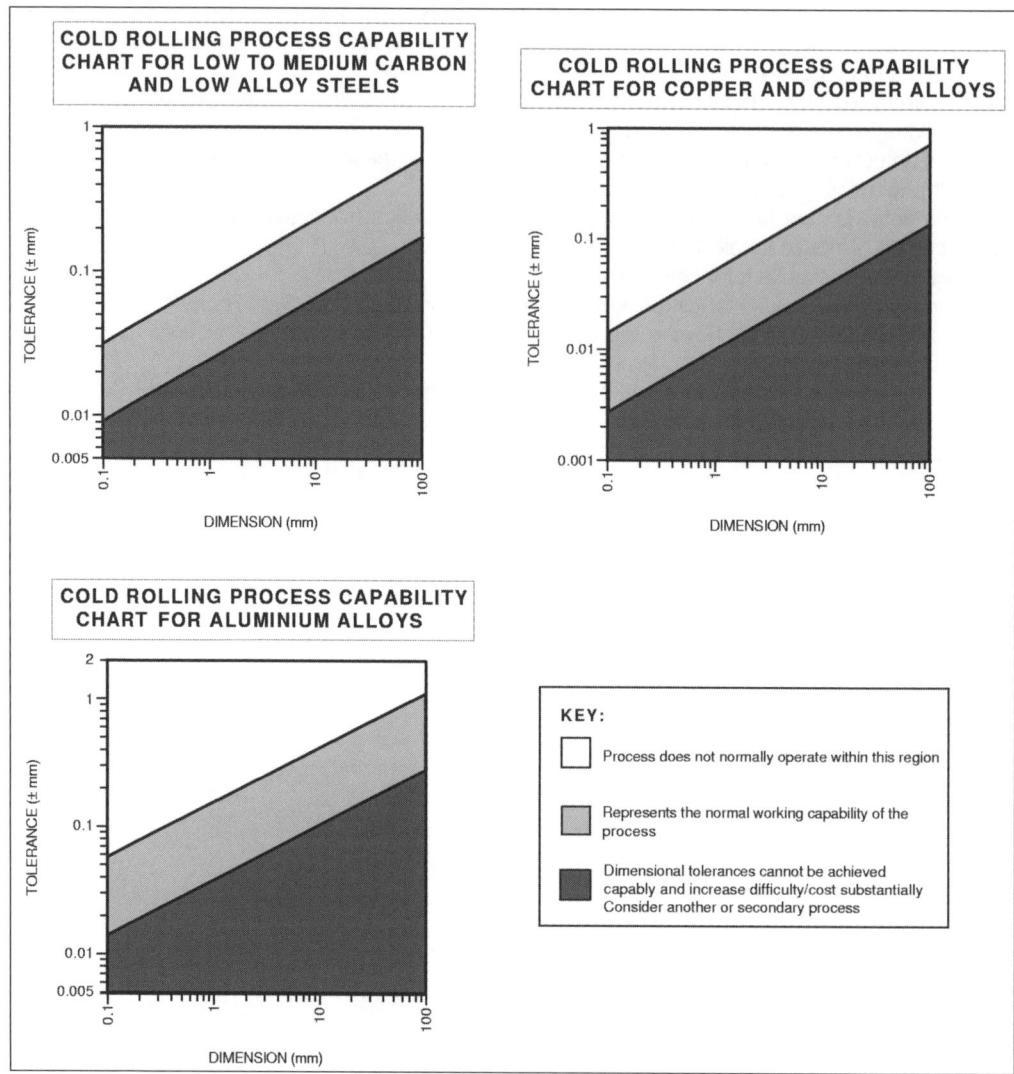

3.2CC Rolling process capability chart.

3.3 Drawing

Process description

- A number of processes where long lengths of rod, tube or wire are pulled through dies to progressively reduce the original cross-section through plastic deformation. The process is performed cold (see 3.3F).

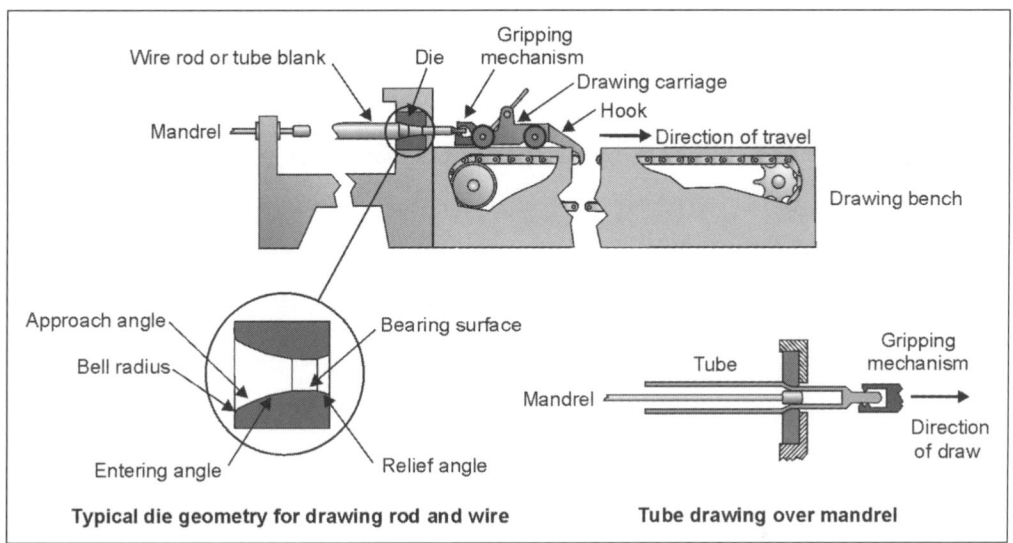

3.3F Drawing process.

Materials

- Any ductile metal at ambient temperatures.

Process variations

- Rod or bar drawing: reduction of the diameter of rod or bar through a single die or progressive reduction through a number of dies.
- Wire drawing: performed on multiple wire drawing machine where the wire is wrapped around blocks before being pulled through the next die to successively reduce diameter. Wire diameters that cannot be wrapped around blocks are drawn out on long benches at low speeds, but give lower production rates.
- Tube drawing: reduction of either the diameter of a tube or simultaneous reduction of diameter and thickness using mandrel.
- Can use rollers in place of dies for plastic deformation.
- Sizing is a low deformation operation sometimes used to finish the drawn section giving closer dimensional accuracy and surface roughness.

Economic considerations

- Production rates from 10 (rod, tube) to 2000 m/min (wire).
- Lead time typically days.
- Material utilization excellent. Some scrap may be generated when cutting to length.
- High degree of automation possible.
- Economical for high production runs (1000 m+).
- Tooling costs low.
- Equipment costs moderate.
- Direct labor costs low to moderate.
- Finishing costs very low. Cutting long lengths of rod, bar and tube to length only.

Typical applications

- Drawing is an important process for producing the stock material for many other processes, e.g. machining and cold heading.
- Rod, bar, wire, tubes
- Fabrication and machine construction
- Spring wire, musical instrument wire

Design aspects

- Simple shapes with rotational symmetry only.
- No draft angles required.
- Rod drawing:
    Minimum diameter $= \varnothing 0.1$ mm.
    Maximum diameter $= \varnothing 50$ mm.
- Wire drawing:
    Minimum diameter $= \varnothing 0.1$ mm.
    Maximum diameter $= \varnothing 20$ mm.
- Tube drawing:
    Minimum diameter $= \varnothing 6$ mm.
    Maximum diameter $= \varnothing 600$ mm.
    Minimum section $= 0.1$ mm.
    Maximum section $= 25$ mm.

Quality issues

- Strain hardening occurs in material during cold working, giving high strength.
- High directionality (anisotropy) due to nature of plastic deformation and grain orientation in direction of drawing.
- High friction between work and die causes high temperatures which must be reduced through external cooling.
- Surface detail excellent.
- Surface roughness ranging 0.2–6.3 μm Ra.
- Finer surface roughness values obtained with finer grit grades.
- Process capability charts showing the achievable dimensional tolerances for cold drawing various materials are provided (see 3.3CC).

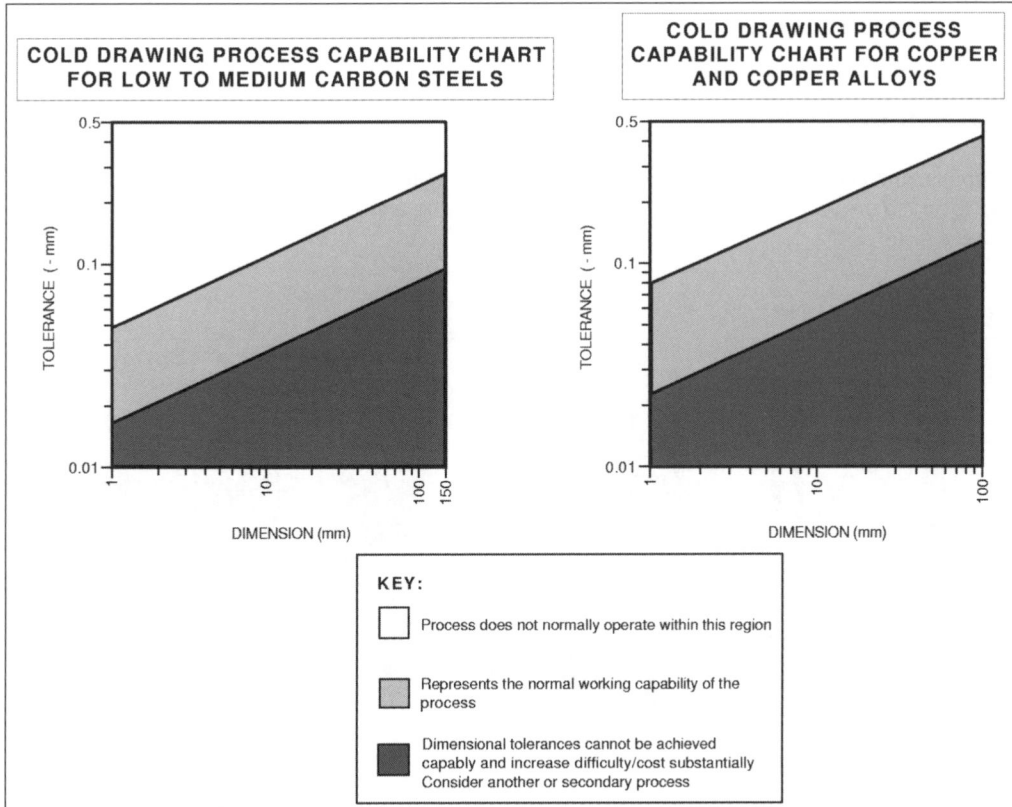

3.3CC Drawing process capability chart.

3.4 Cold forming

Process description

- Various processes under the heading of cold forming tend to combine forward and backward extrusion to produce near-net shaped components by the application of high pressures and forces (see 3.4F).

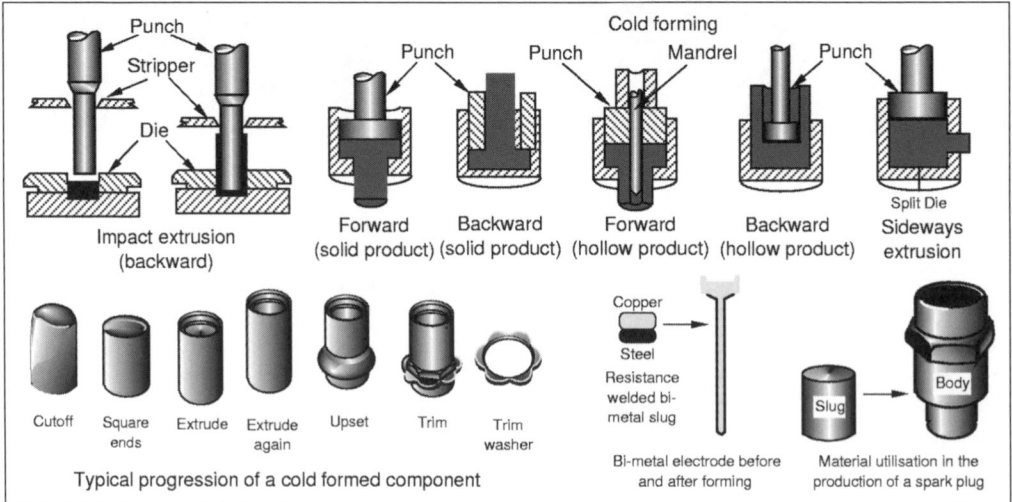

3.4F Cold forming process.

Materials

- Any ductile material at ambient temperature, including: aluminum, copper, zinc, lead and tin alloys, and low carbon steels. Also alloy and stainless steels, nickel and titanium alloys processed on a more limited basis.

Process variations

- Impact extrusion: similar to cold extrusion, but cold billet is plastically deformed by a single blow of the tool. Can be forward or backward extrusion (Hooker process).
- Cold forming: can be forward, backward or a combination of both.
- Hydrostatic extrusion: metal forced through die by high fluid pressure. Used for high strength, brittle and refractory alloys.
- Can incorporate other processes such as: cold heading, drawing, swaging, sizing and coining to produce complex parts at one station.

Economic considerations

- Production rates up to 2000/h.
- Lead times usually weeks.
- High utilization of material (95 per cent). Possible material cost savings over machining can be high. Near elimination of heat treatment and machining requirements.
- Can be economical for quantities down to 10 000, depending on complexity of part. More suited for high production volumes (100 000+).
- Most applications in the formation of symmetrical parts with solid or hollow cross sections.
- Tooling costs high.
- Equipment costs high.
- Direct labor costs low.
- Finishing costs very low.

Typical applications

- Fasteners
- Tool sockets
- Spark plug bodies
- Gear blanks
- Collapsible tubes
- Valve seats

Design aspects

- Complexity limited. Symmetry of the part is important: concentric, round or square cross-sections typical. Limited asymmetry possible.
- To avoid mismatch of dies, every effort should be made to balance the forces, especially on unsymmetrical parts.
- Length to diameter ratios of secondary formed back-extruded parts may approach 10:1; forward extrusion unlimited.
- Any parting lines should be kept in one plane and placement across critical dimensions should be avoided.
- Can be used to process two materials simultaneously to produce parts such as steel-coated copper electrodes.
- Inserts not recommended.
- Undercuts not possible.
- Draft angles not required.
- Maximum section ranging 0.25–22 mm, depending on material for impact extrusion. No limit for cold forming.
- Minimum section ranging 0.09–0.25 mm, depending on material.
- Sizes ranging $\varnothing$1.3–$\varnothing$150 mm, depending on cold formability of material being processed.

Quality issues

- Inside shoulders require secondary processing to ensure flatness.
- Cold working offers valuable increase in mechanical properties, including extended fatigue life.
- Concentricity of blank and punch is important in providing uniform section thickness.

- Supply of lubrication (commonly phosphate based) to the die surfaces is important in providing uniform material flow and reduce friction.
- Small quantities of sulfur, lead, phosphorus, silicon, etc. reduce the ability of ferrous metals to withstand cold working.
- Surface cracking: tearing of the surface of the part, especially with high temperature alloys, aluminum, zinc, magnesium. Control of the billet temperature, extrusion speed and friction are important.
- Pipe or fishtailing: metal flow tends to draw surface oxides and impurities towards center of part. Governing factors are friction, temperature gradients and amount of surface impurities in billets.
- Internal cracking or chevron cracking: similar to the necked region in a tensile test specimen. Governing factors are the die angle and amount of impurities in the billet.
- Surface detail is excellent.
- Surface roughness ranging 0.1–1.6 μm Ra.
- Process capability charts showing the achievable dimensional tolerances for impact extrusion and cold forming are provided (see 3.4CC).
- Dimensional tolerances for non-circular components are at least 50 per cent greater than those shown on the charts.

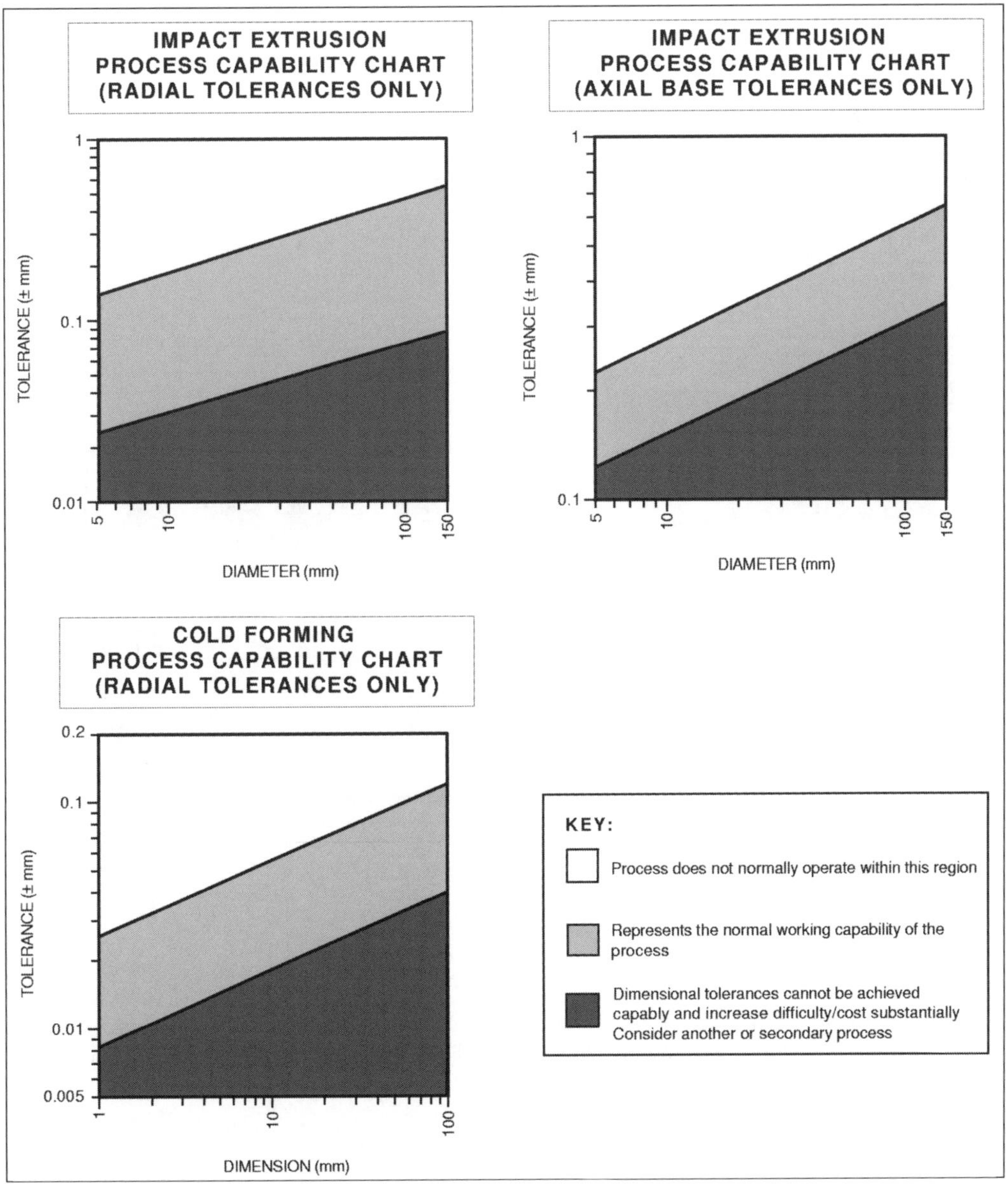
IMPACT EXTRUSION
PROCESS CAPABILITY CHART
(RADIAL TOLERANCES ONLY)
TOLERANCE (± mm)
1
0.1
0.01
5
10
100
150
DIAMETER (mm)
IMPACT EXTRUSION
PROCESS CAPABILITY CHART
(AXIAL BASE TOLERANCES ONLY)
TOLERANCE (± mm)
1
0.1
5
10
100
150
DIAMETER (mm)
COLD FORMING
PROCESS CAPABILITY CHART
(RADIAL TOLERANCES ONLY)
TOLERANCE (± mm)
0.2
0.1
0.01
0.005
1
10
100
DIMENSION (mm)
KEY:
Process does not normally operate within this region
Represents the normal working capability of the process
Dimensional tolerances cannot be achieved capably and increase difficulty/cost substantially
Consider another or secondary process

3.4CC Cold forming process capability chart.

3.5 Cold heading

Process description

- Wire form stock material is gripped in a die with usually one end protruding. The material is subsequently formed (effectively upset) by successive blows into the desired shape by a punch or a number of progressive punches. Shaping of the shank can be achieved simultaneously (see 3.5F).

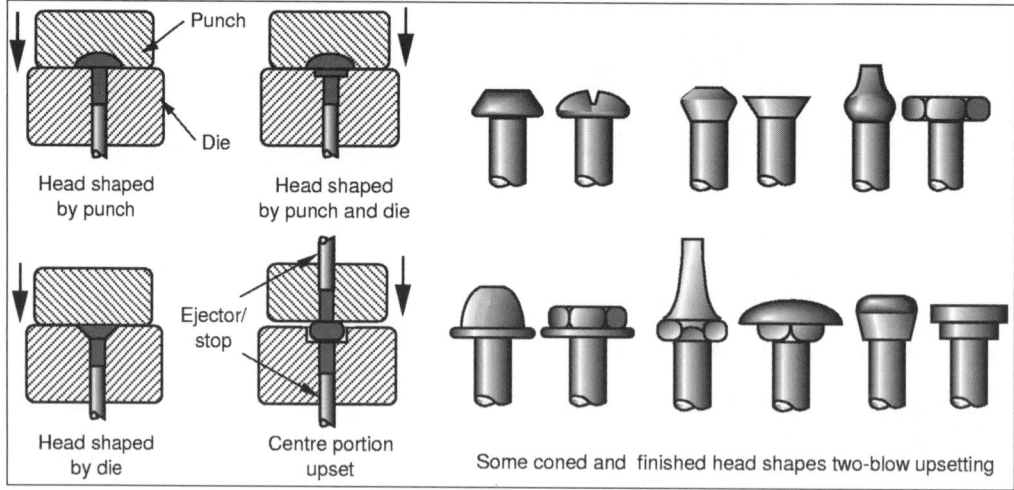

3.5F Cold heading process.

Materials

- Suitable for all ductile metals: principally, carbon steels, aluminum, copper and lead alloys.
- Alloy and stainless steels, zinc, magnesium, nickel alloys and precious metals are also processed.

Process variations

- Usually performed with stock material at ambient temperature (cold), but also with stock material warm or hot.
- Solid die: single stroke, double stroke, three blow, two die, progressive bolt makers, cold or hot formers – the choice is determined by the length to diameter ratio of the raw material.
- Open die: parts made by this process have wide limits and are too long for solid dies.
- Continuous rod or cut lengths of material can be supplied to the dies.
- Can incorporate other forming processes, for example: knurling, thread rolling and bending to produce complex parts at one machine.
- Upset forging: heated metal stock gripped by dies and end pressed into desired shape, i.e. increasing the diameter by reducing height.

Economic considerations

- Production rates between 35 and 120/min common.
- Lead times relatively short due to simple dies.
- High material utilization. Virtually no waste.
- Flexibility moderate. Tooling tends to be dedicated.
- Production quantities typically very high, 100 000+, but can be as low as 10 000.
- Tooling costs moderate.
- Equipment costs moderate.
- Direct labor costs low. Process highly automated.
- Finishing costs low: normally no finishing is required.

Typical applications

- Electronic components
- Electrical contacts
- Nails
- Bolts and screws
- Pins
- Small shafts

Design aspects

- Complexity limited to simple cylindrical forms with high degree of symmetry.
- Significant asymmetry difficult.
- Minimization of shank diameter and upset volume important.
- Radii should be as generous as possible.
- Threads on fasteners should be rolled wherever possible.
- Head volumes limited due to amount of deformation possible.
- Inserts possible at added cost.
- Undercuts produced via secondary operations.
- Machining usually not required.
- Draft angles not required.
- Minimum diameter $= \varnothing 0.8$ mm.
- Maximum diameter $= \varnothing 50$ mm.
- Minimum length $= 1.5$ m.
- Maximum length $= 250$ mm.

Quality issues

- Cold working process gives improved mechanical properties.
- Fatigue, impact and surface strength increased giving a tough, ductile, crack resistance structure.
- Small quantities of sulfur, lead, phosphorus, silicon, etc. reduce the ability of ferrous metals to withstand cold working.
- Length to diameter ratio of protruding shank to be formed should be below 2:1 to avoid buckling.
- Residual stresses may be left at critical points.
- Sharp corners reduce tool life.
- Surface detail is good to excellent.

- Surface roughness ranging 0.8–6.3 µm Ra.
- Process capability charts showing the achievable dimensional tolerances for cold heading are provided (see 3.5CC).

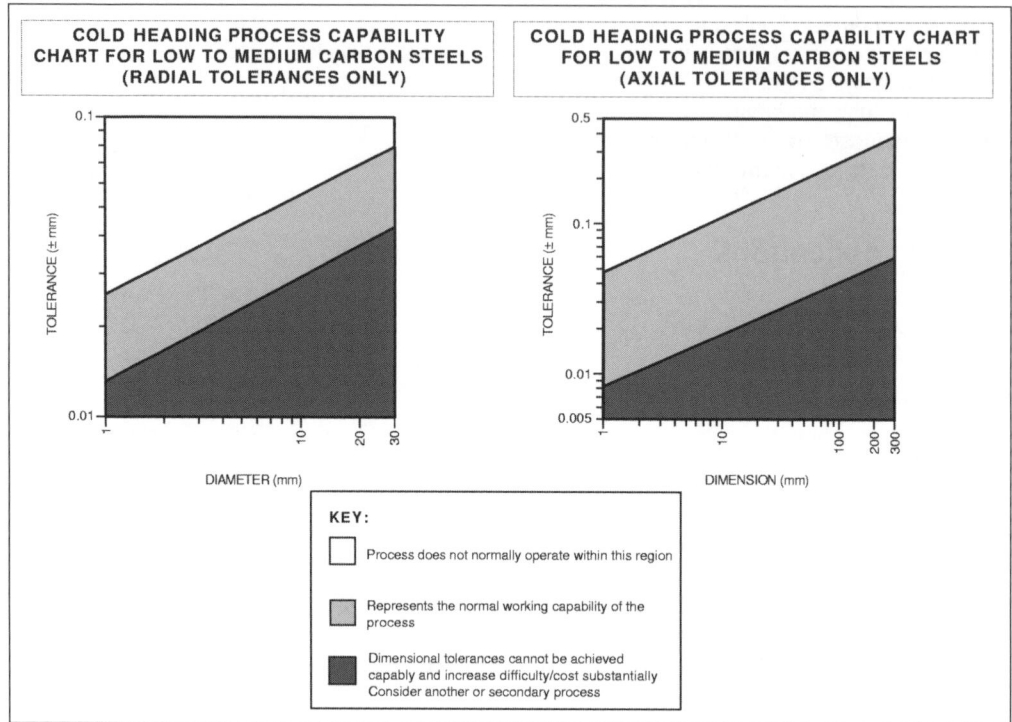

3.5CC Cold heading process capability chart.

3.6 Swaging

Process description

- Process of gradually shaping and reducing the cross section of tubes, rods and wire using successive blows from hard dies rotating around the material (on a mandrel if necessary for tubular sections). Operation performed at ambient temperature (see 3.6F).

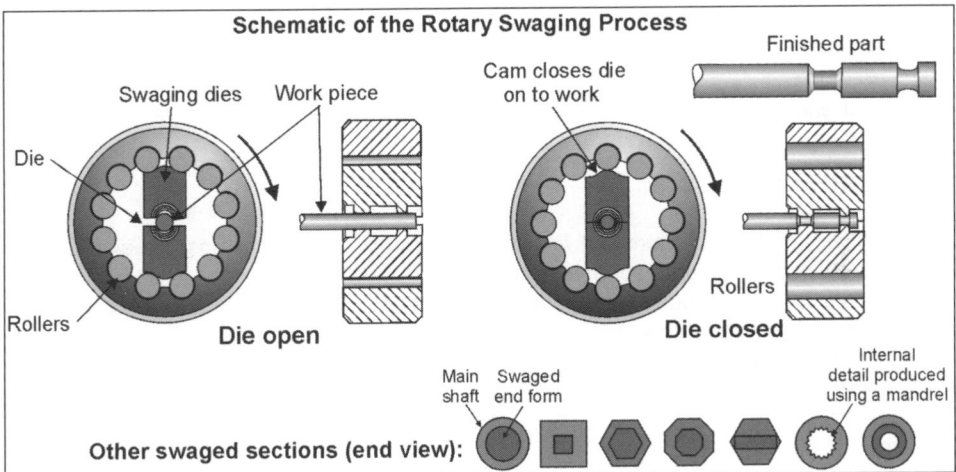

3.6F Swaging process.

Materials

- Carbon, low alloy and stainless steels, aluminum, magnesium, nickel and their alloys. Copper, zinc, lead and their alloys less commonly.

Process variations

- Using a shaped mandrel can generate inner section profiles different to outer.
- Hand forging: hot material reduced, upset and shaped using hand tools and an anvil, commonly associated with the blacksmith's trade. For decorative and architectural work.

Economic considerations

- Production rates moderate to high (100–300/h).
- Lead time typically days depending on complexity of tool.
- Special tooling not necessarily required for each job.
- Material utilization excellent. No scrap generated.
- Some automation possible.
- Economical for low production runs. Can be used for one-offs.
- Tooling costs high.

- Equipment costs generally moderate.
- Direct labor costs low to moderate.

Typical applications

- Used to close tubes, produce tapering, clamping and steps in sections
- Many section types possible either parallel or tapered
- Tool shafts
- Punches
- Chisels
- Handles
- Exhaust pipes
- Cable assemblies
- Architectural work

Design aspects

- Complexity fairly high. Round, square, rectangular and polygon sections possible either parallel or tapered. Splines and contoured surfaces also possible.
- Holes possible, but only through the length of the part.
- No undercuts or inserts possible.
- Draft angles ranging 0–3.5°.
- Minimum section = 2.5 mm.
- Maximum section = 50 mm.
- Minimum solid diameter = $\varnothing$2.5 mm.
- Maximum solid diameter = $\varnothing$150 mm.
- Maximum tube diameter = $\varnothing$350 mm.
- Minimum length = 1.5 mm.
- Maximum length = 250 mm.

Quality issues

- Cold working of material gives good mechanical properties and compressive surface stresses improve fatigue life.
- Surface finish of stock material is markedly improved.
- Surface detail is good to excellent.
- Surface roughness values ranging 0.8–3.2 µm Ra.
- A process capability chart showing the achievable dimensional tolerances for swaging is provided (see 3.6CC).

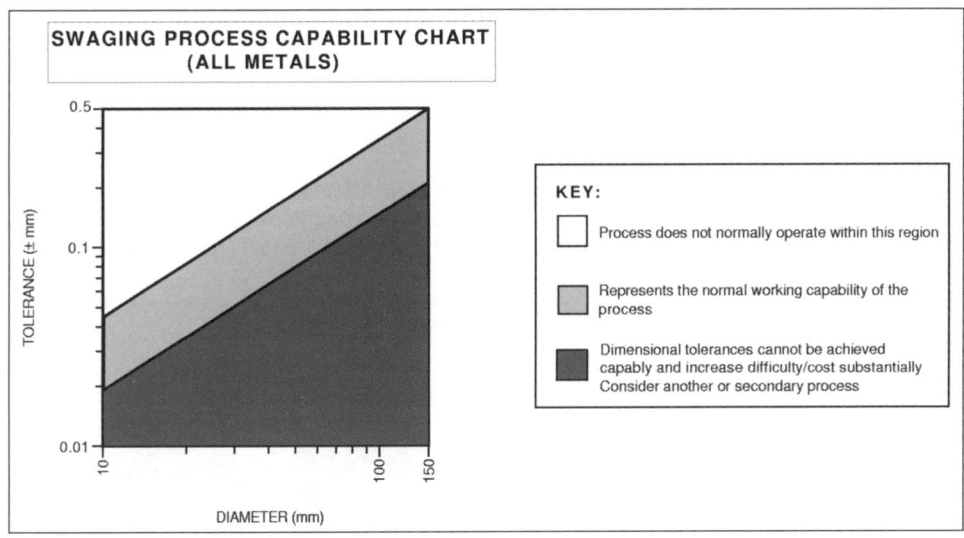

3.6CC Swaging process capability chart.

3.7 Superplastic forming

Process description

- Sheet metal is clamped over a male or female form tool and heated to a high enough temperature to give the material high ductility at low strain rates. Pressurized gas (typically argon) on the back face of the sheet forms the material into cavity or over the surface of the tool (see 3.7F).

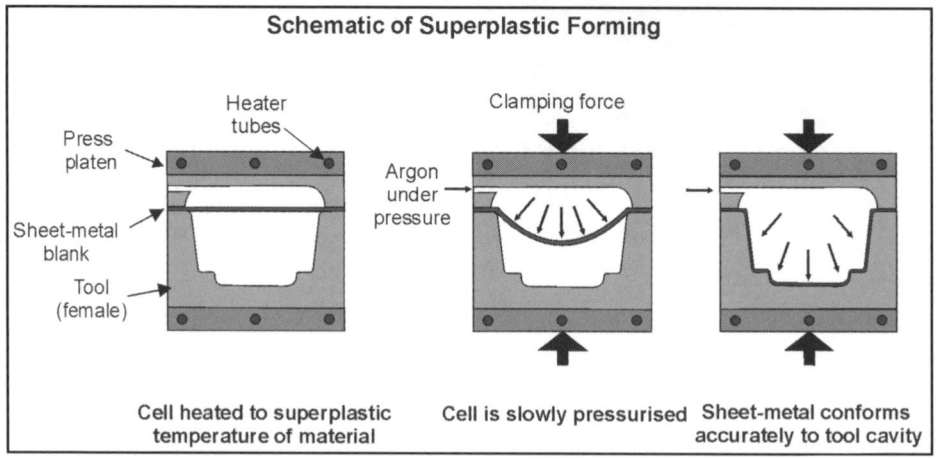

Schematic of Superplastic Forming

Heater tubes

Press platen

Sheet-metal blank

Tool (female)

Argon under pressure

Clamping force

Cell heated to superplastic temperature of material

Cell is slowly pressurised

Sheet-metal conforms accurately to tool cavity

3.7F Superplastic forming process.

Materials

- For stainless steels, aluminum and titanium alloys typically.
- Material must be able to deform at low strain rates and high temperatures and possess a stable microstructure.

Process variations

- Either male or female tool: male forming is more complex, but offers greater design freedom and more uniform material distribution.
- Additional use of tool movement with gas pressure gives deeper parts with more uniform wall thickness.
- Diaphragm forming: uses additional diaphragm sheet behind material to give better control of thickness distribution.
- Can be used in conjunction with Diffusion Bonding (Welding) (DFW) to create complex parts (see 7.9, Solid State Welding).

Economic considerations

- Production rates low. Long cycle times.
- Slower than conventional deep drawing.

- Lead times moderate, typically weeks, depending on complexity of mold.
- Material utilization good. Some waste may be generated during subsequent trimming operations. Scrap not recyclable directly.
- Some aspects can be automated.
- Economically viable for low to moderate production volumes (10–10 000). Can be used for one-offs.
- Tooling costs high.
- Equipment costs high.
- Direct labor costs low to moderate.
- Finishing costs low to moderate. Trimming typically required.

Typical applications

- Used to generate deep and intricate forms in sheet metal
- Aerospace fuselage panels
- Containers
- Casings
- Architectural and decorative work
- Can also be used to clad other materials

Design aspects

- Complexity limited to shape of female or male tool and constant thickness parts.
- Ribs, bosses and recesses possible.
- Re-entrant features not possible.
- Radii should be greater than five times the wall thickness.
- Sharp radii at extra cost can be produced.
- Draft angles ranging 2 to 3°.
- Maximum drawing ratio (height to width) = 0.6.
- Maximum dimension = 2.5 m.
- Maximum thickness = 4 mm.
- Minimum thickness = 0.8 mm.

Quality issues

- No spring back exhibited after processing.
- No residual stresses.
- Creep performance poor due to small grain sizes produced.
- Cavitation and porosity can occur in some alloys at high temperatures and low strain rates.
- Graphite coating on the blank sheet used to reduce friction.
- Surface detail good.
- Secondary operations such as heat treatment, paint, powder coating and anodizing commonly used to improve finish.
- Surface roughness ranging 0.4–6.3 μm Ra.
- Achievable dimensional tolerances ranging ±0.13 mm–±0.25 mm up to 25 mm, ±0.45 mm–±0.78 mm up to 150 mm. Wall thickness tolerances are typically ±0.25 mm.

3.8 Sheet-metal shearing

Process description

- Various shearing processes used to cut cold rolled sheet metal with hardened punch and die sets. The most common shearing processes are: cutting, piercing, blanking and fine blanking (see 3.8F).

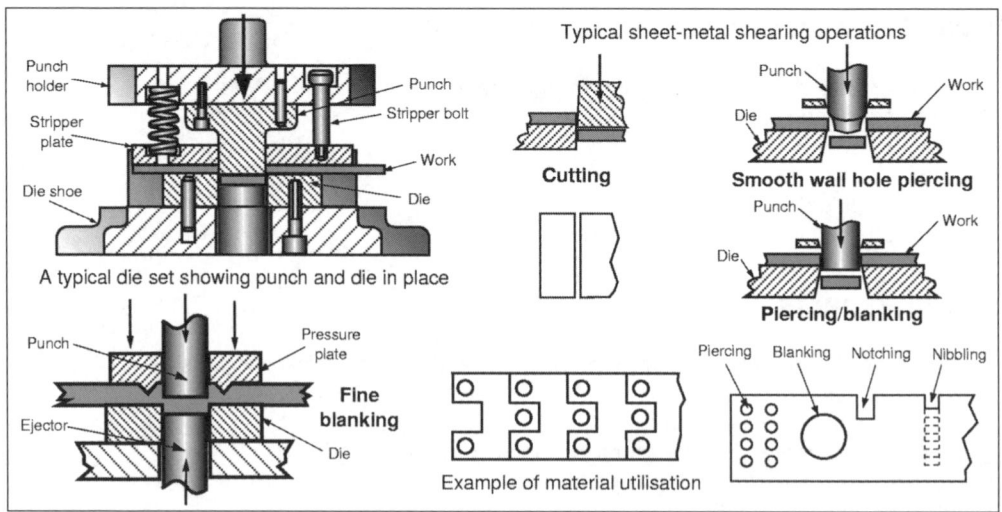

3.8F Sheet-metal shearing process.

Materials

- All ductile metals available in cold rolled sheet form, supplied flat or coiled.
- Most commonly used metals are: carbon steels, low alloy steels, stainless steels, aluminum alloys and copper alloys. Also, nickel, titanium, zinc and magnesium alloys are processed to a lesser degree.

Process variations

- Mechanical drives: faster action and more positive displacement control.
- Hydraulic drives: greater forces and more flexibility.
- Cutting: large sheets of metal are clamped and cut along a straight line.
- Piercing: removal of material from a blank, for example, a hole.
- Blanking: parts are blanked to obtain the final outside shape.
- Fine blanking: uses special clamping tooling to produce a smooth and square-edged contoured blank or hole.
- Smooth wall hole piercing: special punch profiles are used to produce crack-free holes.
- Other operations include: nibbling, notching, trimming and shaving.
- Computer Numerical Control (CNC) common on piercing and blanking machines.

Economic considerations

- Production rates high, 10 000+/h for small components.
- High degree of automation possible.
- Cycle time usually determined by loading and unloading times for stock material.
- Progressive dies can incorporate shearing and forming processes.
- Lead times can be several weeks depending on complexity and degree of automation, but more typically, several days.
- Material utilization moderate to high, however, substantial amounts of scrap can be produced in piercing and blanking.
- Production quantities should be high for dedicated tooling, 10 000+. Economical quantities can range from 1 for blanking and piercing to 2000 for fine blanking.
- Tooling cost moderate to high, depending on process and degree of automation.
- Equipment costs vary greatly. Low for simple guillotines to high for high speed, precision CNC presses.
- Labor costs low to moderate depending on degree of automation.
- Finishing costs low to moderate. Deburring and cleaning usually required.

Typical applications

- Blanks for forming work
- Cabinet panels
- Domestic appliance components
- Machine parts
- Gears and levers
- Washers

Design aspects

- Complex patterns of contours and holes possible in two dimensions.
- Material used dictates press forces and die clearances.
- Blanked parts should be designed to make the most use of the stock material.
- Pierced holes with their diameter greater than the material thickness should be drilled.
- Fine blanked holes with diameters 60 per cent of the material thickness possible.
- Holes should be spaced at least 1.5 times the thickness of the material away from each other.
- Maximum sheet thickness $= 13$ mm.
- Minimum sheet thickness $= 0.1$ mm.
- Maximum sheet dimension for cutting is 3 m; 1 m for fine blanking.

Quality issues

- Conventional hole piercing, blanking and cutting do not result in a perfectly smooth and parallel cut.
- Acceptable hole wall and blank edge quality may be achieved with fine blanking and piercing processes.
- Holes placed too close to a bend line can be distorted subsequently in forming operations.
- Inspection and maintenance of die wear and breakage is important.
- Variations in stock material thickness and flatness should be controlled.
- Surface detail is good.
- Surface roughness values ranging 0.1–12.5 μm Ra.
- Process capability charts showing the achievable dimensional tolerances for several sheet-metal shearing processes are provided (see 3.8CC).

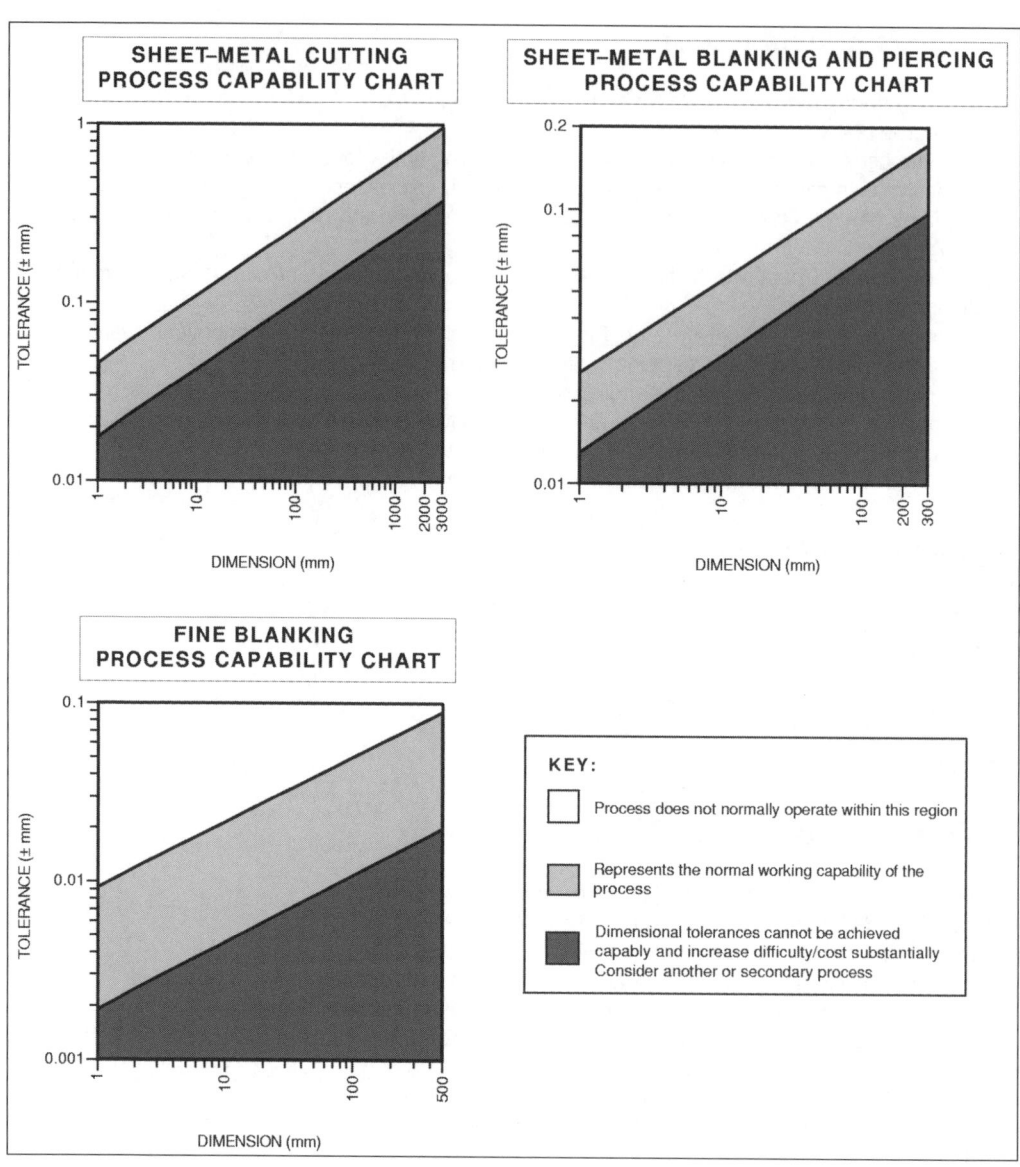

3.8CC Sheet-metal shearing process capability chart.

3.9 Sheet-metal forming

Process description

- Various processes are used to form cold rolled sheet metal using die sets, formers, rollers, etc. The most common processes are: deep drawing, bending, stretch forming and roll forming (see 3.9F).

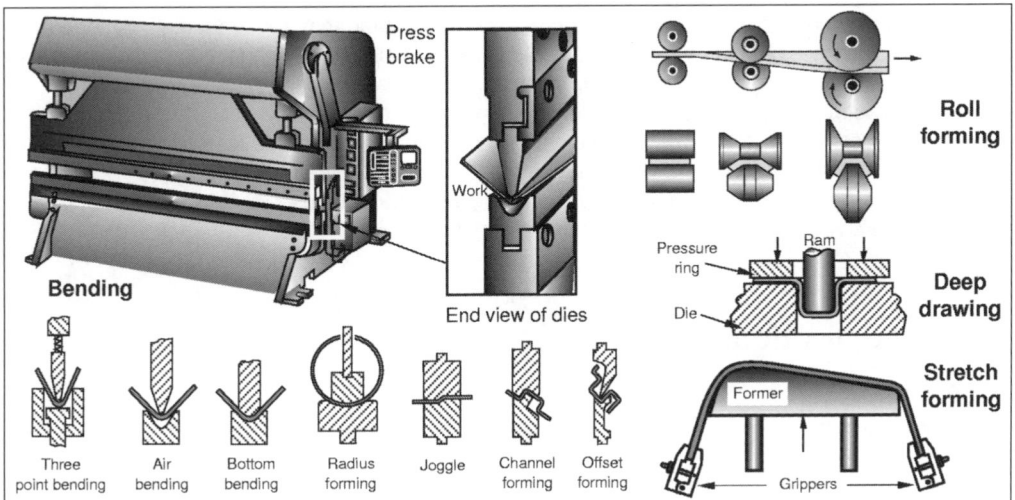

3.9F Sheet-metal forming process.

Materials

- All ductile metals available in cold rolled sheet form, supplied as blanks, flat or coiled.
- Most commonly used metals are: carbon steels, low alloy steels, stainless steels, aluminum alloys and copper alloys. Also, nickel, titanium, zinc and magnesium alloys are processed to a lesser degree.
- Also coated materials, such as galvanized sheet steel.

Process variations

- Mechanical drives: faster action and more positive displacement control.
- Hydraulic drives: greater forces and more flexibility.
- Deep drawing: forming of a blank into a closed cylindrical or rectangular shaped die. Incorporating an ironing operation improves dimensional tolerances.
- Bending: deformation about a linear axis to form an angled or contoured profile.
- Stretch forming: sheet metal is clamped and stretched over a simple form tool.
- Roll forming: forming of long lengths of sheet metal into complex profiles using a series of rolls.
- Beading: edge of sheet bent into cavity of die. May be used to remove sharp edges.
- Hemming: edge of sheet folded over. May be used to remove sharp edges.
- Can incorporate initial sheet metal shearing operations.

Economic considerations

- Production rates vary, up to 3000/h for small components using automated processes.
- Deep drawing punch speeds a function of material; high to low – brass, aluminum, copper, zinc, steel, stainless steel (typically 800/h).
- High degree of automation is possible.
- Cycle time is usually determined by loading and unloading times for the stock material.
- Lead times vary, up to several weeks for deep drawing and stretch forming; could be less than an hour for bending.
- Material utilization is moderate to high (10–25 per cent scrap generated). Bending and roll forming do not produce scrap directly. Deep drawing and stretch forming may require a trimming operation.
- Production quantities should be high for dedicated tooling, 10 000+. Minimum economical quantities range from 1 for bending to 1000 for deep drawing.
- Tooling cost is moderate to high, depending on component complexity.
- Equipment costs vary greatly; low for simple bending machines, moderate for roll forming machines and high for automated deep drawing, sheet metal presses and stretch forming.
- Labor costs are low to moderate, depending on degree of automation.
- Finishing costs are low. Trimming and cleaning may be required.

Typical applications

- Cabinets
- Mounting brackets
- Electrical fittings
- Cans
- Machine frames
- Automotive body panels
- Aircraft fuselage panels
- Light structural sections
- Domestic appliances
- Kitchen utensils

Design aspects

- Complex forms possible: several processes may be combined to produce one component, or a series of operations used to progressively form the part.
- Working envelope of machine and uniform thickness of sheet can restrict design options.
- No inserts or re-entrant angles.
- Draft angles maybe required (0.25°).
- Minimum bend radii are a function of material and sheet thickness, but typically four times the sheet thickness.
- Radii kept as large as possible, particularly if parallel with grain of material.
- Square or rectangular boxes limited by sharpness of corner detail required.
- Minimum sheet thickness = 0.1 mm.
- Maximum sheet thickness: deep drawing = 12 mm, bending = 25 mm, roll/stretch forming = 6 mm.
- Sizes ranging $\varnothing$2–$\varnothing$600 mm for deep drawing; 10 mm–1.5 m width for roll forming; 2 mm–3.6 m width for bending.

Quality issues

- Bending and stretch forming are limited by the onset of necking.
- The limiting drawing ratio (blank diameter/punch diameter) is between 1.6 and 2.2 for most materials. This should be observed where drawing takes place without progressive dies, otherwise excessive thinning and tearing could occur.
- Variations in stock material thickness and flatness should be controlled.
- Other problems include: spring-back (metal returns to original form) and wrinkling during drawing (comparable with forcing a circular piece of paper into a drinking glass), eliminated by adjustment of blank holder force.
- Spring-back can also be compensated for by over-bending, coining and stretch-bending operations.
- High residual stresses can be generated. Subsequent heat treatment may be necessary.
- Surface detail is good.
- Surface roughness is approximately that of the sheet-material used.
- Process capability charts showing the achievable dimensional tolerances are provided (see 3.9CC).

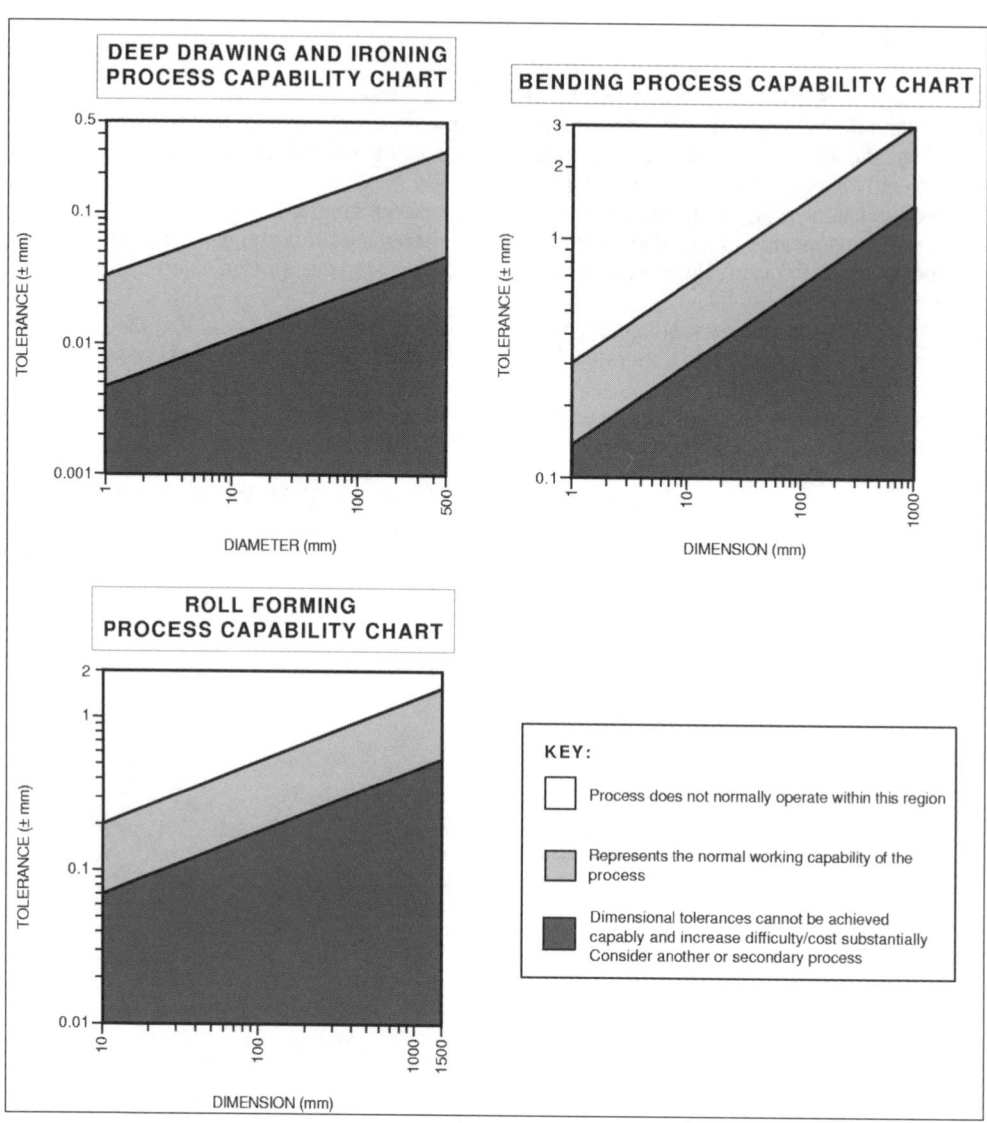

3.9CC Sheet-metal forming process capability chart.

3.10 Spinning

Process description

- Forming of sheet-metal or thin tubular sections using a roller or tool to impart sufficient pressure for deformation against a mandrel while the work rotates (see 3.10F).

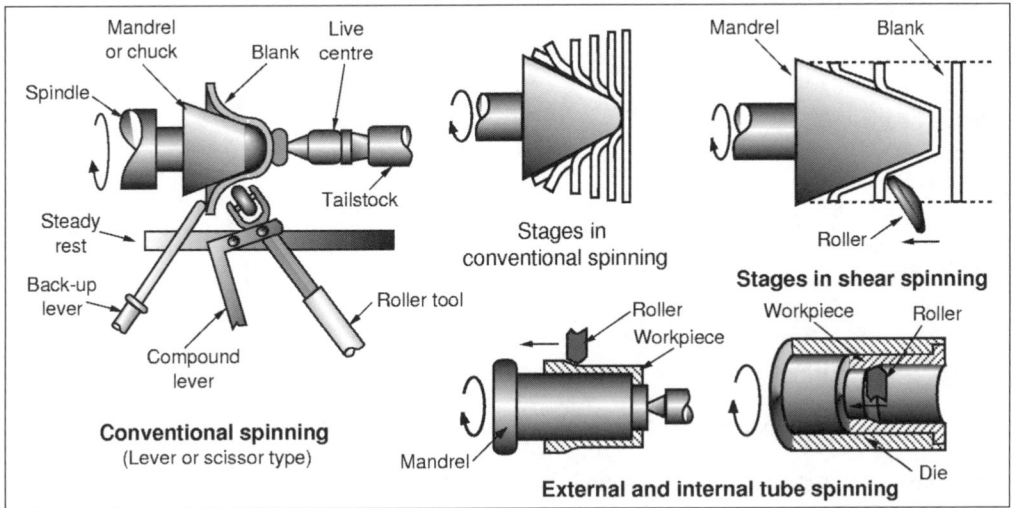

3.10F Spinning process.

Materials

- All ductile metals that are available in sheet form. The most common metals used are carbon steels, stainless steels, aluminum alloys, copper alloys and zinc alloys.
- Used on a more limited basis are: magnesium, tin, lead, titanium and nickel alloys.

Process variations

- Tube spinning: thickness reduction of a cylindrical section on a mandrel, either internally or externally.
- Flame spinning: oxyacetylene flame heats material prior to forming. Permits rapid forming of parts with thick sections.
- Shear spinning: point extrusion process reduces thickness of starting blank or shape to produce the final form.

Economic considerations

- Production rates are low, typically 10–30/h.
- Lead times are short. Simple mandrels made quickly.

- Material utilization is moderate. Main losses occur in cutting blanks.
- Flexibility is high: formers are changed quickly and setup times are low.
- Production volumes are viable from 10 to 10 000. Can be used for one-offs.
- Tooling costs are low.
- Equipment costs are moderate.
- Labor costs are high. Skilled labor is needed.
- Finishing costs are low. Cleaning and trimming required.

Typical applications

- Flanged, dished, spherical and conical shapes
- Nose cones
- Missile heads
- Bells
- Light shades
- Cooking utensils
- Funnels
- Reflectors

Design aspects

- Complexity is limited to thin-walled, conical, concentric shapes. Typically, the diameter is twice the depth.
- Cylindrical or cup-shaped pieces are the most difficult of the simple shapes.
- Oval or elliptical parts are possible, but expensive.
- Material thickness, bend radii, depth of spinning, diameter, steps in diameter and workability of the material are important issues in spinning.
- Radii should be at least 1.5 times the material thickness.
- Stiffening beads should be formed externally rather than internally.
- No draft angle is required.
- Undercuts are possible, but at added cost.
- Maximum section is 75 mm for automated spinning, but approximately 6 mm for hand spinning.
- Minimum section $= 0.1$ mm.
- Sizes ranging $\varnothing 6$ mm–$\varnothing 7.5$ m.

Quality issues

- Skill and experience are required to cause the metal to flow at the proper rate avoiding wrinkles and tears.
- Streamlined or smooth curves and large radii are an aid both to manufacture and improved appearance.
- Associated problems are blank development and proper feed pressure.
- Grain flow and cold working give good mechanical properties.
- Surface detail is good.
- Surface roughness ranging 0.4–3.2 μm Ra.
- A process capability chart showing the achievable dimensional tolerances is provided (see 3.10CC).

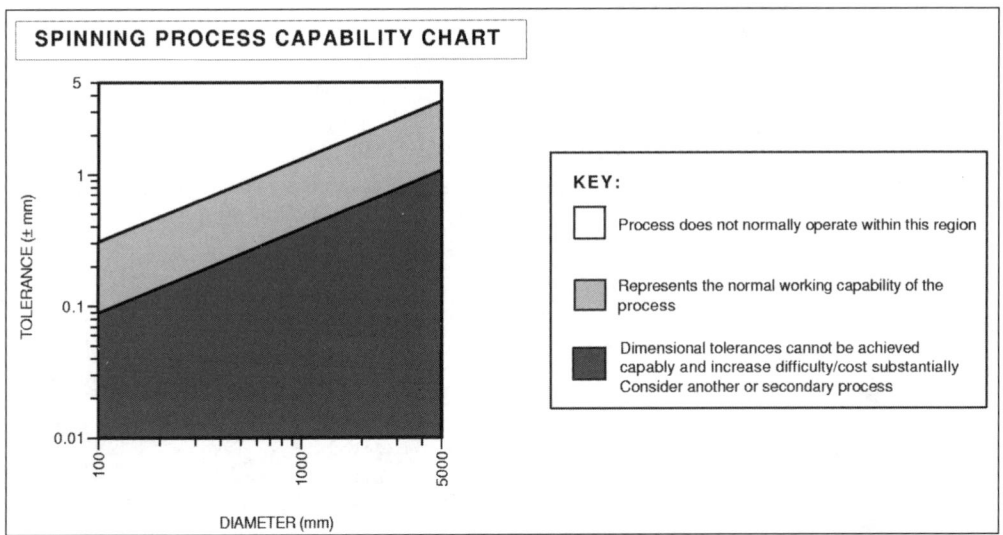

3.10CC Spinning process capability chart.

3.11 Powder metallurgy

Process description

- Die compaction of a blended powdered material into a 'green' compact which is then sintered with heat to increase the bond strength. Usually secondary operations are performed to improve dimensional accuracy, surface roughness, strength and/or porosity (see 3.11F).

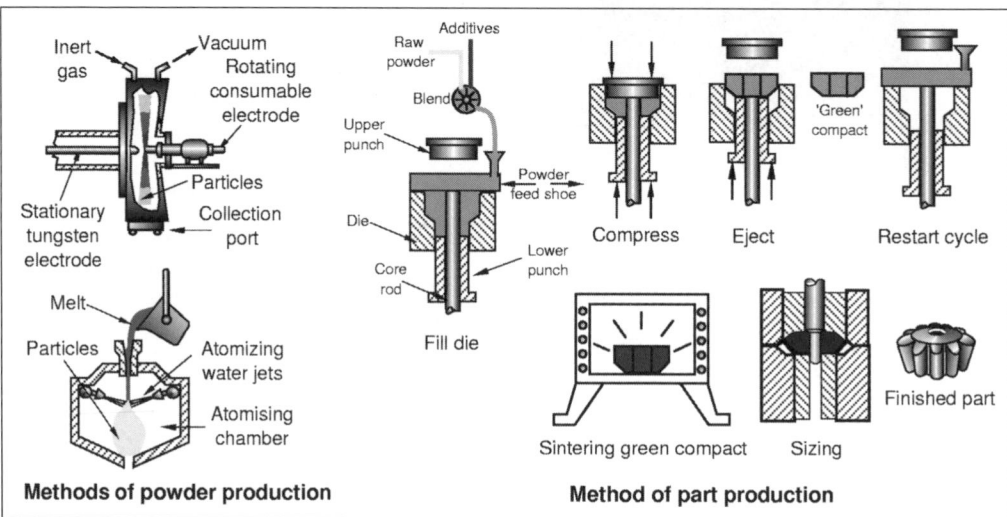

Methods of powder production   **Method of part production**

3.11F Powder metallurgy process.

Materials

- All materials, typically metals and ceramics. Iron, copper alloys and refractory metals most common.
- Can process materials not formable by other methods.
- Powder production: atomization, electrolysis and chemical reduction methods.

Process variations

- Cold die compaction: performed at room temperature. Gives high porosity and low strength.
- Hot forging: deformation of reheated sintered compact to final density and shape.
- Continuous compaction: for strip or sheet product. Slower than conventional rolling.
- Isostatic compaction (hot or cold): compaction of powder in a membrane using pressurized fluid (oil, water) or gas. Permits more uniform compaction and near-net shapes. Undercuts and reverse tapers possible, but not transverse holes. Used for ceramics mainly.
- Extrusion: high pressure ram forces powder through an orifice determining the section profile.
- Injection molding: fine powder coated with thermoplastic injected into dies. Relatively complex shapes with thin walls achievable.
- Spark sintering: gives magnetic and electrical properties.

- Pressureless compaction: for porous components.
- Secondary operations include: repressing, sizing and machining.

Economic considerations

- High production rates, small parts up to 1800 pieces/hour.
- Cycle times dictated by sintering mechanisms.
- Lead times several weeks. Dies must be carefully designed and made.
- Production quantities of 20 000+ preferred, but may be economic for 5000 for simple parts.
- Material utilization very high. Less than 5 per cent lost in scrap.
- Powders expensive to produce.
- Automation of process common.
- New set of die and punches required for each new product, i.e. flexibility low.
- Tooling costs very high. Dedicated tooling.
- Equipment costs high. Sintering equipment not dedicated though.
- Labor costs low to moderate. Some skilled labor may be required.
- Finishing costs generally low.
- Final grinding may be more economical than sizing for very close tolerances.

Typical applications

- Cutting tools
- Small arms parts
- Bearings
- Filters (porous)
- Lock components (keys, barrels)
- Machine parts (ratchets, pawls, cams, gears)

Design aspects

- Complexity and part size limited by powder flow through die space (powders do not follow hydro-dynamic laws) and pressing action.
- Near-net shapes generated.
- Concentric, cylindrical shapes with uniform, parallel walls preferred.
- Multiple-action tooling can be used to create complex parts.
- Complex profiles on one side only.
- Parts can be quenched, annealed and surface treated like wrought products to alter mechanical properties.
- Inert plastics can be impregnated for pressure sealing or a low melting point metal for powder forging.
- Densities typically between 90 per cent and 95 per cent of original material.
- Density can be controlled for special functional properties, e.g. porosity for filters.
- Spheres approximated. Complicated radial contours possible.
- Marked changes in section thickness should be avoided.
- Narrow slots, splines, long thin section, knife-edges and sharp corners should be avoided. Use secondary processing operations.
- Threads not possible.
- Tapered, blind and non-circular holes, vertical knurls possible in direction of powder compaction.
- Grooves, cutouts and off-axis holes perpendicular to the pressing direction can not be produced directly.

- Undercuts perpendicular to compaction direction not possible. Better to secondary process.
- Radii should be as generous as possible.
- Chamfers preferred to radii on part edges.
- Maximum length to diameter ratio = 4:1.
- Maximum length to wall thickness ratio = 8:1.
- Inserts possible at extra cost.
- Draft angles can be zero.
- Minimum section can be as low as 0.4 mm, but 1.5 mm typically.
- Sizes ranging 10 g–15 kg in weight or 4 mm^2–0.016 m^2 in projected area.

Quality issues

- Density and strength variations in product can occur with asymmetric shapes. Can be minimized by die design.
- High densities required for subsequent welding of sintered parts.
- Porosity in sintered parts means excessive absorption of braze and solder fillers.
- Product strength determinable by powder size, compacting pressure, sintering time and temperature, but generally, lower mechanical properties than wrought materials.
- Can give a highly porous structure, but can be controlled and used to advantage, e.g. filters and bearing lubricant impregnation (10–30 per cent oil by volume). Also, resin impregnation can greatly improve machinability of sintered products.
- Generally, lower mechanical properties than wrought materials.
- Sharp edges on tools should be avoided. Causes excessive tool wear.
- Remnants of contaminates at grain boundaries may act as crack initiators.
- Oxide film may impair properties of finished part, for example: chromium and high temperature superalloys.
- Surface detail good.
- Surface roughness in the range 0.2–3.2 µm Ra.
- Process capability charts showing the achievable dimensional tolerances are provided (see 3.11CC).
- Repressing, coining and sizing improves surface finish, density and dimensional accuracy. Also for embossing.

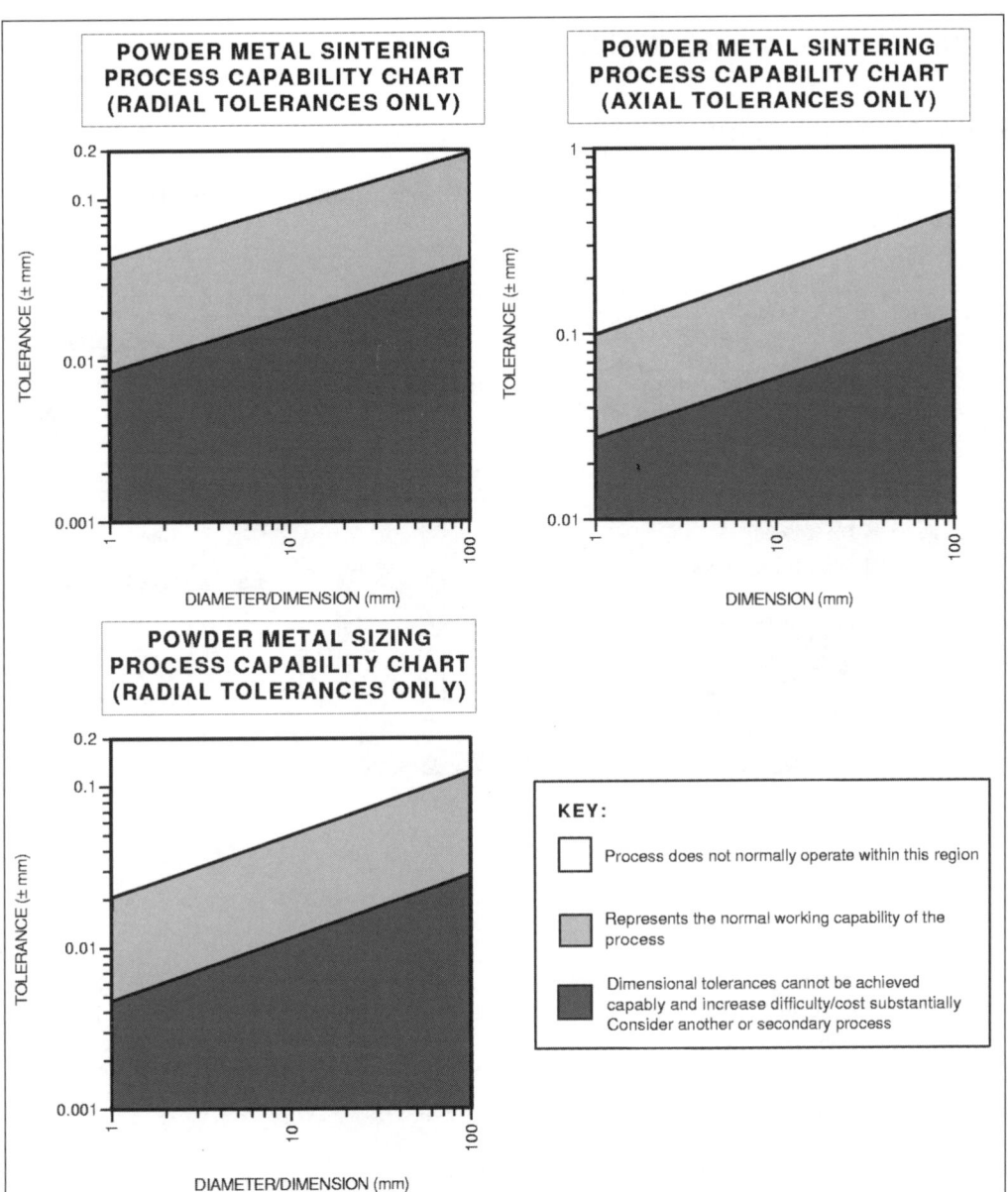

3.11CC Powder metallurgy process capability chart.

3.12 Continuous extrusion (metals)

Process description

- A billet of the raw material, either hot or cold, is placed into a chamber and forced through a die of the required profile with a ram (see 3.12F).

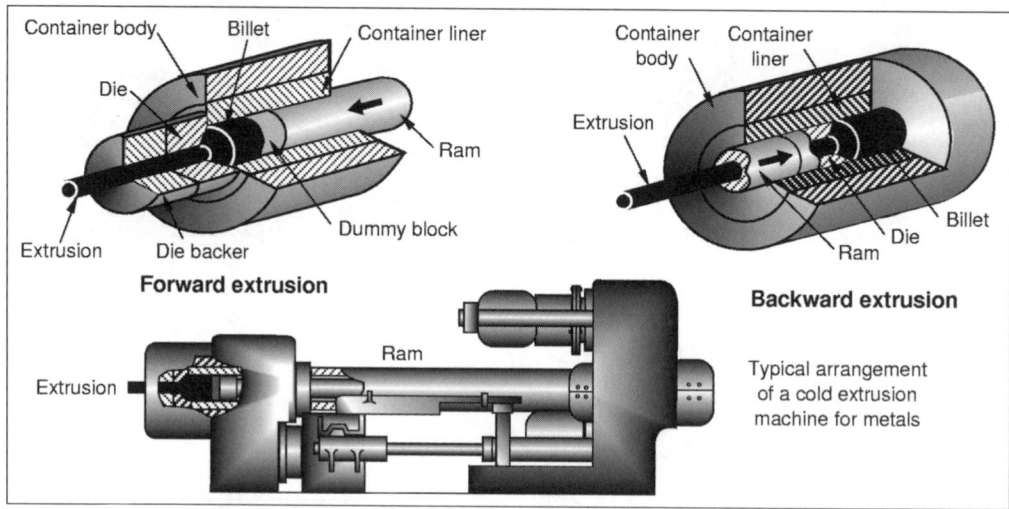

3.12F Continuous extrusion (metals) process.

Materials

- Most ductile metals, for example: aluminum, copper and magnesium alloys. To a lesser degree: zinc, lead, tin, nickel and titanium alloys, refractory metals, and carbon, low alloy and stainless steels are processed.

Process variations

- Forward extrusion: billet extruded by ram from behind.
- Backward extrusion: billet displaced by advancing ram with die attached to front and extrudate travels through the center of the ram. Limited to short lengths only. Similar to Hooker process.
- Cold extrusion: increases friction and therefore processing energy/forces, but increases dimensional accuracy. May be performed warm. Viable for materials possessing adequate cold-working ability.

Economic considerations

- Production rates high but are dependent on size and complexity. Continuous lengths up to 12 m/min.
- Cut extruded length up to 1000/h is possible.

- Extruders often run below their maximum speed for trouble-free production.
- Can have multiple holes in die for increased production rates and lower wear rates.
- Extruder costs increasing steeply at the higher range of output.
- Lead times dependent on the complexity of the 2-dimensional die, but normally weeks.
- Material utilization high (less than 1.5 per cent scrap). Waste is only produced when cutting continuous section to length.
- Process flexibility is moderate: tooling is dedicated, but changeover and setup times are short.
- Short production runs viable if section designed with part consolidation and integral fastening in mind.
- Minimum billet size = 250 kg (equivalent to 500 m).
- Tooling costs moderate.
- Equipment costs high.
- Direct labor costs low.
- Finishing costs low. Deburring cost can be high for small cut lengths.

Typical applications

- Profiles cut to required length, e.g. gear blanks
- Wrought bar and sections for other processing methods
- Window frames
- Structural sections, corner and edge members
- Decorative trim
- Railings

Design aspects

- Dedicated to long products with uniform, symmetrical or varying cross section.
- Cross section may be fairly complex (round, tube, square, T-section, L-section).
- Difficult to control internal dimensions of hollow sections which use complex mandrels or spiders held in the die.
- Can eliminate or minimize secondary processing operations. Section profile designed to decrease amount of machining required and/or increase assembly efficiency by integrating part consolidation features.
- Solid forms including re-entrant angles.
- Grooves and holes not parallel to the axis of extrusion must be produced by a secondary operation.
- Too greater variation in adjacent section thicknesses (less than 2:1 if possible) should be avoided.
- Very small holes through the profile should be avoided.
- Use of materials other than aluminum and copper alloys can cause shape restrictions.
- Radii should be as generous as possible. Concave radii greater than 0.5 mm.
- No draft angle required.
- Maximum extrusion ratios: 40:1 for aluminum alloys, 5:1 for carbon steel.
- Section thickness should be greater than 1.5 per cent of the maximum dimension.
- Maximum diameter = $\emptyset$250 mm.
- Minimum section = 1 mm for aluminum and magnesium alloys; 3 mm for steel.
- Sizes ranging 8–500 mm sections, but dependent on complexity and material used.

Quality issues

- Cold extrusion eliminates oxidation and gives a better surface finish.
- Knife-edges and long, thin sections should be avoided.

- Warp and twist can be troublesome.
- Plastic working in cold extrusion produces favorable grain structure and directional properties (anisotropy).
- The rate and uniformity of cooling are important for dimension control because of shrinkage and distortion in hot extrusion.
- Die swell, where the extruded product increases in size as it leaves the die, may be compensated for by:

  - Increasing haul-off rate compared with extrusion rate
  - Decreasing extrusion rate
  - Increasing the length of the die land
  - Decreasing the melt temperature.

- High die wear is reduced by selection of appropriate lubricant to reduce friction.
- Stretching post extrusion eliminates bowing and twisting (2°/m not untypical).
- Surface detail is good to excellent. Surface roughness is in the range 0.4–12.5 μm Ra.
- Process capability charts showing the achievable dimensional tolerances are provided (see 3.12CC).
- Wall thickness tolerances are between ±0.15 and ±0.25 mm.
- Straightness tolerance is 0.3 mm/m, typically.

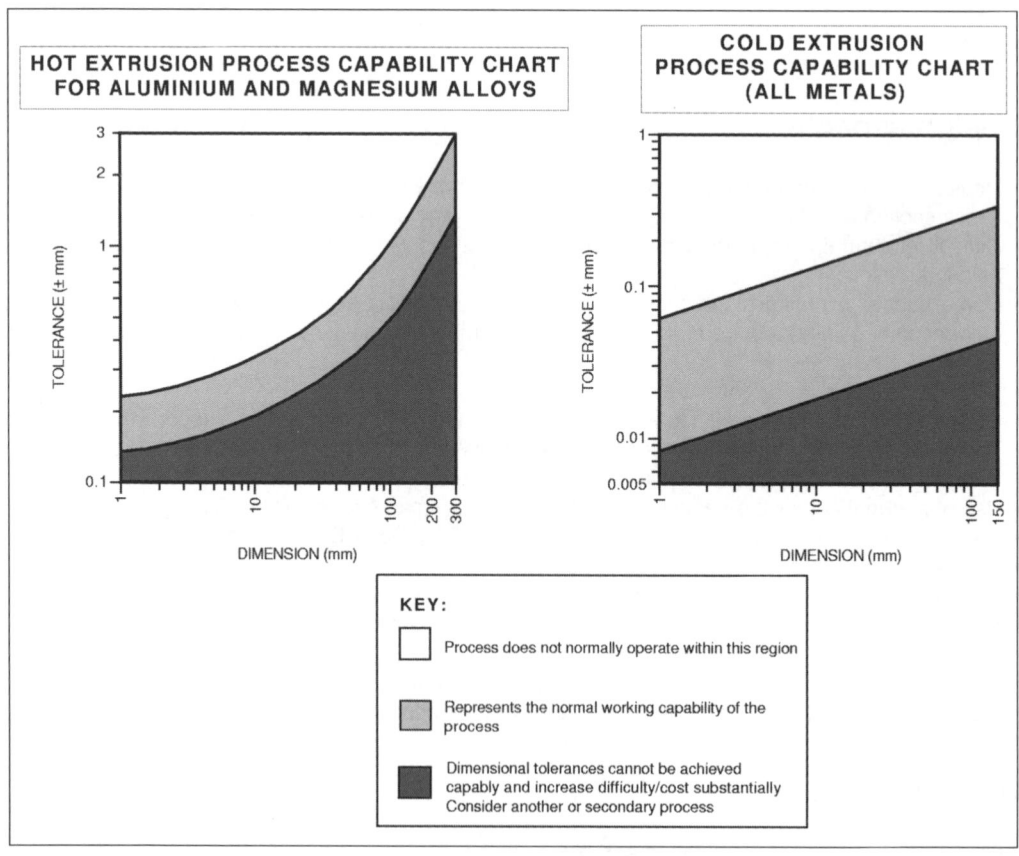

3.12CC Continuous extrusion (metals) process capability chart.

4 Machining processes

4.1 Automatic and manual turning and boring

Process description

- The removal of material by chip processes using sequenced or simultaneous machining operations on cut to length bar or coiled bar stock. The stock can be automatically or manually fed into the machine (see 4.1F).

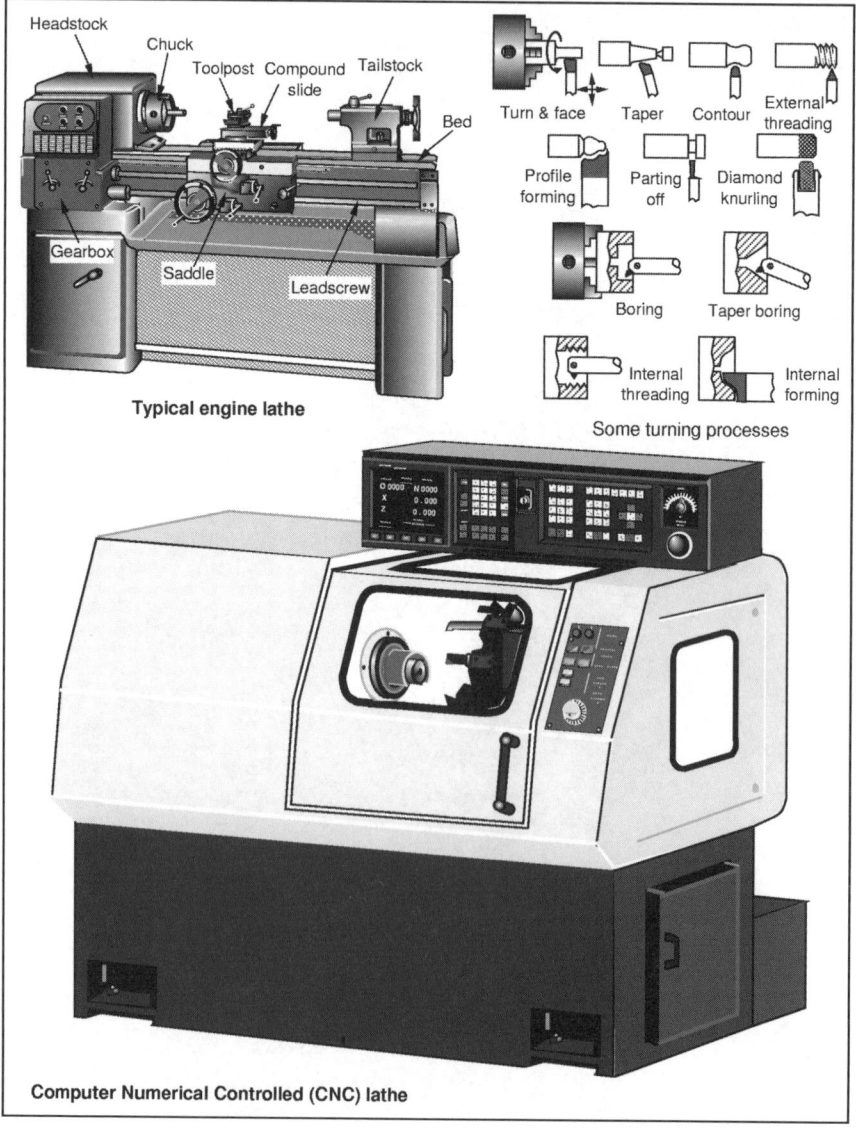

4.1F Automatic and manual turning and boring process.

Materials

- All metals (mostly free machining), some plastics, elastomers and ceramics.

Process variations

- Manually operated machines include: bench lathes (can machine non-standard shape parts) and turret lathes (limited to standard stock material).
- Automatic machines: fully or semi-automated. Follow operations activated by mechanisms on the machine.
- Automatic bar machines: used mainly for the production of screws and similar parts. Single spindle, multiple spindle and Swiss-types are available.
- CNC machines: movement and control of tool, headstock and saddle are performed by a computer program via stepper motors.
- Machining centers: fully automated, integrated turning, boring, drilling and milling machines capable of performing a wide range of operations.
- Extensive range of cutting tool geometries and materials available.

Economic considerations

- Production rates ranging 1–60/h for manual machining, 10 to 1000/h for automatic machining.
- Lead times vary from short to moderate.
- Material utilization is poor to moderate depending on specific operation (10–60 per cent scrap generated typically). Large quantities of chips generated which can be recycled.
- Flexibility is low to moderate for automatic machines: changeover and setup times can be many hours. Manual machines very flexible.
- Economical quantities are 1000+ for automatic machines. Production volumes of 100 000+ are common. Manual and CNC machining commonly used for small production runs, but can also be economic for one-offs.
- Tooling costs are moderate to high for automatic machines, low for manual.
- Equipment costs are high for automatic/CNC machines. Moderate for manual machining.
- Direct labor costs are high for manual machining, low to moderate for automatic/CNC machining.
- Finishing costs are low. Only cleaning and deburring required.

Typical applications

- Any component with rotational symmetrical elements requiring close tolerances
- Non-standard shapes requiring secondary operations
- Shafts
- Screws and fasteners
- Transmission components
- Engine parts

Design aspects

- Complexity limited to elements with rotational symmetry.
- Little opportunity for part consolidation.
- Can perform many different operations in a logical sequence on the same machine.
- Potential for linking with CAD very high.

- Machining operations should be reduced to a minimum (for simplicity and lower cycle time).
- Fillet corners and chamfer edges where possible to increase tool life.
- Holes should be drilled with a standard drill point at the bottom for economy.
- Required number of full threads should always be specified.
- Leading threads on both male and female work should be chamfered to assure efficient assembly.
- Auxiliary operations made possible by special attachments, for example, drilling and milling perpendicular to the length of the work.
- Some special machines allow larger pieces but then operations restricted.
- Sizes ranging $\varnothing$0.5 mm–$\varnothing$2 m+ for manual and CNC machining. Automatic machines usually have a capacity of less than $\varnothing$60 mm.

Quality issues

- Machinability of the material to be processed is an important issue with regard to: surface roughness, surface integrity, tool life, cutting forces and power requirements. Machinability is expressed in terms of a 'machinability index'* for the material.
- Multiple setups can be a source of variability.
- Selection of appropriate cutting tool, coolant/lubricant, feed rate, depth of cut and cutting speed with respect to material to be machined is important.
- Coolant also helps flush swarf from cutting area.
- Regular inspection of cutting tool condition and material specification is important for minimum variability.
- Surface detail is good to excellent.
- Surface roughness values ranging 0.05–25 μm Ra are obtainable.
- Process capability charts showing the achievable dimensional tolerances for turning/boring (using conventional and diamond tipped cutting tools) are provided (see 4.1CC). Note, the tolerances on these charts are greatly influenced by the machinability index for the material used and the part geometry.

---

* Machinability index for a material is expressed as a percentage based on the relative ease of machining a material with respect to free cutting mild steel which is 100 per cent and taken as the standard.

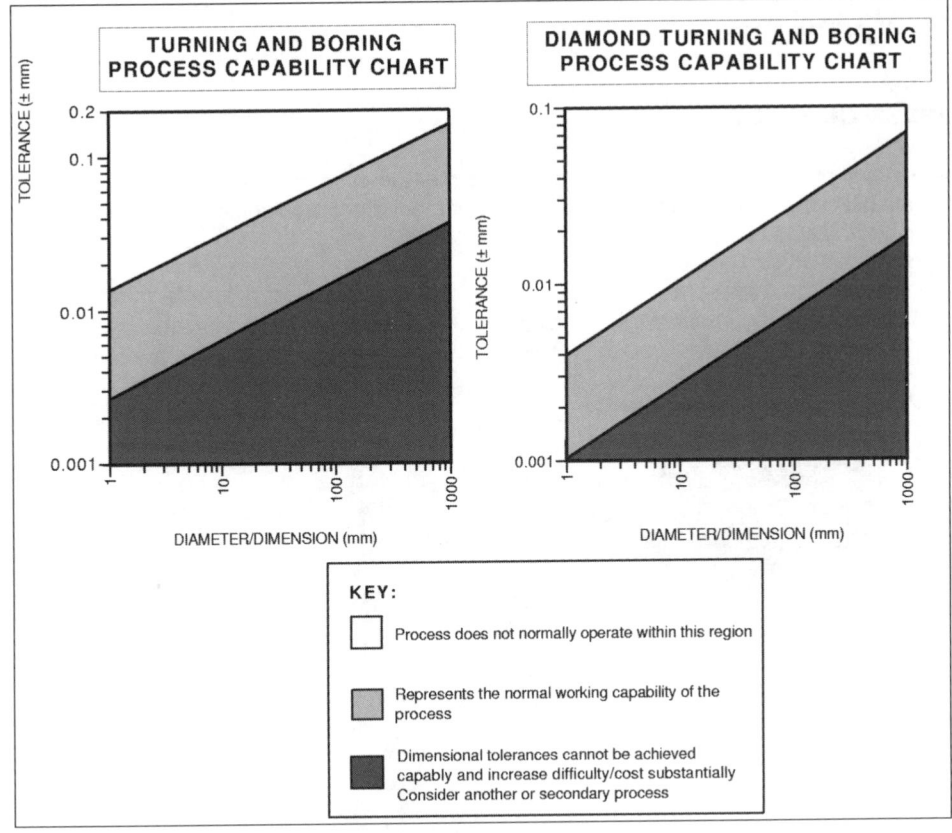

4.1CC Automatic and manual turning and boring process capability chart.

4.2 Milling

Process description

- The removal of material by chip processes using multiple-point cutting tools of various shapes to generate flat surfaces or profiles on a workpiece of regular or irregular section (see 4.2F).

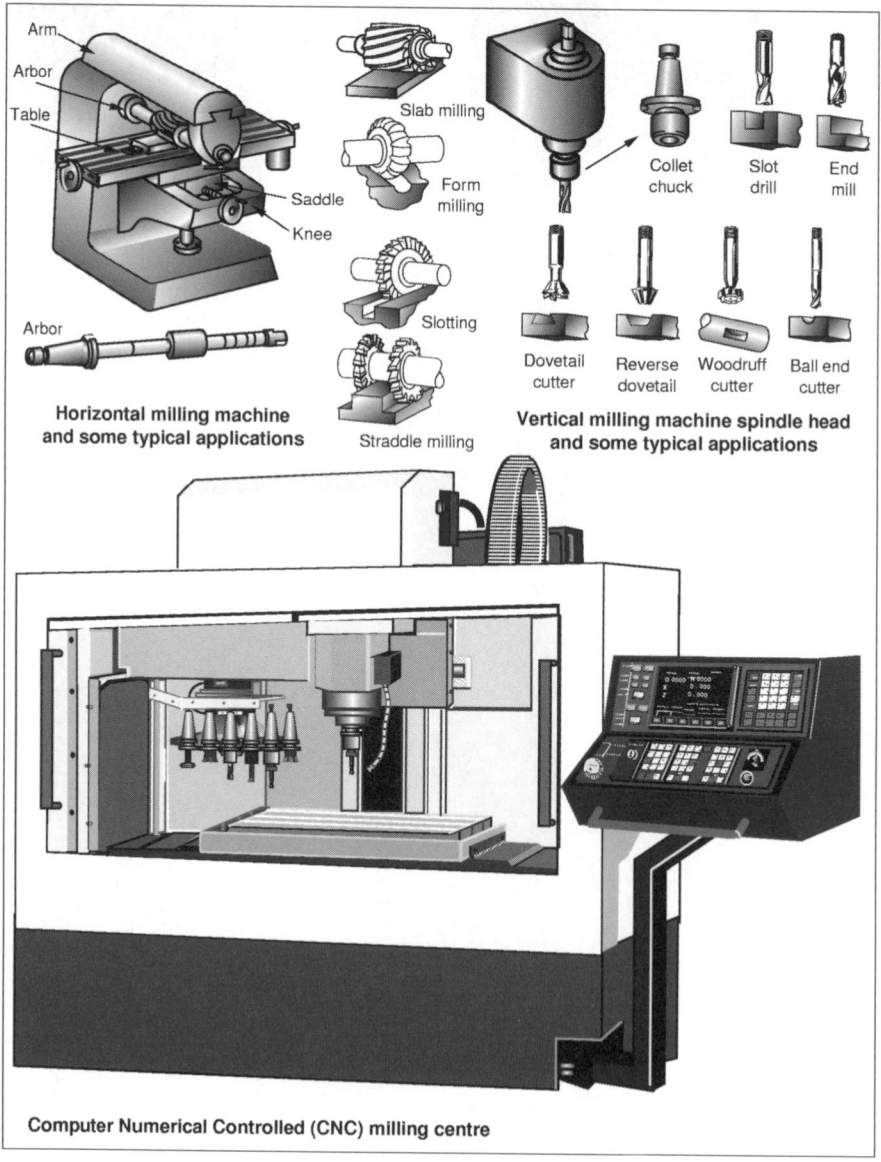

4.2F Milling process.

Materials

• All metals (mostly free machining) and some plastics and ceramics.

Process variations

• Horizontal milling: axis of cutter rotation is parallel to surface of workpiece. Includes slab milling, form milling, slotting, gang milling and slitting. Can be either up-cut or down-cut milling.
• Vertical milling: axis of cutter rotation is perpendicular to surface of workpiece. Includes face milling, slotting, dovetail and woodruff milling.
• CNC machines: movement and control of tool, headstock and bed are performed by a computer program via stepper motors.
• Extensive range of cutting tool geometries and tool materials available.

Economic considerations

• Production rates ranging 1–100/h.
• Lead times vary from short to moderate. Reduced by CNC.
• Material utilization is poor. Large quantities of chips generated.
• Recycling of waste material is possible but difficult.
• Flexibility is high. Little dedicated tooling.
• Production volumes are usually low. Can be used for one-offs.
• Tooling costs are moderate to high depending on degree of automation (tool carousels, mechanized tool loading, automatic fixturing, etc.).
• Equipment costs are moderate to high.
• Direct labor costs are moderate to high. Skilled labor required.
• Finishing costs are low. Cleaning and deburring required.

Typical applications

• Any standard or non-standard shapes requiring secondary operations
• Aircraft wing spars
• Engine blocks
• Pump components
• Machine components
• Gears

Design aspects

• Complexity limited by cutter profiles and workpiece orientation.
• Potential for linking with CAD very high.
• Chamfered edges preferred to radii.
• Standard sizes and shapes for milling cutter used wherever possible.
• Auxiliary operations made possible by special attachments, for example gear cutting using an indexing head.
• Minimum section less than 1 mm, but see below.
• Minimum size limited by ability to clamp workpiece to milling machine bed, typically $1.5\,m^2$, but length 5 m have been milled on special machines.

Quality issues

- Machinability of the material to be processed is an important issue with regard to: surface roughness, surface integrity, tool life, cutting forces and power requirements. Machinability is expressed in terms of a 'machinability index'* for the material.
- Rigidity of milling cutter, workpiece and milling machine is important in preventing deflections during machining.
- Selection of appropriate cutting tool, coolant/lubricant, depth of cut, feed rate and cutting speed with respect to material to be machined is important.
- Coolant also helps flush swarf from cutting area.
- Regular inspection of cutting tool condition and material specification is important for minimum variability.
- Surface detail is good.
- Surface roughness values ranging 0.2–25 µm Ra are obtainable.
- A process capability chart showing the achievable dimensional tolerances for milling and a chart for positional tolerance capability of CNC milling centers are provided (see 4.2CC). Note, the tolerances on the milling process capability chart are greatly influenced by the machinability index for the material used.

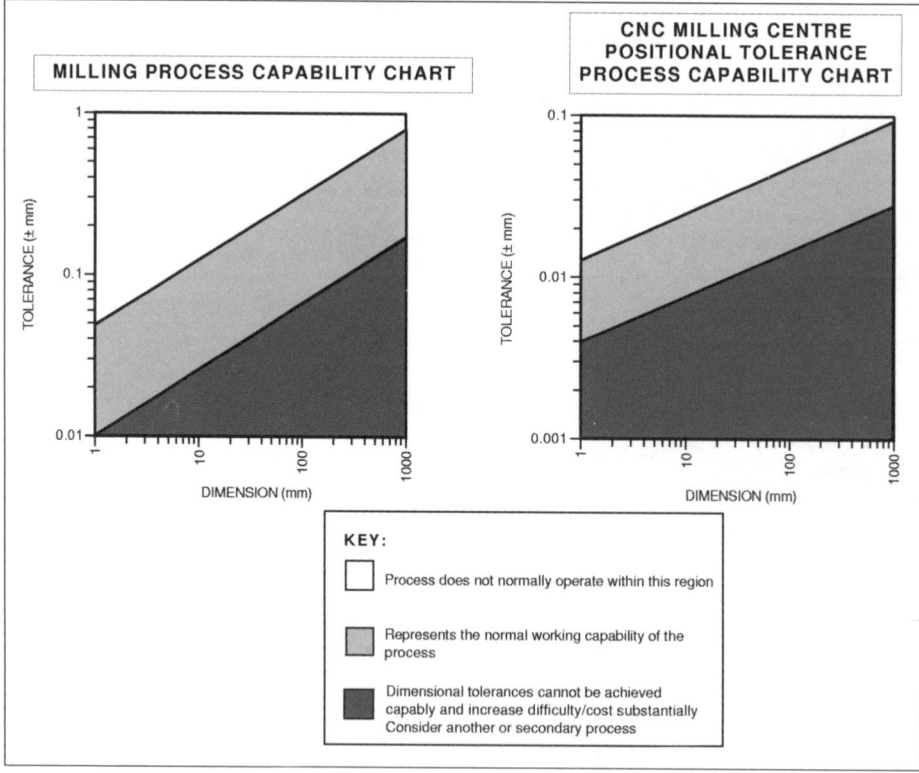

4.2CC Milling process capability chart.

---

* Machinability index for a material is expressed as a percentage based on the relative ease of machining a material with respect to free cutting mild steel which is 100 per cent and taken as the standard.

4.3 Planing and shaping

Process description

- The removal of material by chip processes using single-point cutting tools that move in a straight line parallel to the workpiece surface with either the workpiece reciprocating, as in planing, or the tool reciprocating, as in shaping. Simplest of all machining processes (see 4.3F).

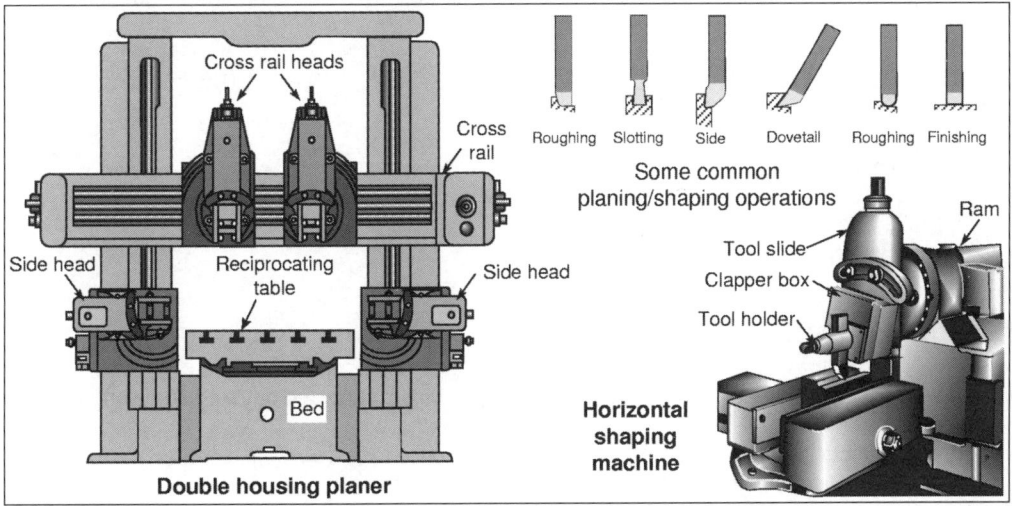

4.3F Planing and shaping process.

Materials

- All metals (mostly free machining).

Process variations

- Double housing planer: closed gantry carrying several tool heads.
- Open side planer: open gantry to accommodate large workpieces carrying usually one tool-head.
- Horizontal shaping: includes push-cut and pull-cut.
- Vertical shaping: includes slotters and key-seaters.
- Wide range of cutting tool geometries and tool materials available.

Economic considerations

- Production rates ranging 1–50/h.
- Lead times vary from short to moderate.
- Material utilization is poor. Large quantities of chips are generated, which can be recycled.
- Flexibility is high. Little dedicated tooling and setup times are generally short.

- On larger parts, the elapsed time between cutting strokes can be long making the process inefficient. Can be improved by having the cutting stroke in both directions, using several cutting tools and/or machining several parts at once.
- Other processes, for example, milling or broaching, may be more economical for larger production runs of smaller parts.
- Planing machines are usually integrated with milling machines to make them more flexible.
- Least economical quantity is one. Production volumes are usually very low.
- Tooling costs are low.
- Equipment costs are moderate to high, depending on machine size and requirements.
- Direct labor costs are high to moderate. Skilled labor may be required.
- Finishing costs are moderate. Normally requires some other machining operations for finishing.

Typical applications

- Machine tool beds
- Large castings
- Die blocks
- Key-seats, slots and notches
- Large gear teeth

Design aspects

- Complexity limited by nature of process, i.e. straight profiles, slots and flat surfaces along length of workpiece.
- As many surfaces as possible should lie in the same plane for machining.
- Rigidity of workpiece design important in preventing vibration.
- Minimum section less than 2 mm, but see below.
- Minimum size limited by ability to clamp workpiece to machine bed.
- Maximum size approximately 25 m long in planing; 2 m long in shaping.

Quality issues

- Machinability of the material to be processed is an important issue with regards to: surface roughness, surface integrity, tool life, cutting forces and power requirements. Machinability is expressed in terms of a 'machinability index'* for the material.
- Adequate clearance should be provided for to prevent rubbing and chipping of the cutting tool on return strokes.
- Cutting tools require chip breakers for ductile materials, because the strokes can be long during machining and the swarf may tangle and pose a safety hazard.
- Selection of appropriate cutting tool, coolant/lubricant, depth of cut, feed rate and cutting speed with respect to material to be machined is important.
- Coolant also helps flush swarf from cutting area.
- It can produce large, accurate, distortion free surfaces due to low cutting forces and low local heat generation.
- Surface detail is fair.

---

* Machinability index for a material is expressed as a percentage based on the relative ease of machining a material with respect to free cutting mild steel which is 100 per cent and taken as the standard.

- Surface roughness values ranging 0.4–25 µm Ra are obtainable.
- A process capability chart showing the achievable dimensional tolerances is provided (see 4.3CC). Note, the tolerances on this chart are greatly influenced by the machinability index for the material used.

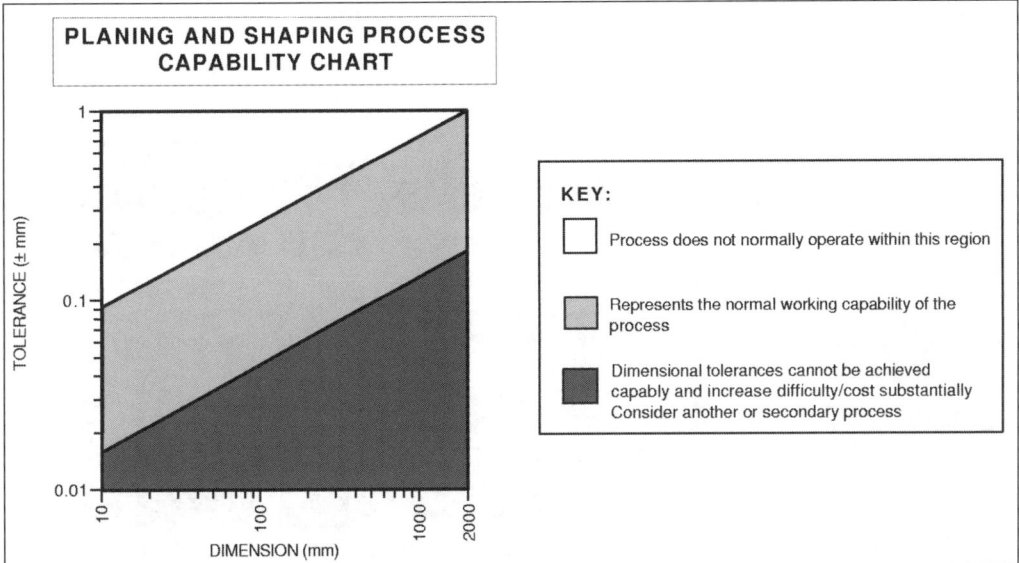

4.3CC Planing and shaping process capability chart.

4.4 Drilling

Process description

- The removal of material by chip processes using rotating tools of various types with two or more cutting edges to produce cylindrical holes in a workpiece (see 4.4F).

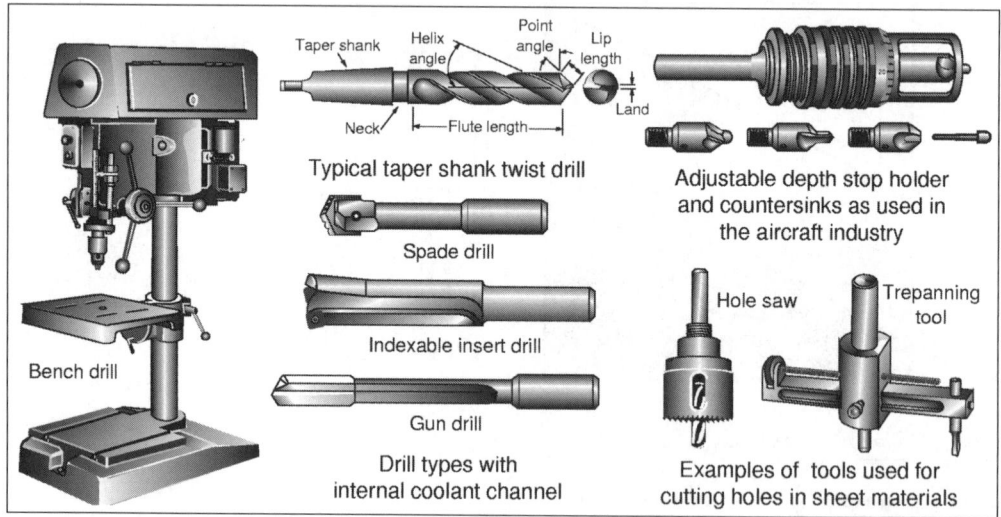

4.4F Drilling process.

Materials

- All metals (mostly free machining) and some plastics and ceramics.

Process variations

- Variations on the basic drilling machine include: bench, column, radial arm, gang, multiple spindle, turret and CNC controlled turret.
- Variations on the basic drill types include: twist drill (either three flute, taper shank, bit shank and straight flute), gun drills, spade drill, indexable insert drill, ejector drill, hole saw, trepanning and solid boring drill.
- Variations on conventional drill point geometry are aimed at reducing cutting forces and self-centering capability and include: four facet, helical, Racon, Bickford and split point.
- Wide range of cutting tool materials are available. Titanium nitride coatings are also used to increase tool life.
- Drilling can also be performed on lathes, milling machines and machining centers.
- Spot facing, counterboring and countersinking are related drilling processes.

Economic considerations

- Production rates ranging 10–500/h.
- Lead times vary from short to moderate. Reduced by automation.
- Material utilization is very poor. Large quantities of chips generated which can be recycled.
- Flexibility is high. Little dedicated tooling and generally short setup times.
- Drill jigs facilitate the reproduction of accurate holes on large production runs.
- Production volumes are usually low to moderate. Can be used for one-offs.
- Production costs are significantly reduced with multiple spindle machines when used on large production runs.
- Tooling costs are low.
- Equipment costs are low to moderate, depending on degree of automation and simultaneous drilling heads.
- Direct labor costs are low to moderate. Low operator skill required.
- Finishing costs are low. Cleaning and deburring required.

Typical applications

- Any component requiring cylindrical holes, either blind or through
- Engine blocks
- Pump components
- Machine components

Design aspects

- Complexity limited to cylindrical blind or through hole.
- Standard sizes used wherever possible.
- Faces to be drilled usually required to be perpendicular to the drilling direction unless spot faced, and adequate clearance should be provided for.
- Exit surfaces should be perpendicular to hole.
- Through holes preferred to blind holes.
- Allowances should be made for drill point depths in blind holes.
- Flat-bottomed holes should be avoided.
- Center drilling usually required before drilling unless special drill point geometry used.
- Holes with a length to diameter ratio of greater than 70 have been produced, but problems with hole straightness, coolant supply and chip removal may cause drill breakage.
- Sizes ranging from $\varnothing$0.1 mm for twist drills to $\varnothing$250 mm for trepanning.

Quality issues

- Machinability of the material to be processed is an important issue with regards to: surface roughness, surface integrity, tool life, cutting forces and power requirements. Machinability is expressed in terms of a 'machinability index'* for the material.
- Hard spots, oxide layers and poor surfaces can cause drill point to blunt or break.

---

* Machinability index for a material is expressed as a percentage based on the relative ease of machining a material with respect to free cutting mild steel which is 100 per cent and taken as the standard.

- Accurate re-grinding of the drill point geometry is required to maintain correct hole size and balance cutting forces to avoid drill breakage.
- Rigidity of drilling machine, workpiece and drill holder and concentricity of drill spindle are important in preventing oversize holes, chatter and poor surface finish.
- Selection of appropriate drill geometry (including relief and rake angles), coolant/lubricant, size of cut/hole, feed rate and cutting speed with respect to material to be machined is important.
- Drills may require chip breakers for ductile materials to efficiently remove swarf from cutting area.
- Coolant also helps flush swarf from cutting area in long through holes, and blind holes.
- Surface detail is fair.
- Surface roughness values ranging 0.4–12.5 µm Ra are obtainable.
- A process capability chart showing the achievable dimensional tolerances is provided (see 4.4CC). Note, the tolerances on this chart are greatly influenced by the machinability index for the material used.

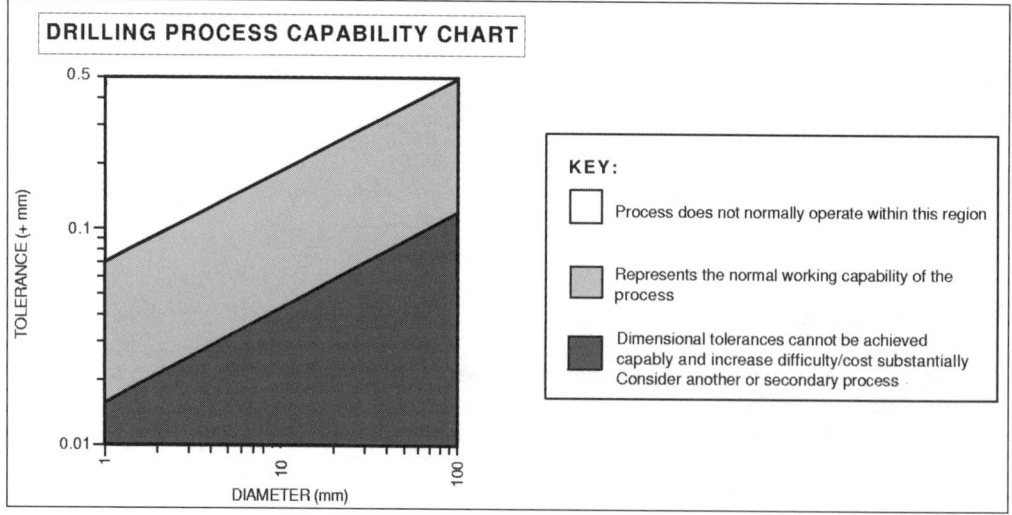

4.4CC Drilling process capability chart.

4.5 Broaching

Process description

- The removal of material by chip processes using a multiple-point cutting tool, which is pushed or pulled across the workpiece surface. With successively deeper cuts, the desired profile is gradually generated in a single pass (see 4.5F).

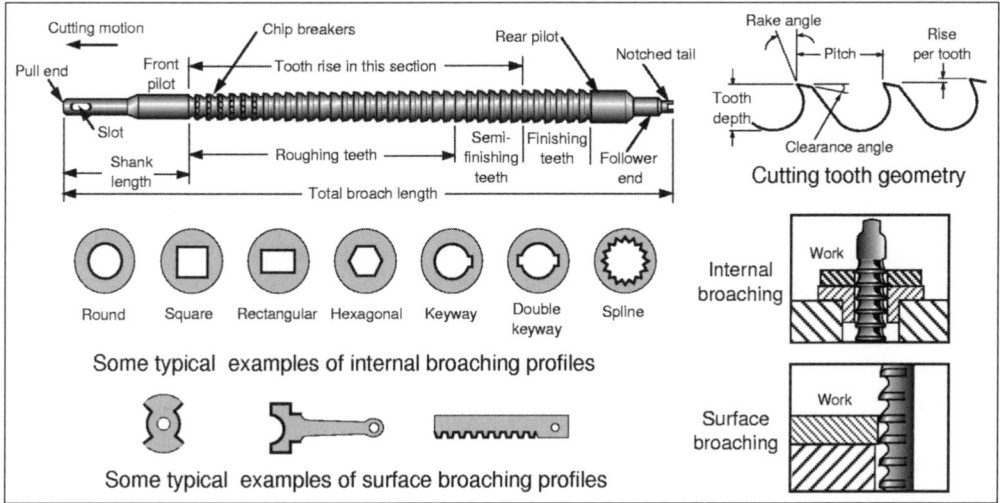

4.5F Broaching process.

Materials

- All metals (mostly free machining).

Process variations

- Horizontal, vertical or rotary broaching machines with push and/or pull capability.
- Broaching tools can be single or combination types, internal or external, performing either roughing or finishing operations.
- Some indexable insert broaches are available for surface broaching and titanium nitride coatings are also used to increase tool life.

Economic considerations

- Production rates up to 400/h.
- To improve production rates, many parts can be machined at once, called stacking. Stacking is best suited to internal features.
- Automation possible to improve production rates.
- Lead times moderate.

- Material utilization poor. Large quantities of chips are generated, which can be recycled.
- Flexibility high. Little dedicated tooling and setup times are generally short.
- Accurate re-grinding of the broaching tool required on large production runs, which uses expensive fixtures and grinding machines.
- Production volumes usually very high, 10 000–100 000.
- Tooling costs high. Broaching tools are very expensive due to their complexity and the economics of this process must be carefully studied on this basis.
- Equipment costs low to moderate.
- Direct labor costs low to moderate. Some skilled labor may be required.
- Finishing costs low. Some deburring may be required.

Typical applications

- Many regular or irregular, internal or external profiles
- Turbine blade root forms
- Connecting rod ends
- Rifling on gun barrels
- Flat surfaces
- Key seats and slots
- Splines, both straight and helical
- Gear teeth

Design aspects

- Complexity is limited by nature of process, i.e. straight, curved and complex profiles, slots and flat surfaces along length of workpiece.
- Part design should allow for sufficient clamping area and clearance for broaching tool.
- A hole is initially required for internal broaching for broaching tool access. This can be achieved by either punching, boring or drilling the blank.
- Ideally, between 0.5 and 6 mm should be removed by the broaching tool on any one surface.
- More than one surface can be cut simultaneously.
- Workpiece must be strong enough to withstand the pressure of continuous cutting action of broach.
- Large surfaces, blind holes and sharp corners should be avoided.
- Chamfers are preferred to radiused corners.
- Minimum stroke = 25 mm.
- Maximum stroke = 3 m.

Quality issues

- Machinability of the material to be processed is an important issue with regards to: surface roughness, surface integrity, tool life, cutting forces and power requirements. Machinability is expressed in terms of a 'machinability index'* for the material.
- For materials with high surface hardness, the first tooth on the broach should cut beneath this layer to improve tool life.
- Soft or non-uniform materials may tear during machining.

---

* Machinability index for a material is expressed as a percentage based on the relative ease of machining a material with respect to free cutting mild steel which is 100 per cent and taken as the standard.

- Adequate clearance should be provided for to prevent rubbing and chipping of the broaching tool on return strokes.
- Broaching tools may require chip breakers for very ductile materials to efficiently remove swarf from cutting area.
- Selection of appropriate cutting tool material, coolant/lubricant, depth of cut per tooth and cutting speed with respect to material to be machined is important.
- Coolant also helps flush swarf from cutting area.
- Surface detail is excellent.
- Surface roughness values ranging 0.4–6.3 µm Ra are obtainable.
- A process capability chart showing the achievable dimensional tolerances is provided (see 4.5CC). Note, the tolerances on this chart are greatly influenced by the machinability index for the material used and geometry complexity.

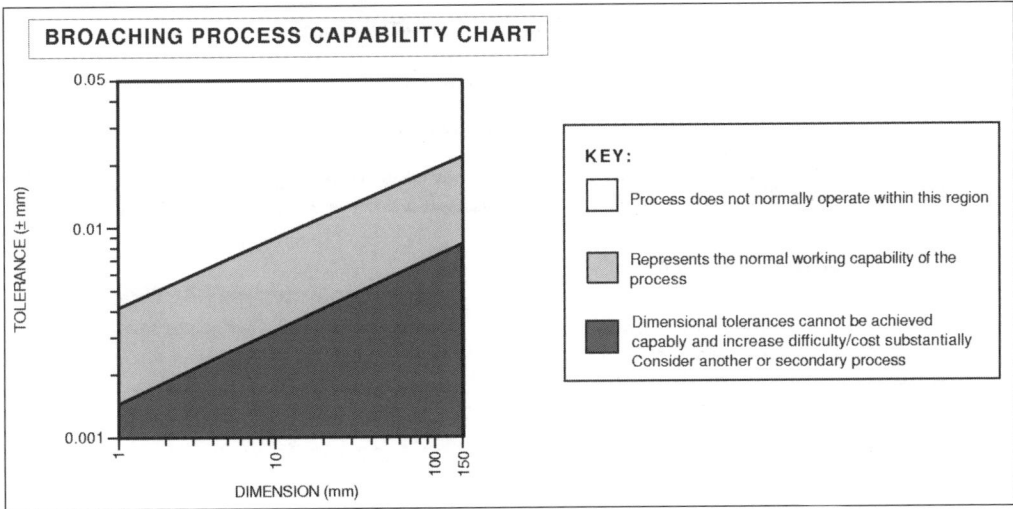

4.5CC Broaching process capability chart.

4.6 Reaming

Process description

- The removal of small amounts of material by chip processes using tools of various types with several cutting edges to improve the accuracy, roundness and surface finish of existing cylindrical holes in a workpiece. The tool or the work can rotate relative to each other (see 4.6F).

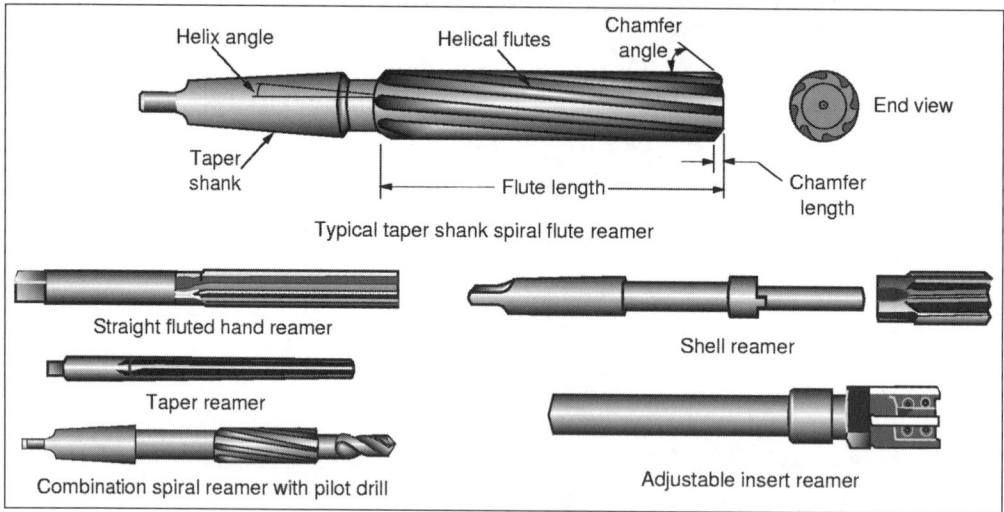

4.6F Reaming process.

Materials

- All metals (mostly free machining).

Process variations

- No special machines are used for reaming. Reaming can be performed on drilling machines, lathes, milling machines and machining centers, or by hand.
- Basic reamer types include: hand (straight and tapered), machine (rose and fluted), shell, expansion, adjustable and indexible insert reamers. Titanium nitride coatings are sometimes used to increase tool life. Combination drills and reamers are also available.

Economic considerations

- Production rates ranging from 10–500/h.
- Lead times varying from short to moderate. Reduced by automation.
- Minimum amount of material removed.
- Flexibility high. Little dedicated tooling and generally short setup times.

- Production volumes usually low to moderate.
- Can be used for one-offs.
- Production costs significantly reduced with multiple spindle machines.
- Tooling costs low.
- Equipment costs low.
- Direct labor costs low to moderate. Low operator skill required.
- Finishing costs low. Cleaning and deburring required.

Typical applications

- Any component requiring accurate, cylindrical or tapered holes with good surface finish, either blind or through after a primary hole making operation, typically drilling.

Design aspects

- Complexity limited to straight or tapered cylindrical blind or through holes.
- Ideally, reaming allowances should be 0.1 mm per 5 mm of diameter, i.e. for a finished reamed hole $\varnothing$20 mm, the pilot hole should be approximately $\varnothing$19.6 mm. However, drilled holes prior to reaming should be standard size, wherever possible.
- Allowances should be made for reamer-end chamfers and the slight taper on some reamers when machining blind holes, although more suited to through holes.
- Standard sizes used wherever possible.
- Through holes preferred to blind holes.
- Sizes ranging $\varnothing$3–$\varnothing$100 mm.

Quality issues

- Machinability of the material to be processed is an important issue with regard to: surface roughness, surface integrity, tool life, cutting forces and power requirements. Machinability is expressed in terms of a 'machinability index'* for the material.
- Any misalignment between workpiece and reamer will cause chatter, oversize holes and bell-mouthing of hole entrance. Piloted reamers ensure alignment of the workpiece and reamer.
- Most accurate holes are center drilled, drilled, bored and reamed to finished size.
- Proper maintenance and reconditioning of reamers is required to maintain correct hole size and surface finish requirements. To work efficiently, a reamer must have all its teeth cutting.
- Pick-up or galling is caused by too much material being removed by the reamer.
- Selection of appropriate reamer geometry (including relief and rake angles), coolant/lubricant (if required), size of hole, feed rate and cutting speed with respect to material to be machined is important.
- Reaming is performed at one-third the speed and two-thirds the feed rate of drilling for optimum conditions.
- Coolant also helps flush swarf from cutting area in long through holes, and blind holes.
- Surface detail is good.
- Decreasing feed rate improves surface finish.
- Surface roughness values ranging 0.4–6.3 μm Ra are obtainable.
- A process capability chart showing the achievable dimensional tolerances is provided (see 4.6CC). Note, the tolerances on this chart are greatly influenced by the machinability index for the material used.

---

* Machinability index for a material is expressed as a percentage based on the relative ease of machining a material with respect to free cutting mild steel which is 100 per cent and taken as the standard.

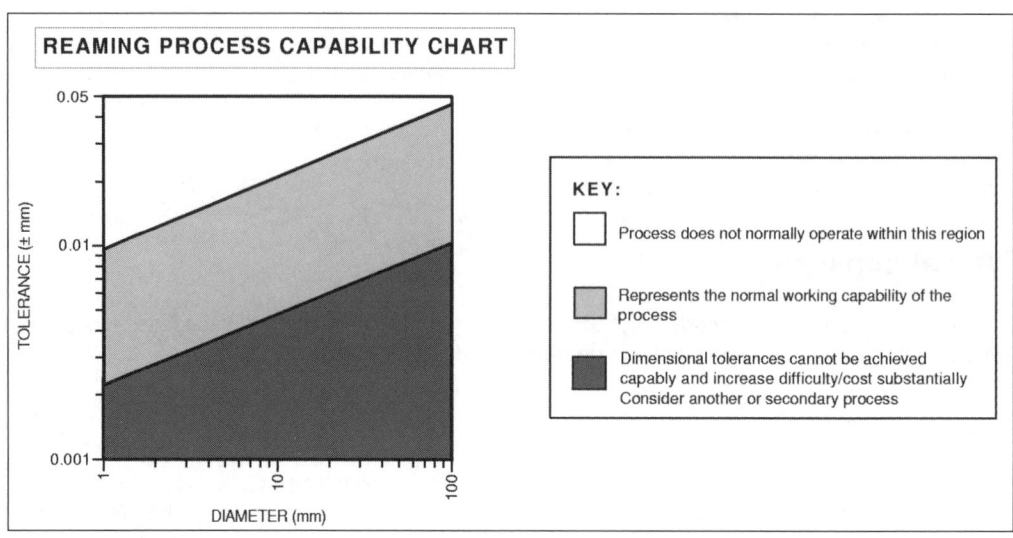

4.6CC Reaming process capability chart.

4.7 Grinding

Process description

- The removal of small layer's material by the action of an abrasive spinning wheel on a rotating or reciprocating workpiece (see 4.7F).

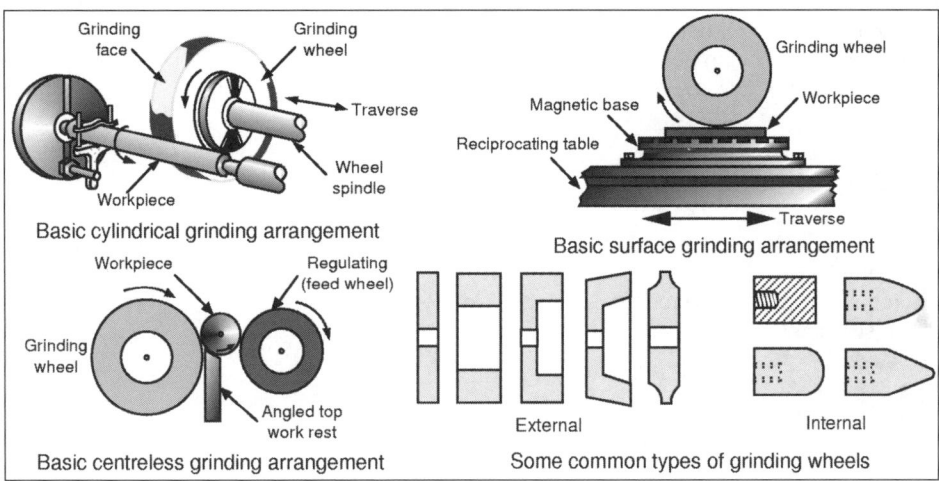

4.7F Grinding process.

Materials

- All hard materials. Not suitable for soft or flexible materials.

Process variations

- Surface grinding: workpiece is mounted on a reciprocating or rotating bed and a rotating abrasive wheel (either horizontal or vertical axis of rotation) is fed across the surface.
- Cylindrical grinding: rotating abrasive wheel is fed along the periphery of a slower rotating cylindrical workpiece. Also includes: thread, form and plunge grinding.
- Internal grinding: small rotating abrasive wheel is fed into the bore of a cylindrical rotating workpiece.
- Centreless grinding: workpiece is supported on a work rest blade and ground between two wheels, one of which is a regulating wheel operating at 5 per cent the speed of the other.
- Tool grinder: precision bench grinding unit for tool dressing.
- Off-hand grinding: a fixed grinding machine (either bench or pedestal) where the work is manually presented to the grinding wheel.
- Portable grinding: a hand held unit used for fettling and cutting.
- CNC machines: movement and control of abrasive wheel and workpiece are performed by a computer program via stepper motors.
- Extensive ranging of abrasive wheel geometries, abrasive materials (aluminum oxide, emery, corundum, diamond, Cubic Boron Nitride (CBN)), grain size, hardness grading and bond types (resin, vitrified glass, rubber, metal) are available.

Economic considerations

- Production rates range from 1 to 1000/h.
- Lead times vary from short to moderate, depending on degree of automation and geometry.
- Material utilization is poor. Recycling of waste material difficult.
- Flexibility of grinding is high.
- Turning can compete with grinding in some situations.
- Suitable for all quantities.
- Tooling costs are low to moderate.
- Equipment costs are moderate to high, depending on degree of automation.
- Direct labor costs ranging from high to low, depending on degree of automation and part complexity.
- Finishing costs are very low. Cleaning required.

Typical applications

- Grinding is used for the generation of basic geometric surfaces and finishing of a wide range of components
- Parts requiring fine surface roughness and/or close tolerances
- Bearing surfaces
- Valve seats
- Gears
- Cams

Design aspects

- Complexity is limited to nature of workpiece surface, i.e. cylindrical or flat, unless profiled wheels and/or special machines are used.
- Grinding should be used to remove the minimum amount of material.
- Surface features should be kept simple to avoid frequent dressing of the wheel.
- Fillets and corner radii should be as liberal as possible.
- Deep holes and recesses should be avoided.
- Parts should be mounted securely to avoid deflections as high forces can be generated during the grinding process.
- May not be suitable for delicate workpieces.
- For best results use the largest wheel possible for the relevant workpiece.
- Minimum section = 0.5 mm
- Sizes ranging $\varnothing 0.5\,mm$–$\varnothing 2\,m+$ for cylindrical grinding. Maximum size for surface grinding is approximately 6 m in length. Less than $\varnothing 1\,m$ for centreless grinding.

Quality issues

- Interruptions on the workpiece surface, for example key seats and recesses, may cause vibration and chatter.
- Unit pressures vary with area of contact. High pressures use hard grade, fine grit abrasive wheels.
- Surface tensile residual stresses remain in the workpiece due to localized high-temperature gradients. This may be critical in heat sensitive applications or when fatigue strength is important. Low stress grinding can impart beneficial compressive stresses.

- The final size of the workpiece is determined by the speed of response of the gauging system and the forces built up in machine as a result of cutting loads.
- Gauging may be contact or non-contact, this will probably be dictated by the part.
- The properties of the wheel may change in the course of the process. Grinding wheels require occasional dressing to ensure uniform cutting properties.
- Use of grinding fluid is important for chip removal and cooling of the workpiece.
- Grinding wheels need careful storage and to be visually inspected for cracks before use.
- Grinding wheels require balancing before use, to minimize vibration because of the high rotational speeds.
- Surface roughness is controlled by the wheel grading, wheel condition, feed rate at finish size and cleanliness of the cutting fluid.
- Surface detail is excellent.
- Surface roughness values ranging 0.025–6.3 µm Ra are obtainable.
- Process capability charts showing the achievable dimensional tolerances for surface and cylindrical grinding are provided (see 4.7CC).

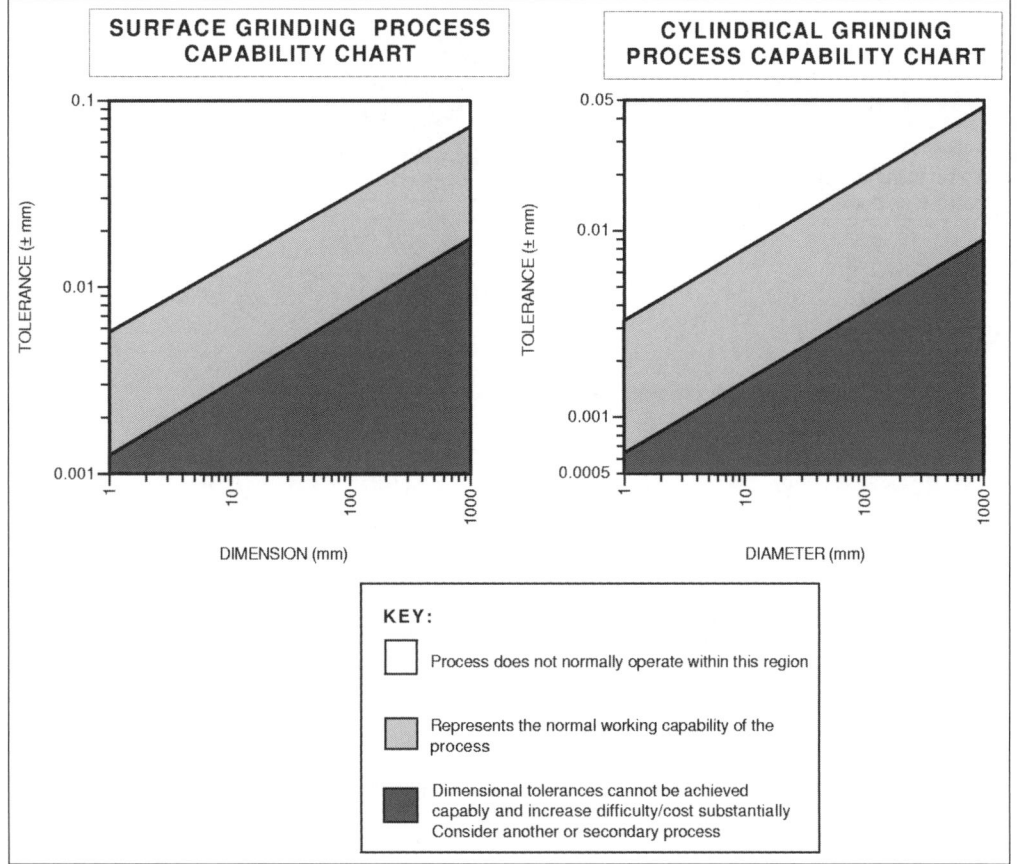

4.7CC Grinding process capability chart.

4.8 Honing

Process description

- The removal of small amounts of material by floating segmented abrasive stones mounted on an expanding mandrel, which rotates with low rotary speed and reciprocates along the surface of the workpiece (see 4.8F).

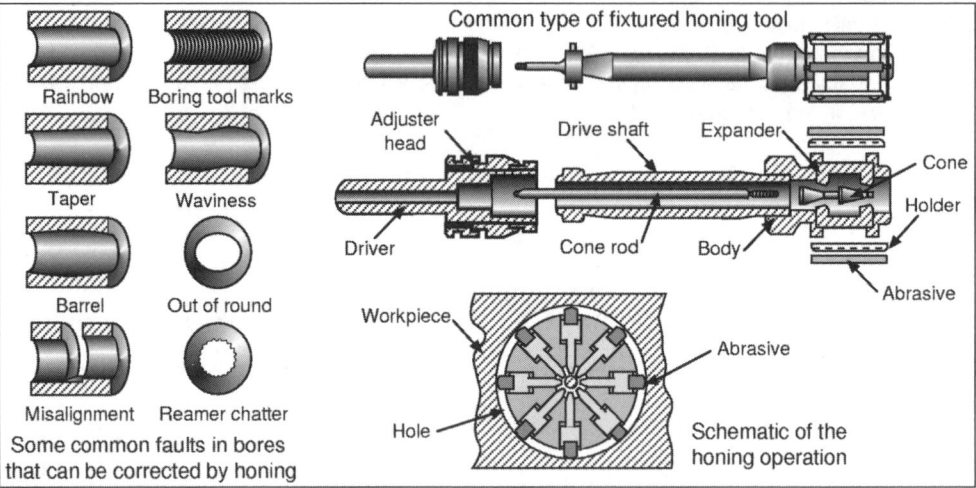

4.8F Honing process.

Materials

- All materials, including some ceramics and plastics.

Process variations

- Horizontal and vertical honing machines with single or multiple spindles, are with either short or long stroke capability.
- Honing can also be performed on lathes and drilling machines.
- Internal and external cylindrical surfaces are honed commonly. Also, spherical, toroidal and flat surfaces can be honed, but are less common applications.
- Single-stroke bore finishing and superfinishing using 'superabrasives' such as diamond and CBN, are related processes. Superfinishing is similar to honing, but performed on outside surfaces previously ground or lapped to improve finish.
- Laser honing: for precise surface topographies.
- Large range of stone geometries, abrasive materials, grain size, hardness grading, bond types and methods (co-axial and match honing) are available.
- Automation aspects include in-process gauging and adaptive control to optimize cutting conditions and control accuracy.
- Workpieces can be manually presented to honing mandrel.

Economic considerations

- Production rates ranging 10–1000/h depending on number of spindles. Typically 60/h for single spindle machines.
- Lead times short.
- Very little material removed.
- Suitable for all quantities.
- Tooling costs varying, depending on degree of automation and size.
- Equipment costs moderate.
- Direct labor costs moderate. Skill level required is moderate to high (manual).
- Finishing costs very low. Cleaning only required.

Typical applications

- Any component where superior accuracy, surface finish and/or improvement of geometric features required on cylindrical features
- Bearing surfaces
- Pin and dowel holes
- Engine cylinder bores
- Rifle bores

Design aspects

- Honing is performed to remove the minimum amount of material, usually between 0.02 and 0.2 mm.
- Complexity is limited to nature of workpiece surface, i.e. cylindrical (internal and external), spherical, flat or toroidal.
- Honing logically follows the grinding process to produce precision surfaces.
- Surface features should be kept simple.
- Chamfers are required on entrance to bores to facilitate easy access of honing tool.
- Blind holes should have undercuts.
- For small holes less than $\varnothing$15 mm, the maximum length that can be honed is 20 times the diameter of the hole. For best results, a length to diameter ratio of 1 is recommended.
- Maximum length for large holes is 12 m.
- Sizes ranging $\varnothing$6–$\varnothing$750 mm+ for cylindrical honing.

Quality issues

- Interruptions on the workpiece surface, for example key seats and holes, reduce the quality of finish. Can be offset by increasing rotary speed of honing stone.
- The process has the ability to correct geometrical inaccuracies, for example, bell-mouthing, barreling, tapers and waviness in holes, as well as removing machining marks.
- Surface finish and accuracy is controlled by the stone grain size, feed pressure, area of contact, coolant access, stroke length, rotary speed and stone reciprocation speed, which when optimized ensure breakdown of the stone and good self-dressing characteristics.
- Coolant also helps flush swarf from cutting area in long through holes, and blind holes.
- Little heat is generated at surface, therefore, original surface characteristics of the component not altered.
- Surface detail is excellent.
- Surface roughness values ranging 0.025–1.6 μm Ra are obtainable.
- Relatively soft and flexible materials tend to give inferior surface finish to hard materials.
- A process capability chart showing the achievable dimensional tolerances is provided (see 4.8CC).

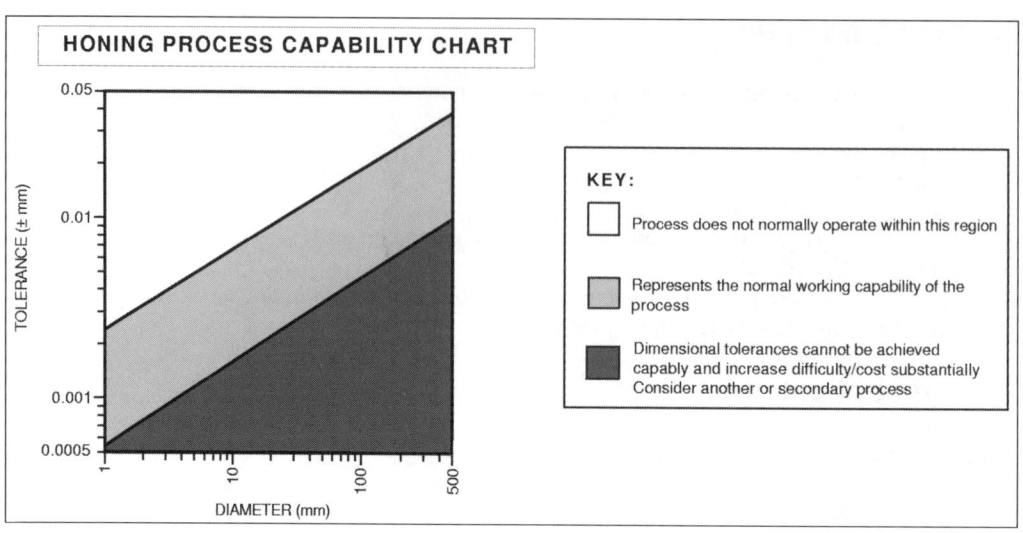

4.8CC Honing process capability chart.

4.9   Lapping

Process description

- The removal of very small amounts of material by the relative motion of fine abrasive particles, embedded in a soft material (the lap), with the aid of a lubricating and carrier fluid (see 4.9F).

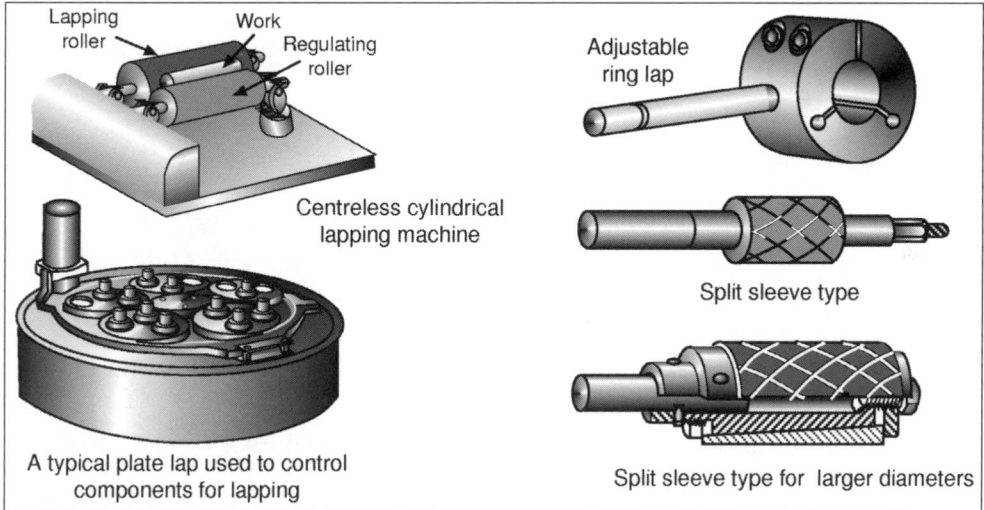

4.9F Lapping process.

Materials

- All materials, but materials of low hardness or high flexibility present problems.

Process variations

- Hand lapping: operator moves the workpiece over a grooved surface plate in an irregular rotary motion, turning the part frequently to ensure uniformity.
- Machine lapping: horizontal and vertical lapping machines with variety of floating work holding devices that can carry many parts at once over the rotating plate lap.
- Centreless lapping: used for internal and external cylindrical, spherical and contoured surfaces.
- Pressure jet lapping: uses a low viscosity mix of abrasive grit and water applied at high speed to the surface using compressed air. Similar to Abrasive Jet Machining (AJM) (see 5.7).
- Range of lap materials, abrasive materials, grain size and carrier fluids are available for different materials.

Economic considerations

- Production rates ranging 10–3000/h, depending on level of automation.
- Lead times are short.

- Very little material is removed.
- Suitable for all quantities.
- Tooling costs vary depending on degree of automation and size.
- Equipment costs are moderate.
- Direct labor costs are low to moderate. Operator skill required for hand lapping.
- Finishing costs are very low. Cleaning only required.

Typical applications

- Any component where superior surface finish is required on flat, cylindrical or contoured surfaces
- Bearing surfaces
- Gauge blocks
- Piston rings
- Balls for ball bearings
- Piston pins
- Valve seats
- Glass lenses
- Pump gears

Design aspects

- Complexity is limited to nature of workpiece surface, i.e. flat, cylindrical (internal and external) or spherical.
- Lapping is performed to remove the minimum amount of material, usually between 0.005 and 0.01 mm.
- Lapping should not be specified if the surface finish on the component is not critical and can be produced by other processes.
- Lapping logically follows the grinding or honing process to produce precision surfaces.
- Parts required to provide lapping pressure under their own weight should have a low center of gravity and be stable.
- Surface features should be kept simple.
- Sizes ranging 1–500 mm for flat lapping.
- Centerless lapping sizes ranging $\varnothing$0.75–$\varnothing$300 mm. Maximum lengths are 4+ for up to $\varnothing$75 mm.

Quality issues

- Soft materials difficult to lap due to abrasive particles becoming embedded in workpiece material.
- Low lapping speeds can introduce beneficial compressive residual stresses into the surface of workpiece to improve fatigue resistance.
- Choice of abrasive, lap and carrier important for specific material types.
- Surface detail excellent.
- Surface roughness values in the range 0.012–0.8 μm Ra obtainable.
- A process capability chart showing the achievable dimensional tolerances is provided (see 4.9CC).

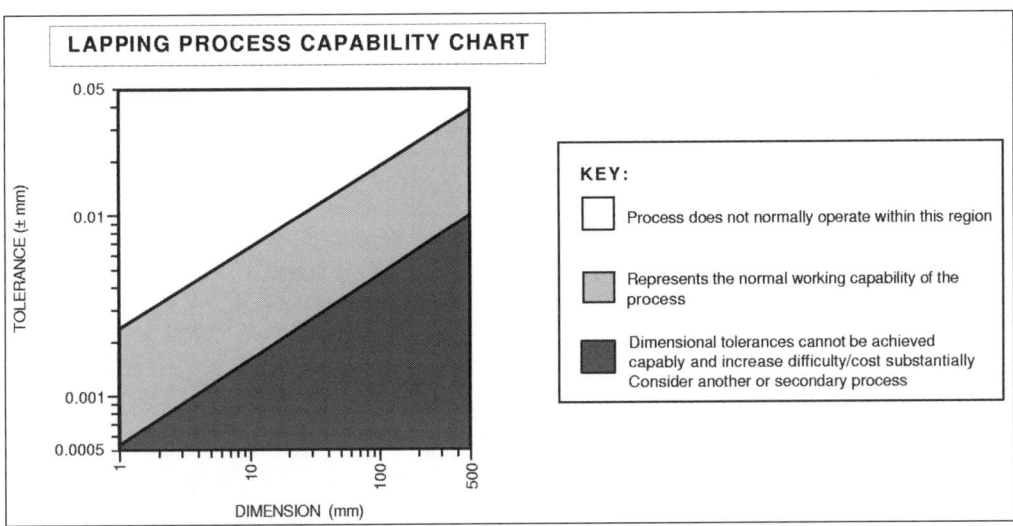

4.9CC Lapping process capability chart.

5 Non-Traditional Machining (NTM) processes

5.1 Electrical Discharge Machining (EDM)

Process description

- The tool, usually graphite, and the workpiece are essentially electrodes, the tool being the negative of the cavity to be produced. The workpiece is vaporized by spark discharges created by a power supply. The gap between the workpiece and tool is kept constant and a dielectric fluid is used to cool the vaporized 'chips' and then flush them away from the workpiece surface (see 5.1F).

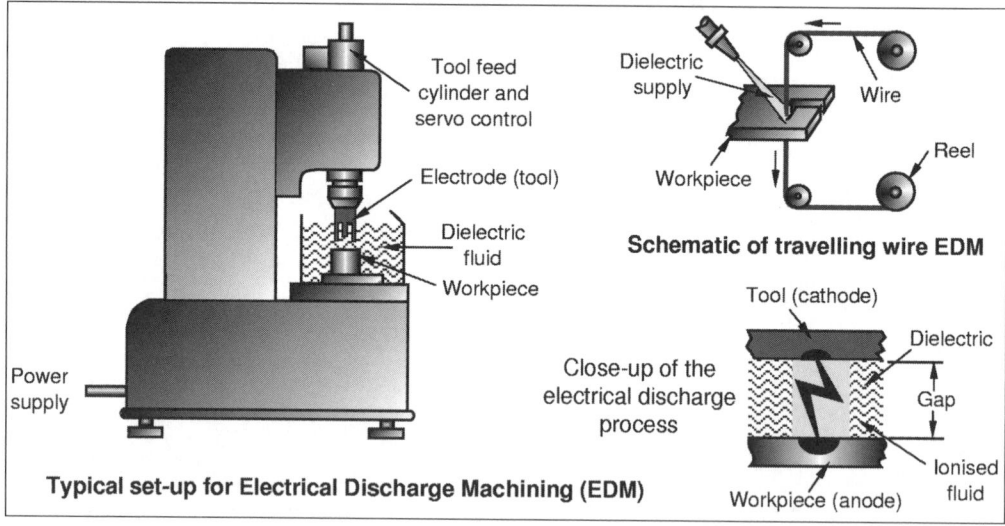

Typical set-up for Electrical Discharge Machining (EDM)

Schematic of travelling wire EDM

Close-up of the electrical discharge process

5.1F Electrical discharge machining process.

Materials

- Any electrically conductive material irrespective of material hardness, commonly, tool steels, carbides, Polycrystalline Diamond (PCD) and ceramics, but not cast iron.
- Melting point and latent heat of melting are important properties, partially determining the material removal rate.

Process variations

- Traveling wire EDM: wire moves slowly along the prescribed path on the workpiece and cuts the metal with sparks creating a slot of 'kerf'. CNC control is common.
- No-wear EDM: minimizing tool wear of steels by reversing the polarity and using copper tools.
- Electrical Discharge Grinding (EDG): graphite or brass grinding wheel rotates relative to the rotating workpiece and removes material by spark erosion (no abrasives involved).
- Ultrasonic EDM: increases production rate and gives less surface damage.

Economic considerations

- Production rates very low.
- Material removal rates up to 1.6 mm^3/min.
- Cutting rate for traveling wire EDM approximately 0.635 mm/s.
- Material removal/cutting rates a function of the current rate and material properties.
- Lead time days to several weeks depending on complexity of electrode tool.
- Tools can be of segmented construction for high complexity work.
- Material utilization very poor. Scrap material cannot be recycled.
- Disposal of sludge and chemicals used can be costly.
- High degree of automation possible.
- Economical for low production runs. Can be used for one-offs.
- Tooling costs high. High tool wear rates mean period changing.
- Equipment costs generally high.
- Direct labor costs low to moderate.

Typical applications

- Tool and die blocks for forging, extrusion, casting, punching, blanking, etc.
- Honeycomb structures and irregular shapes
- Prototype parts
- Burr free parts

Design aspects

- High degree of shape complexity possible, limited only by ability to produce tool shape.
- Traveling wire EDM limited to 2-dimensional profiles.
- Suitable for small diameter, deep holes with length to diameter ratios up to 20:1. Can be up to 100:1 for special applications.
- Undercuts possible with specialized tooling.
- No mechanical forces used for cutting, therefore simple fixtures can be used.
- Possible to machine thin and delicate sections due to minimal machining forces.
- Minimum radius = 0.025 mm.
- Minimum hole/slot size = 0.05 mm.
- Traveling wire EDM can cut sections up to 150 mm.

Quality issues

- Burr free part production.
- Produces slightly tapered holes, especially if blind, and some overcut.
- Optimum tool to workpiece gap ranges from 0.012 to 0.51 mm.
- Surface layer is altered metallurgically and chemically due to high thermal energies.
- A hard skin, or recast layer, produced may offer longer life, lower friction and lubricant retention for dies, but can be removed if undesirable.
- Beneath the recast layer is a heat affected zone which may be softer than the parent material.
- Finishing cuts made at lower removal rates.
- Tool wear related to the melting points of the materials involved, and this affects accuracy. May require changing periodically.
- Being a thermal process, residual stresses and fine cracks may form.

- Removal rate can be increased with the expense of a poorer surface finish.
- Surface detail good.
- Surface roughness values ranging 0.4–25 μm Ra. Dependent on current density, material being machined and rate of removal.
- Achievable tolerances ranging ±0.01–±0.125 mm. (Process capability charts have not been included. Capability is not primarily driven by characteristic dimension but by the material being processed.)

5.2 Electrochemical Machining (ECM)

Process description

- Workpiece material is removed by electrolysis. A tool, usually copper (−ve electrode), of the desired shape is kept a fixed distance away from the electrically conductive workpiece (+ve electrode), which is immersed in a bath containing a fast flowing electrolyte and connected to a power supply. The workpiece is then dissolved by an electrochemical reaction to the shape of the tool. The electrolyte also removes the 'sludge' produced at the workpiece surface (see 5.2F).

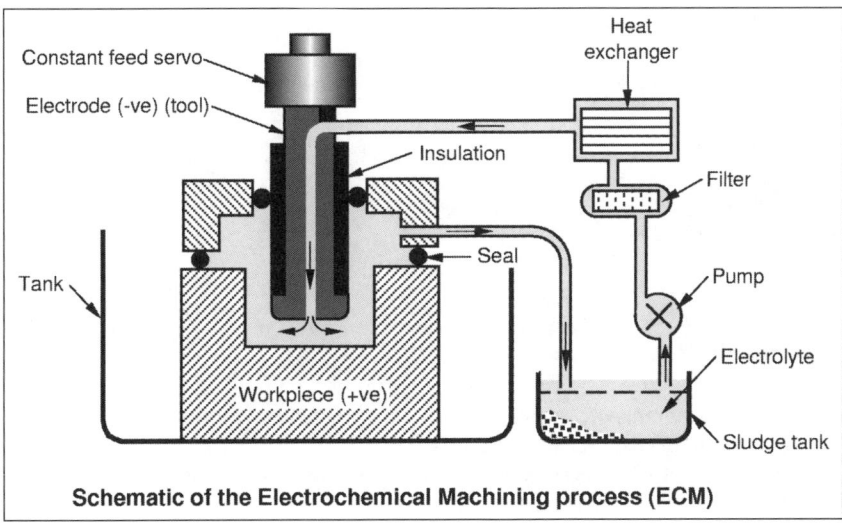

Schematic of the Electrochemical Machining process (ECM)

5.2F Electrochemical machining process.

Materials

- Any electrically conductive material irrespective of material hardness, commonly, tool steels, nickel alloys and titanium alloys. Ceramics and copper alloys are also processed occasionally.

Process variations

- Electrochemical Grinding (ECG): combination of electrochemical reaction and abrasive machining of workpiece.
- Electrochemical drilling: for the production of deep, small diameter holes.
- Electrochemical polishing: for deburring and honing.

Economic considerations

- Production rates moderate.
- Material removal rates typically 50–250 mm^3/s.
- Linear penetration rates up to 0.15 mm/s.

- Dependent on current density, electrolyte and gap between tool and workpiece.
- High power consumption.
- Lead time can be several weeks. Tools are very complex.
- Setup times can be short.
- Material utilization very poor. Scrap material cannot be recycled.
- Disposal of sludge and chemicals used can be costly and hazardous.
- High degree of automation possible.
- Economical for moderate to high production runs.
- Tooling costs very high. Dedicated tooling.
- Equipment costs generally high.
- Direct labor costs low to moderate.

Typical applications

- Hole (circular and non-circular) production, profiling and contouring of components
- Engine casting features
- Turbine blade shaping
- Dies for forging
- Gun barrel rifling
- Honeycomb structures and irregular shapes
- Burr free parts
- Deep holes

Design aspects

- High degree of shape complexity possible, limited only by ability to produce tool shape.
- Can be used for material susceptible to heat damage.
- Suitable for small diameter, deep holes with length to diameter ratios up to 50:1.
- Suitable for parts affected by thermal processes.
- Undercuts possible with specialized tooling.
- Possible to machine thin and delicate sections due to no processing forces.
- Cannot produce perfectly sharp corners.
- Minimum radius = 0.05 mm.
- Minimum hole size = $\varnothing$0.1 mm.

Quality issues

- Burr free part production.
- Produces slightly tapered holes, especially if deep, and some overcut possible.
- Finishing cuts are made at low material removal rates.
- Deep holes will have tapered walls.
- No stresses introduced, either, thermal or mechanical.
- Virtually no tool wear.
- Arcing may cause tool damage.
- Some electrolyte solutions can be corrosive to tool, workpiece and equipment.
- Surface detail good.
- Surface roughness values ranging 0.2–12.5 µm Ra. Dependent on current density and material being machined.
- Achievable tolerances ranging ±0.013–±0.5 mm. (Process capability charts have not been included. Capability is not primarily driven by characteristic dimension but by the material being processed.)

5.3   Electron Beam Machining (EBM)

Process description

- An electron gun bombards the workpiece with electrons up to 80 per cent the speed of light generating localized heat and evaporating the workpiece surface. Magnetic lenses focus the electron beam, and electromagnetic coils control its position. The workpiece is contained within a vacuum chamber typically (see 5.3F).

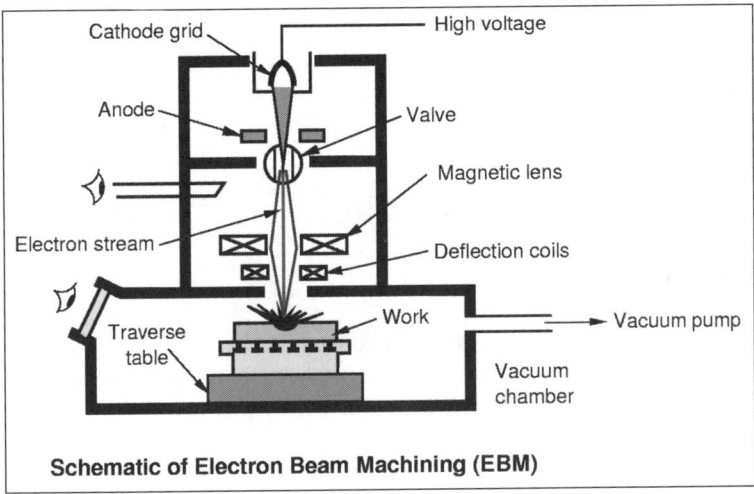

Schematic of Electron Beam Machining (EBM)

5.3F   Electron beam machining process.

Materials

- Any material regardless of its type, electrical conductivity and hardness.

Process variations

- Electron Beam Welding (EBW) (see 7.5): used to weld a range of material of varying thicknesses giving a small weld area and heat affected zone, with no flux or filler.
- The electron beam process can also be used for cutting, profiling, slotting and surface hardening, using the same equipment by varying process parameters.

Economic considerations

- Production rates dependent on size of vacuum chamber and by the ability to process a number of parts in batches at each loading cycle (less than 1 s per hole cycle time on thin workpieces).
- Parts should closely match size of chamber.
- Material removal rates low, typically 10 mm^3/min. Penetration speeds up to 600 mm/min possible.
- Lead times can be several weeks.

- Setup times can be short, but the time to create a vacuum in the chamber at each loading cycle is an important consideration.
- Material utilization good.
- High degree of automation possible.
- High energy consumption process.
- Economical with low to moderate production runs for thin parts requiring small cuts.
- Tooling costs very high.
- Equipment costs very high.
- Direct labor costs high. Skilled labor required.
- Finishing costs very low.

Typical applications

- Multiple small diameter holes in very thin and thick materials
- Injector nozzle holes
- Small extrusion die holes
- Irregular shaped holes and slots
- Engraving
- Features in silicon wafers for the electronics industry

Design aspects

- Electron beam path can be programmed to produce the desired pattern.
- Suitable for small diameter, deep holes with length to diameter ratios up to 100:1.
- Possible to machine thin and delicate sections due to no mechanical processing forces.
- Sharp corners difficult to produce.
- Better to have more small holes requiring less heat than a few large holes requiring considerable heat.
- Maximum thickness = 150 mm.
- Minimum hole size = $\emptyset$0.01 mm.

Quality issues

- Localized thermal stresses giving very small heat affected zones, small recast layers and low distortion of thin parts possible.
- Integrity of vacuum important. Beam dispersion occurs due to electron collision with air molecules.
- The reflectivity of the workpiece surface important. Dull and unpolished surfaces are preferred.
- Hazardous X-rays produced during processing which require lead shielding.
- Produces slightly tapered holes, especially if deep holes are required.
- Critical parameters to control during process: voltage, beam current, beam diameter and work speed.
- The melting temperature of the material may also have a bearing on quality of surface finish.
- Surface roughness values ranging 0.4–6.3 µm Ra.
- Achievable tolerances ranging ±0.013–±0.125 mm. (Process capability charts have not been included. Capability is not primarily driven by characteristic dimension.)

5.4   Laser Beam Machining (LBM)

Process description

- A pulsed beam of coherent monochromatic light of high power density, commonly known as a laser (Light Amplification by Stimulated Emission of Radiation), is focused on to the workpiece surface causing it to vaporize locally. The material then leaves the surface in the vaporized or liquid state at high velocity (see 5.4F).

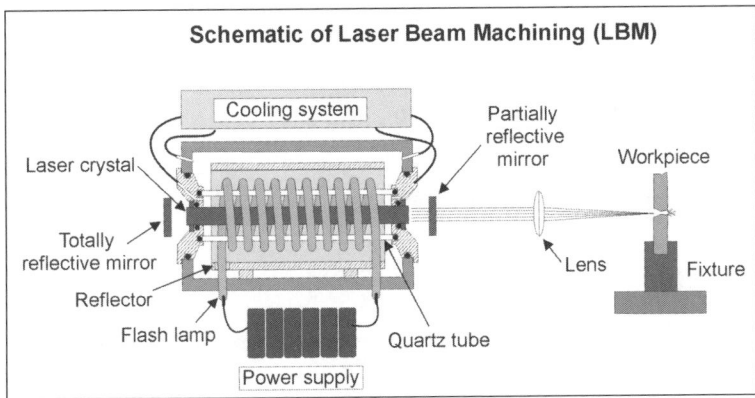

Schematic of Laser Beam Machining (LBM)

Cooling system

Partially reflective mirror

Workpiece

Laser crystal

Totally reflective mirror

Lens

Fixture

Reflector

Flash lamp

Quartz tube

Power supply

5.4F Laser beam machining process.

Materials

- Most materials, but dependent on thermal diffusivity and to a lesser extent the optical characteristics of material, rather than chemical composition, electrical conductivity or hardness.

Process variations

- Many types of laser are available, used for different applications. Common laser types available are: CO_2, Nd:YAG, Nd:glass, ruby and excimer. Depending on economics of process, pulsed and continuous wave modes are used.
- High pressure gas streams are used to enhance the process by aiding the exothermic reaction process, keeping the surrounding material cool and blowing the vaporized or molten material and slag away from the workpiece surface.
- Laser beam machines can also be used for cutting, surface hardening, welding (LBW) (see 7.6), drilling, blanking, honing, engraving and trimming, by varying the power density.

Economic considerations

- Production rates are moderate to high; 100 holes/s possible for drilling.
- Higher material removal rate than conventional machining.
- Material removal rates typically 5 mm^3/s and cutting speeds 70 mm/s.
- High power consumption.

- Lead times can be short, typically weeks.
- Setup times short.
- Material utilization good.
- High degree of automation possible.
- High flexibility. Integration with CNC punching machines is popular giving greater design freedom.
- Possible to perform many operations on same machine by varying process parameters.
- Economical for low to moderate production runs.
- Tooling costs very high.
- Equipment costs very high.
- Direct labor costs medium to high. Some skilled labor required.

Typical applications

- For holes, profiling, scribing, engraving and trimming
- Non-standard shaped holes, slots and profiling
- Prototype parts
- Small diameter lubrication holes
- Features in silicon wafers in the electronics industry

Design aspects

- Laser can be directed, shaped and focused by reflective optics permitting high spatial freedom in 2-dimensions and 3-dimensions with special equipment.
- Suitable for small diameter, deep holes with length to diameter ratios up to 50:1.
- Special techniques required to drill blind and stepped holes, but not accurate.
- Minimal work holding fixtures required.
- Sharp corners possible, but radii should be provided for in the design.
- Maximum thicknesses: mild steel $= 25\,mm$, stainless steel $= 13\,mm$, aluminum $= 10\,mm$.
- Maximum hole size (not profiled) $= 1.3\,mm$.
- Minimum hole size $= \varnothing 0.005\,mm$.

Quality issues

- Difficulty of material processing is dictated by how close the material's boiling and vaporization points are.
- Localized thermal stresses, heat affected zones, recast layers and distortion of very thin parts may be produced. Recast layers can be removed if undesirable.
- No cutting forces, so simple fixtures can be used.
- It is possible to machine thin and delicate sections due to no mechanical contact.
- The cutting of flammable materials is usually inert gas assisted. Metals are usually oxygen assisted.
- Control of the pulse duration is important to minimize the heat-affected zone, depth and size of molten metal pool surrounding the cut.
- The reflectivity of the workpiece surface is important. Dull and unpolished surfaces are preferred.
- Hole wall geometry can be irregular. Deep holes can cause beam divergence.
- Surface detail is fair.
- Surface roughness values ranging 0.4–6.3 μm Ra.
- Achievable tolerances ranging ± 0.015–$\pm 0.125\,mm$. (Process capability charts have not been included. Capability is not primarily driven by characteristic dimension.)

5.5 Chemical Machining (CM)

Process description

- Selective chemical dissolution of the workpiece material by immersion in a bath containing an etchant (usually acid or alkali solution). The areas that are not required to be etched are masked with 'cut and peel' tapes, paints or polymeric materials (see 5.5F).

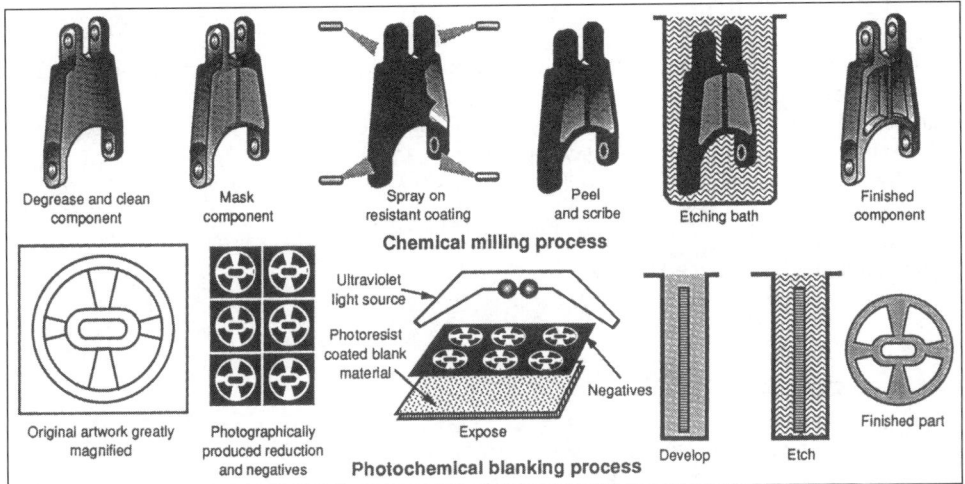

5.5F Chemical machining process.

Materials

- Most materials can be chemically machined with the correct chemical etchant selection, commonly: ferrous, nickel, titanium, magnesium and copper alloys, and silicon.

Process variations

- Chemical milling: chemical removal of material to a specified depth on large areas.
- Chemical blanking: used for thin parts requiring penetration through thickness.
- Photochemical blanking: uses photographic techniques to blank very thin sheets of metal, primarily for the production of printed circuit boards.
- Thermochemical machining: uses a hot corrosive gas.
- Electropolishing: for removal of residual stresses in surfaces.
- Chemical jet machining: uses a single jet of etchant.

Economic considerations

- Production rates low to moderate. Can be improved by machining a large sheet before cutting out the individual parts. Parts can also be etched on both sides simultaneously.
- Linear penetration rate very slow, typically 0.0025–0.1 mm/min, but dependent on material.

- Lead times short.
- Setup times short.
- Material utilization poor. Scrap material cannot be recycled.
- Disposal of chemicals used can be costly.
- Economical for low production runs. Least economical quantity is 1.
- Tooling costs low.
- Equipment costs generally low.
- Direct labor costs low.

Typical applications

- Primarily used for weight reduction in aerospace components, panels, extrusions and forgings by producing shallow cavities
- Printed circuit board tracks
- Features in silicon wafers for the electronics industry
- Decorative panels
- Printing plates
- Honeycomb structures
- Irregular contours and stepped cavities
- Burr free parts

Design aspects

- High degree of shape complexity possible in two-dimensions.
- Suitable for parts affected by thermal processes.
- Undercuts always present. The etch factor for a material is the ratio of the etched depth to the size of undercut.
- Controlling the size of small holes in thin sheet difficult.
- Compensation for the undercut should be taken into account when designing the masking template.
- Inside edges always have radii. Outside edges have sharp corners.
- Possible to machine thin and delicate sections due to no processing forces.
- Minimum thickness $= 0.013\,mm$.
- Maximum depth of cut $= 13\,mm$.
- Maximum size $= 3.7\,m \times 15\,m$, but dependent on bath size.

Quality issues

- Residual stresses in the part should be removed before processing to prevent distortion.
- Surfaces need to be clean and free from grease and scale to allow good masking adhesion and uniform material removal.
- Masking material should not react with the chemical etchant.
- Parts should be washed thoroughly after processing to prevent further chemical reactions.
- Porosity in castings/welds and intergranular defects are preferentially attacked by the etchant. This causes surface irregularities and non-uniformities.
- Room temperature and humidity, bath temperature and stirring need to be controlled to obtain uniform material removal.
- Surface detail is good.

- Surface roughness values ranging 0.4–6.3 μm Ra and are dependent on the material being processed.
- Achievable dimensional tolerances for selected process and material combinations are provided (see 5.5CC).

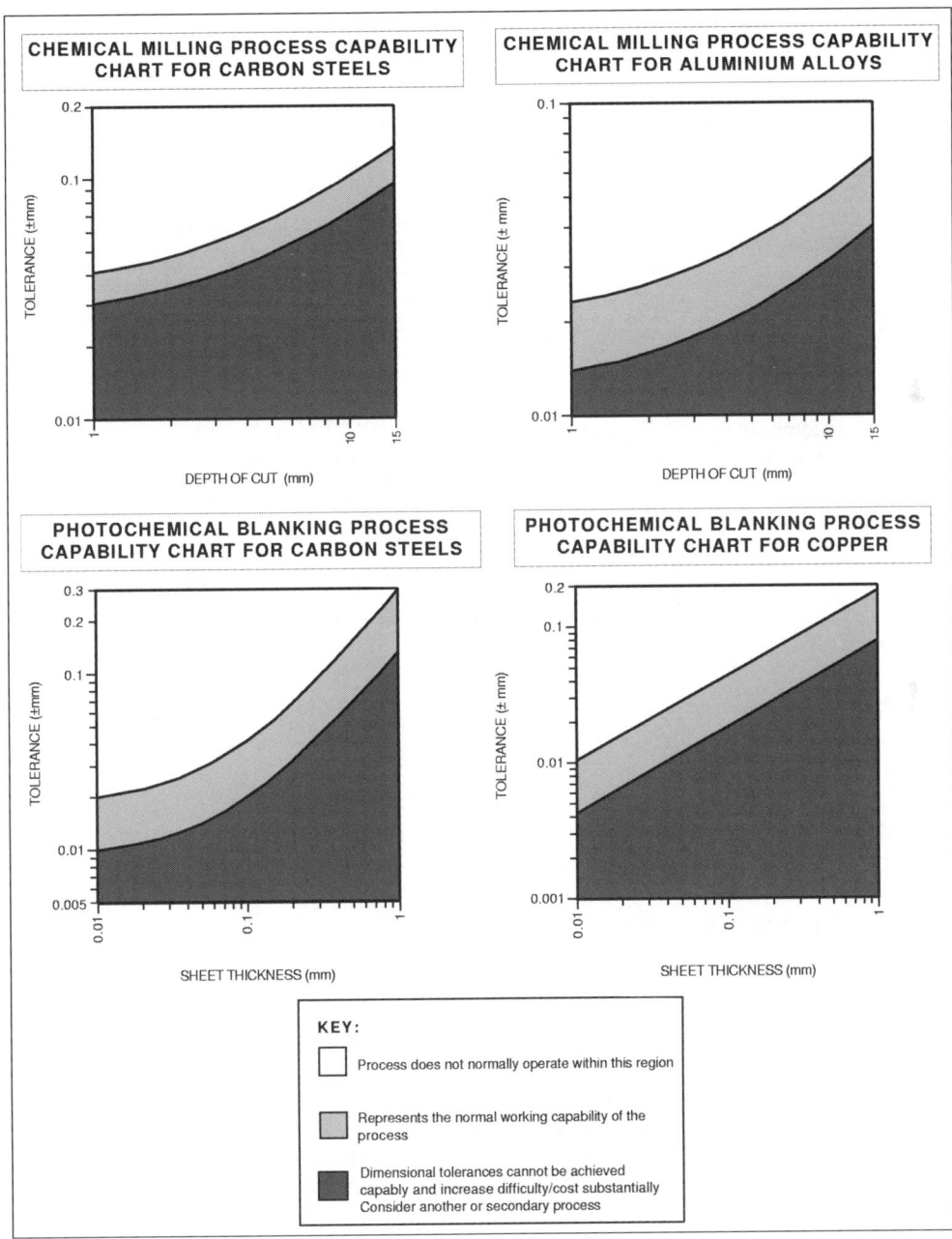

5.5CC Chemical machining process capability chart.

5.6   Ultrasonic Machining (USM)

Process description

- The tool, which is negative of the workpiece, is vibrated at around 20 kHz with an amplitude between 0.013 mm and 0.1 mm in an abrasive grit slurry at the workpiece surface. The workpiece material is removed by essentially three mechanisms: hammering of the grit against the surface by the tool, impact of free abrasive grit particles (erosion) and micro-cavitation. The slurry also removes debris away from the surface. The tool is gradually moved down maintaining a constant gap of approximately between the tool and workpiece surface (see 5.6F).

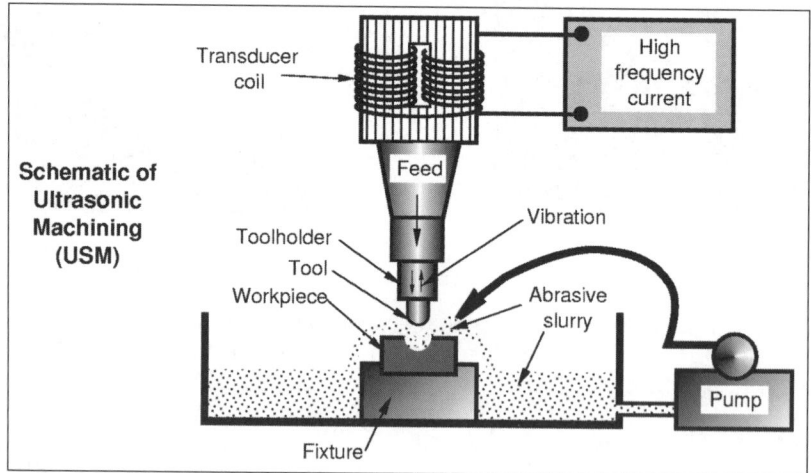

5.6F Ultrasonic machining process.

Materials

- Any material, however, brittle hard materials are preferred to ductile, for example, ceramics, precious stones, tool steels, titanium and glass.

Process variations

- Vibrations are either piezo-electric or magnetostrictive-transducer generated.
- Tool materials vary with application and allowable tool wear during machining. Common tool materials are: mild steel, stainless steel, tool steel, aluminum, brass and carbides (higher wear rates are experienced with aluminum and brass).
- Abrasive grit is available in many grades and material types. Materials commonly used are: boron carbide, aluminum oxide, diamond and silicon carbide.
- Liquid medium can be water, benzine or oil. Higher viscosity mediums decrease material removal rates.
- Rotary USM: a rotating diamond coated tool is used for drilling and threading, but with no abrasive involved.
- Ultrasonic cleaning: uses high-frequency sound waves in a liquid causing cavitation, which cleans the surface of the component, similar to a scrubbing action. Used to remove scale, rust, etc.

Economic considerations

- Production rates very low.
- Material removal rates low, typically $13\,mm^3/s$.
- Linear penetration rates up to $0.4\,mm/s$.
- Lead time typically days depending on complexity of tool. Special tooling required for each job.
- Material utilization poor. Scrap material cannot be recycled.
- High degree of automation possible.
- Economical for low production runs. Can be used for one-offs.
- Tooling costs high.
- Equipment costs generally moderate.
- Direct labor costs low to moderate.

Typical applications

- Burr free holes and slots in hard, brittle materials
- Complex cavities
- Coining operations

Design aspects

- Limited to shape of tool and control in 2-dimensions.
- Tool and tool holder designed with mass, shape and mechanical property considerations.
- Sharp profiles, corners and radii should be avoided as abrasive slurry erodes them away.
- Overcut will be produced which is approximately twice the grit size.
- Suitable for small diameter holes with length to diameter ratios typcially 3:1. Up to 4:1 using special equipment.
- Waste removal limits hole depths.
- Maximum hole size $= 90\,mm$.
- Minimum hole size $= 0.08\,mm$.

Quality issues

- Tapering of slots and holes occurs.
- Through holes in brittle materials should have a backing plate.
- Amplitude and frequency of vibration, tool material, impact force, abrasive grit grade and slurry viscosity and concentration all impact on accuracy, surface roughness and material removal rate.
- Finishing cuts made at lower material removal rates.
- Tool wear problematic. Tool changes can be frequent.
- Part is burr free with no residual stresses, distortion or thermal effects.
- Difference in wear rate between the tool and workpiece materials should be as high as possible.
- Surface detail good.
- Surface roughness values ranging 0.2–$1.6\,\mu m$ Ra.
- Finer surface roughness values obtained with finer grit grades.
- Achievable tolerances ranging ±0.005–$\pm0.05\,mm$. (Process capability charts have not been included. Capability is not primarily driven by characteristic dimension.)

5.7 Abrasive Jet Machining (AJM)

Process description

- Erosive action of an abrasive in a fluid is focused to a high velocity (150–300 m/s) jet through a sapphire nozzle. The abrasive and fractured particles are carried away from the cutting area by the jet (see 5.7F).

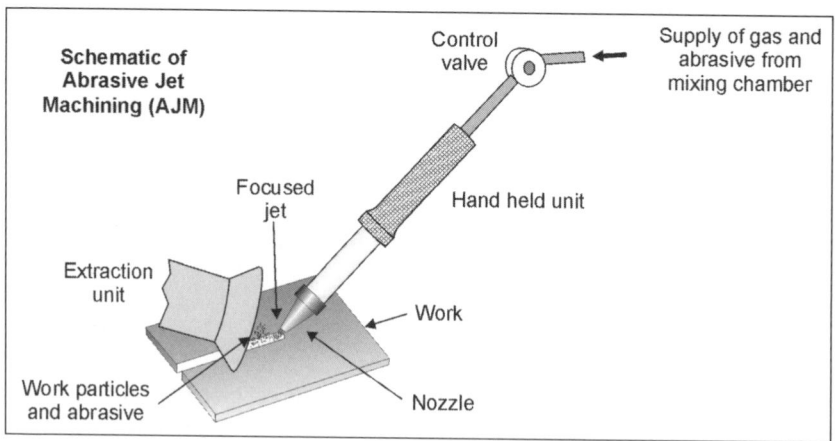

5.7F Abrasive jet machining process.

Materials

- Suitable for brittle and/or fragile materials.
- Refractory metals, titanium alloys, ceramics, metallic honeycomb structural materials, acrylic, composites, glass, silicon and graphite.

Process variations

- Two systems for introducing the abrasive to the jet stream:

  - Entrainment system: pressurized water jet pulls in abrasive particles into the stream, they are mixed in a tube and exit the nozzle.
  - Abrasive slurry system: mixing of fluid medium and abrasive particles takes place prior to pressurization in a separate chamber to create the slurry. Higher wear rates throughout the equipment experienced using this system, but less expensive.

- Fluid medium: either water or a gas (air or CO_2).
- Abrasive types: aluminum oxide and silicon carbide use.
- Tungsten can also be used for the nozzle, but has a higher wear rate than sapphire.
- Nozzle orifice can be round or square.
- Water jet machining: very high pressure focused jet of water used for cutting food, leather, paper and foamed plastics.
- Chemical jet machining: uses a single jet of etchant for deburring.

Economic considerations

- Production rates moderate.
- Material removal rates low, typically $15\,mm^3$/min.
- Penetration rate ranging from 10 to 1200 mm/min.
- Removal rate depends on the hardness of material and process parameters.
- Material utilization poor. Scrap material cannot be recycled.
- Can be fully automated using robots. Added flexibility.
- Small power requirements needed.
- Economical for low production runs.
- Tooling costs high.
- Equipment costs generally high.
- Direct labor costs low to moderate, depending on degree of automation.

Typical applications

- Through-holes, slots and profiles in hard, brittle materials
- For cutting, slitting, drilling, contouring, etching, cleaning, deburring and polishing
- Electronic component etching
- Etching and cutting glass
- Cutting metal foils

Design aspects

- Features limited to profiles, holes and slots.
- Depth of cut can be increased with jet pressure.
- Blind holes not possible.
- Long holes have tapered walls.
- Slot widths ranging from 0.12 to 0.25 mm.

Quality issues

- No heat and therefore no heat affected zone. Part free from metallurgical effects and residual stresses.
- Minimal dust, toxicity and fire hazard, but high noise levels.
- Less than 1 mm focus length from work should be maintained so no loss of definition and stray abrasion occurs.
- Minimal tool dulling.
- Inclination of jet angle to work can be less than $90°$, but at increased jet divergence on work, and therefore less control over material being cut.
- Abrasive size, slurry composition and flow-rate important control variables of the process for consistency.
- Abrasive slurry cannot be recycled due to abrasive grit blunting reducing effectiveness.
- Abrasive can become embedded in work surface.
- Surface detail good to excellent.
- Surface roughness values ranging 0.1–1.6 μm Ra.
- Surface roughness depends on abrasive particle size.
- Achievable tolerances ranging ±0.001–±0.013 mm. (Process capability charts have not been included. Capability is not primarily driven by characteristic dimension.)

6 Assembly systems

6.1 Manual assembly

Process description

- Manual assembly involves the composing of previously manufactured components and/or sub-assemblies into a complete product of unit of a product, primarily performed by human operators using their inherent dexterity, skill and judgment. The operator may be at a workstation (bench) or be part of a transfer system that moves the product as it is being assembled. Manual assembly can be further assisted by mechanized or automated systems for feeding, handling, fitting and checking operations (see 6.1F).

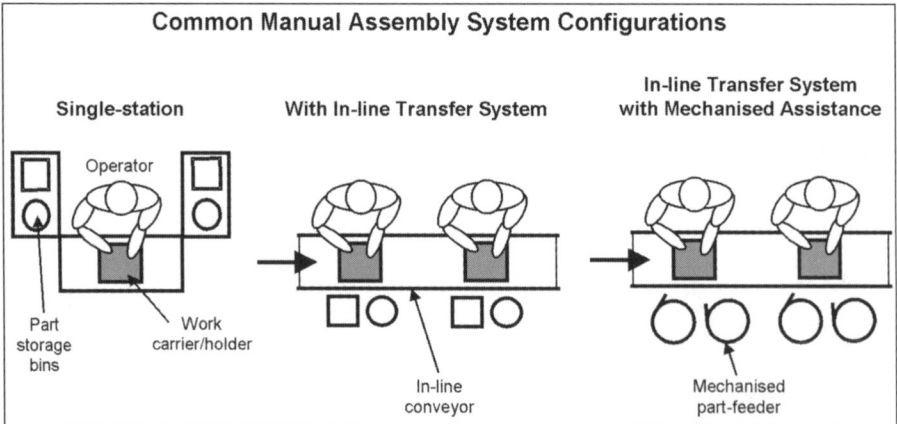

6.1F Manual assembly process.

Process variations

- Feeding: presentation of a component to the handling equipment in the correct orientation by a variety of methods. Parts manually taken from storage bins and then orientated by operator. Orientation can be achieved by vibratory/centrifugal bowl feeders, parts already orientated in pallet/magazine/strip form or use of part-feeding escapement mechanisms.
- Handling: bringing components and/or sub-assemblies together such that later composition can occur. Use of hands, simple lifting aids and jigs and fixtures.
- Fitting: various part placement/location configurations or fastening/joining methods can be utilized, e.g. 'peg in hole', press fit, welding, riveting, adhesive bonding, staking and screwing, using a variety of hand operated or mechanized/electrical tools.
- Checking: detection of missing, incorrect, misshapen or wrongly orientated components by high-level sensing and checking capabilities of operators and mechanical/electrical aids. Also detection of foreign bodies, part-failure and machine in-operation.
- Transfer: typically, the various assembly operations required are carried out at separate stations, usually built up on a work carrier, pallet or holder. Therefore, a system for transferring the partly

completed assemblies from workstation to workstation is required. In general, the various types of workstation/transfer system for manual assembly are:

- Single-station: assembly at one workstation or bench where a specific or a variety of operations are performed.
- Continuous: work carrier flows without stopping, using conveyors (in-line, rotary dial or carousel), overhead rail or towline.
- Intermittent: synchronous/indexing (moved with a fixed cycle time) or non-synchronous/free-transfer (moved as required or when operation/assembly completed by operator) using in-line or rotary systems.

Economic considerations

- Production rates low to moderate, depending on complexity, number and size of component parts. These factors also dictate the degree of mechanized assistance needed.
- Extremely flexible assembly system (many product variants) and therefore most common.
- Lead time typically days, higher if mechanized assistance devices used.
- Economical for low to moderate production runs. Can be used for one-offs.
- Tooling costs low to moderate.
- Equipment costs generally low, except where full-mechanized assistance exists.
- Direct labor costs moderate to high. Relatively easy to train operators.

Typical applications

- Car assembly lines
- Internal combustion engines
- Domestic appliances and office equipment
- Electronic and electrical equipment
- Machine tools
- General fabrication
- Toys, furniture, footwear and clothing

Design aspects

- Use DFA techniques in order to develop assemblies with optimum part-count, improved component geometry for feeding, handling, fitting and checking, and reduce overall assembly costs.
- Use Poka-Yoke (mistake-proofing) techniques to help reduce operator assembly errors by prescribing component features and/or assembly procedures to aid correct assembly.
- Develop an assembly sequence diagram to optimize the assembly line.
- Assess overall assembly tolerance against component tolerances in stack up.

Quality issues

- In general, assembly problems are caused by a number of factors:
  - Components exceeding or being lower than the specified tolerances
  - Component misalignment and adjustment error
  - Gross defects (malformed, missing features, wrong lengths, damage in transit, etc.)
  - Foreign matter causing contamination and blockages
  - Absence of a component due to inefficient feeding or exhausted supply

- Incorrect components caused by wrong supply or instructions
- Inadequate joining technology
- Skill of labor used.

- Manual assembly is not suitable for harsh environments. Also, size and weight of parts to be assembled must be considered for safety handling.
- Operator fatigue, health and relaxation time must be considered, especially for highly repetitive operations.
- Assembly errors increase if components/sub-assemblies are complex, difficult to align, insert or if there is restricted access for insertion.
- Poor quality components can generally be sorted out during the assembly task without difficulty or high loss through the advanced checking capabilities of human operators.
- Repeatable accuracy of component alignment is low to moderate depending on part complexity, typically $\pm 0.5\,$mm.

6.2   Flexible assembly

Process description

• Flexible assembly systems use programmable, robotic devices to compose previously manufactured components and/or sub-assemblies into a complete product of unit of a product. A number of
transfer mechanisms, feeding devices, robot types and end effectors can be utlilized in order to
achieve a general assembly system (see 6.2F).

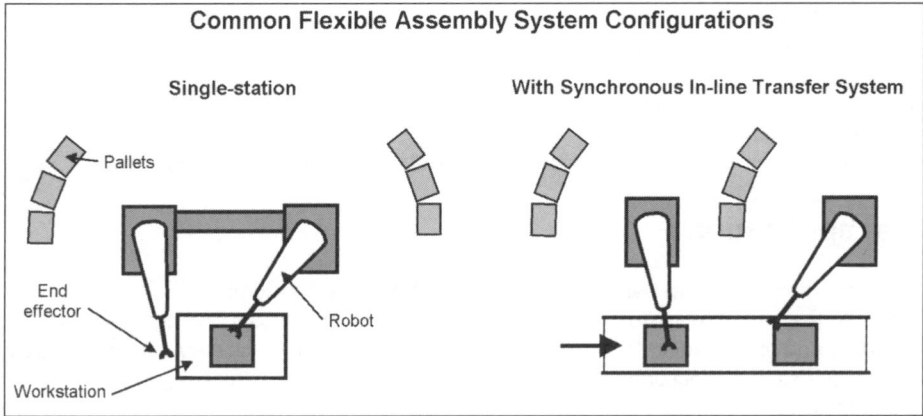

6.2F Flexible assembly process.

Process variations

• Robot types: variety of configurations, loading carrying capacity, working envelope, wrist degrees of
freedom and accuracy/repeatability, e.g. revolute, polar, gantry and pendulum.
• End effectors: variety of arrangements depending on part geometry and feeding direction, flexibility,
fragility and overall part size and weight. Either pneumatic, vacuum, electromechanical or electromagnetic actuation of holding/gripping mechanism.
• Feeding: presentation of a component to the robot arm end effector in the correct orientation.
Orientation can be achieved by vibratory/centrifugal bowl feeders, by receiving parts already
orientated by the supplier in pallet, magazine or by escapement mechanisms for part feeding.
• Handling: bringing components and/or sub-assemblies together such that later composition can
occur using robot arm end effectors.
• Fitting: various part placement/location configurations or fastening/joining methods can be utilized,
e.g. 'peg in hole', adhesive bonding, staking and screwing.
• Checking: detection of missing, incorrect, misshapen or wrongly orientated components by electronic vision systems, tactile/pressure sensors and proximity sensors.
• Transfer: typically, the various assembly operations required are carried out at separate stations,
usually built up on a work carrier, pallet or holder. Therefore, a system for transferring the partly

completed assemblies from workstation to workstation is required. In general, the various types of station/transfer system for flexible assembly are:

- Single-/multi-station: assembly at one or more workstations where a specific or more commonly, a variety of operations are performed. Typically, greater than six components to be assembled requires a multi-station arrangement.
- Synchronous/indexing: moved with a fixed cycle time using in-line, rotary dial or carousel systems.

Economic considerations

- Only moderate flexibility, despite name.
- Systems can be adapted for the assembly of several different products/variants.
- Production rates moderate.
- Lead time weeks to months.
- Economical for moderate to high production volumes.
- Tooling costs high.
- Equipment costs moderate to very high.
- Direct labor costs low.
- Programming/teaching of robot operations and movements is complex and lengthy.

Typical applications

- General assembly, materials handling and transfer of parts and assemblies
- For hazardous environments (to humans), e.g. radioactive, toxic, dusty and high temperatures
- Part loading and/or unloading for manufacturing processes, e.g. machining centers, pressure die casting machines and injection molding machines
- Spot and MIG welding
- Abrasive jet machining
- Surface finishing, grinding, buffing and spray painting operations

Design aspects

- Use DFA techniques in order to develop assemblies with optimum part-count, improved component geometry for feeding, handling, fitting and checking, and reduce overall assembly costs.
- Develop an assembly sequence diagram to optimize the assembly line.
- Assess overall assembly tolerance against component tolerances in stack up.

Quality issues

- In general, assembly problems are caused by a number of factors:
  - Components exceeding or being lower than the specified tolerances
  - Component misalignment and adjustment error
  - Gross defects (malformed, missing features, wrong lengths, damage in transit, etc.)
  - Foreign matter causing contamination and blockages
  - Absence of a component due to inefficient feeding or exhausted supply
  - Incorrect components caused by wrong supply or instructions
  - Inadequate joining technology.

- Approximately 50 per cent of all problems found in automated systems (product defects and down-time) are due to the incoming component quality.

- Robot working envelope must be securely guarded.
- Automated or mechanized systems must be chosen in certain situations, particularly where operator safety is paramount, for example, hazardous or toxic environments, heavy component parts or a high repeatability requirement causing operator fatigue.
- It can use dedicated systems for sterile or clean environment assembly of products.
- Repeatable accuracy of component alignment is high, typically ± 0.1 mm.

6.3 Dedicated assembly

Process description

- Dedicated assembly systems are special purpose, fully mechanized or automated systems for composing previously manufactured components and/or sub-assemblies into a complete product of unit of a product. Typically, a number of workstations comprising automatic part-feeders and fixed work-heads are arranged on an automatically controlled transfer system to compose the product sequentially (see 6.3F).

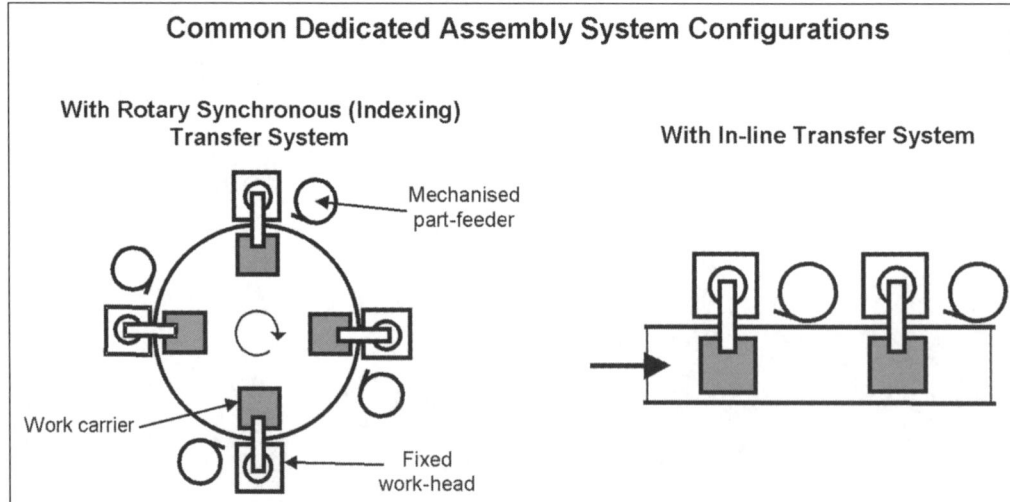

6.3F Dedicated assembly process.

Process variations

- Feeding: presentation of a component to the robot arm end effector in the correct orientation. Orientation can be achieved by vibratory/centrifugal bowl feeders, by receiving parts already orientated by the supplier in pallet, magazine or by escapement mechanisms for part-feeding.
- Handling: bringing components and/or sub-assemblies together such that later composition can occur using fixed work-heads and/or pick and place units.
- Fitting: various part placement/location configurations or fastening/joining methods can be utilized, e.g. 'peg in hole', adhesive bonding, staking and screwing.
- Checking: identification of missing, incorrect, misshapen or wrongly orientated components. Also detection of foreign bodies, part-failure and machine in-operation. Common technologies include vision systems, tactile/pressure sensors, proximity sensors and 'bed of nails'.
- Transfer: typically, the various assembly operations required are carried out at separate stations, usually built up on a work carrier, pallet or holder. Therefore, a system for transferring the partly

completed assemblies from workstation to workstation is required. In general, transfer systems for dedicated assembly are either:

- Synchronous/indexing: moved with a fixed cycle time.
- Non-synchronous/free-transfer: moved as required or when operation/assembly completed using in-line or rotary systems. Can set up a buffer system using this configuration. Typically, greater than ten components to be assembled requires a free-transfer arrangement.

Economic considerations

- Almost totally inflexible. Fixed assembly system for one product type typically, except where variants are based on parts missing from original design.
- Production rates high.
- Lead time typically months.
- Economical for high production volumes.
- Tooling costs high.
- Equipment costs high.
- Direct labor costs very low.

Typical applications

- Electronic and electrical components and devices
- Printed circuit boards
- Small domestic appliances
- Medical products
- Automotive sub-assemblies, e.g. valves, solenoids, relays
- Office equipment

Design aspects

- Use DFA techniques in order to develop assemblies with optimum part-count, improved component geometry for feeding, handling, fitting and checking, and reduce overall assembly costs.
- Develop an assembly sequence diagram to optimize the assembly line.
- Assess overall assembly tolerance against component tolerances in stack up.

Quality issues

- In general, assembly problems are caused by a number of factors:
  - Components exceeding or being lower than the specified tolerances causing interference or location stability problems
  - Component misalignment and adjustment error
  - Gross defects (malformed, missing features, wrong lengths, damage in transit, etc.)
  - Foreign matter causing contamination and blockages
  - Absence of a component due to inefficient feeding or exhausted supply
  - Incorrect components caused by wrong supply or instructions
  - Inadequate joining technology.
- Approximately 50 per cent of all problems found in automated systems (product defects and down-time) are due to the incoming component quality.

- It is difficult and expensive to incorporate insensitivity to component variation and faults in assembly systems to reduce this problem. Sensing capabilities are limited in this capacity.
- Automated or mechanized systems must be chosen in certain situations, particularly where operator safety is paramount, for example, hazardous or toxic environments, heavy component parts or a high repeatability requirement, causing operator fatigue.
- It can use dedicated systems for sterile or clean environment assembly of products.
- Repeatable accuracy of component alignment is high, typically ± 0.1 mm.

7 Joining processes

7.1 Tungsten Inert-Gas Welding (TIG)

Process description

- An electric arc is automatically generated between the workpiece and a non-consumable tungsten electrode at the joint line. The parent metal is melted and the weld created with or without the addition of a filler rod. Temperatures at the arc can reach 12 000 °C. The weld area is shielded with a stable stream of inert gas, usually argon, to prevent oxidation and contamination (see 7.1F).

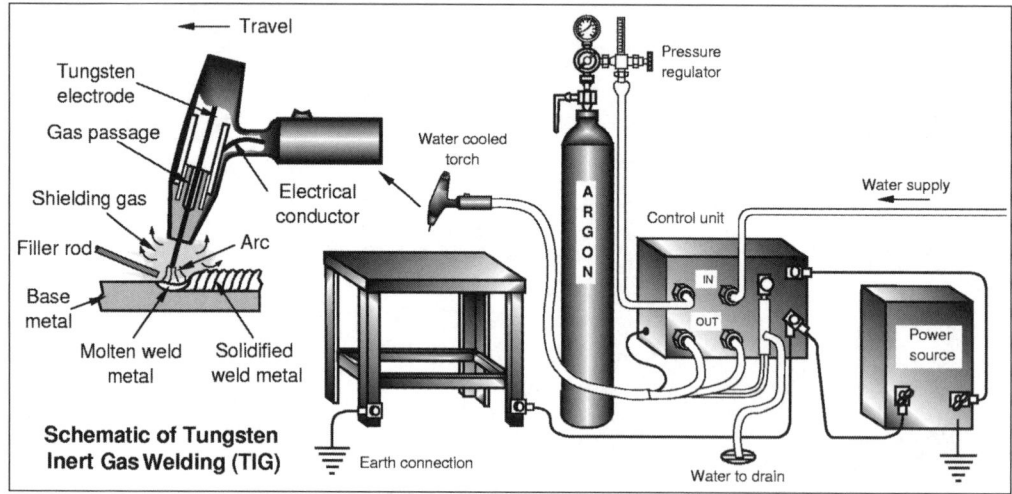

7.1F Tungsten inert-gas welding process.

Materials

- Most non-ferrous metals (except zinc), commonly, aluminum, nickel, magnesium and titanium alloys, copper and stainless steel. Carbon steels, low alloy steels, precious metals and refractory alloys can also be welded. Dissimilar metals are difficult to weld.

Process variations

- Portable manual or automated a.c. or d.c. systems. a.c commonly used for welding aluminum and magnesium alloys.
- Pure helium or more commonly, a helium/argon mix is used as the shielding gas for metals with high thermal conductivity, for example copper, or material thickness greater than 6 mm giving increased weld rates and penetration.
- Pulsed TIG: excellent for thin sheet or parts with dissimilar thickness (low heat input).
- TIG spot welding: used on lap joints in thin sheets.

Economic considerations

- Weld rates vary from 0.2 m/min for manual welding to 1.5 m/min for automated systems.
- Automation is suited to long lengths of continuous weld in the same plane.
- Automation is relatively inexpensive if no filler is required, i.e. use of close fitting parts.
- Process is suited to sheet thickness less than 4 mm, heavier gauges become more expensive due to argon cost and decreased production rate. Helium/argon gas is expensive but may be viable due to increased production rate.
- It is economical for low production runs. Can be used for one-offs.
- Tooling costs are low to moderate.
- Equipment costs are moderate.
- Direct labor costs are moderate to high. Highly skilled labor required for manual welding. Setup costs can be high for fabrications using automated welding.
- Finishing costs are low generally. There is no slag produced at the weld area, however, some grinding back of the weld may be required.

Typical applications

- Chemical plant pipe work
- Nuclear plant fabrications
- Aerospace structures
- Sheet-metal fabrication

Design aspects

- Design complexity is high.
- Typical joint designs possible using TIG are: butt, lap, fillet and edge (see Appendix B – Weld Joint Configurations).
- Design joints using minimum amount of weld, i.e. intermittent runs and simple or straight contours, although TIG is suited to automated contour following.
- Design parts to give access to the joint area, for vision, electrodes, filler rods, cleaning, etc.
- Wherever possible horizontal welding should be designed for, however, TIG welding is suited to most welding positions.
- Sufficient edge distances should be designed for. Avoid welds meeting at end of runs.
- Balance the welds around the fabrication's neutral axis where possible.
- Distortion can be reduced by designing symmetry in parts to be welded along weld lines.
- The fabrication sequence should be examined with respect to the above.
- Provision for the escape of gases and vapors in the design is important.
- Minimum sheet thickness $= 0.2$ mm.
- Maximum thickness, commonly:

  - Copper and refractory alloys $= 3$ mm
  - Carbon, low alloy and stainless steels; magnesium and nickel alloys $= 6$ mm
  - Aluminum and titanium alloys $= 15$ mm.

- Multiple weld runs required on sheet thickness ≥ 5 mm.
- Unequal thicknesses are difficult.

Quality issues

- Clean, high quality welds with low distortion can be produced.
- Access for weld inspection important, e.g. Non-Destructive Testing (NDT).
- Joint edge and surface preparation important. Contaminates must be removed from the weld area to avoid porosity and inclusions.
- A heat affected zone always present. Some stress relieving may be required for restoration of materials' original physical properties.
- Not recommended for site work in wind where the shielding gas may be gusted.
- Control of arc length important for uniform weld properties and penetration.
- Need for jigs and fixtures to keep joints rigid during welding and subsequent cooling to reduce distortion on large fabrications.
- Backing strips can be used for avoiding excess penetration, but at added cost and increased setup times.
- Selection of correct filler rod important (where required).
- Care needed to keep filler rod within the shielding gas to prevent oxidation.
- Workpiece and filler rod must be away from the tungsten electrode to prevent contamination which can cause an unstable arc.
- Shielding gas must be kept on for a second or two to allow tungsten electrode to cool and prevent oxidation.
- Tungsten inclusions can contaminate finished welds.
- Welding variables should be preset and controlled during production.
- Automation reduces the ability to weld mating parts with inherent size and shape variations; reduced by automation however, it does reduce distortion, improve reproduction and produces fewer welding defects.
- 'Weldability' of the material important and combines many of the basic properties that govern the ease with which a material can be welded and the quality of the finished weld, i.e. porosity and cracking. Material composition (alloying elements, grain structure and impurities) and physical properties (thermal conductivity, specific heat and thermal expansion) are some important attributes which determine weldability.
- Surface finish of weld excellent.
- Fabrication tolerances typically ± 0.5 mm.

7.2   Metal Inert-Gas Welding (MIG)

Process description

- An electric arc is manually created between the workpiece and a consumable wire electrode at the joint line. The parent metal is melted and the weld created with the continuous feed of the wire which acts as the filler metal. The weld area is shielded with a stable stream of argon or CO_2 to prevent oxidation and contamination (see 7.2F).

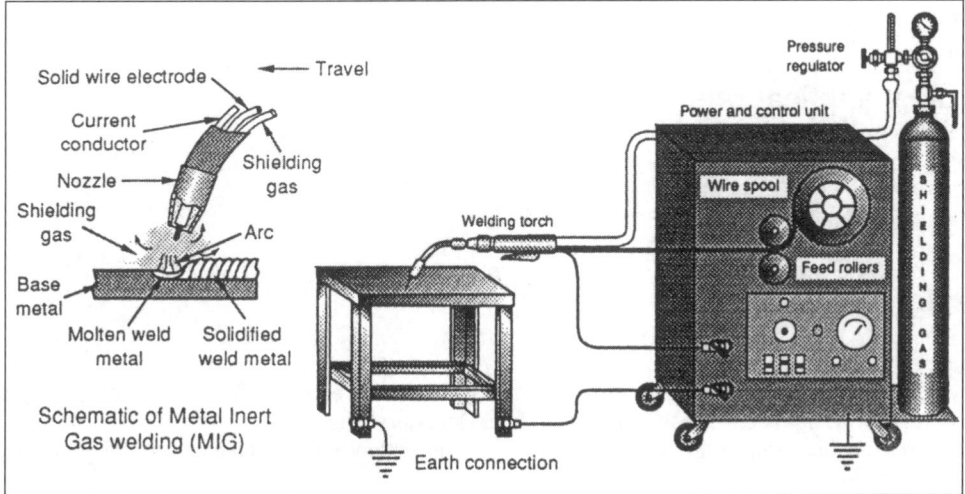

7.2F Metal inert-gas welding process.

Materials

- Carbon, low alloy and stainless steels. Most non-ferrous metals (except zinc) are also weldable; aluminum, nickel, magnesium and titanium alloys and copper. Refractory alloys and cast iron can also be welded. Dissimilar metals are difficult to weld.

Process variations

- Portable semi-automatic (manually operated) or fully automated d.c. systems and robot mounted.
- Three types of metal transfer to the weld area: dip and pulsed transfer use low current for positional welding (vertical, overhead) and thin sheet; spray transfer uses high currents for thick sheet and high deposition rates, typically for horizontal welding.
- Shielding gases: pure CO_2 or argon/CO_2 mix commonly used for carbon and low alloy steels, or a mix of argon/helium, also used for nickel alloys and copper. Pure argon is used for aluminum alloys. High chromium steels use an argon/O_2 mix.
- MIG spot welding: used on lap joints.
- Flux Cored Arc Welding (FCAW): uses a wire containing a flux and gas generating compounds for self-shielding, although flux-cored wire is preferred with additional shielding gas for certain conditions. Limited to carbon steels and lower welding rates.

Economic considerations

- Weld rates from 0.2 m/min for manual welding to 15 m/min for automated setups.
- Production costs reduced by high weld deposition rates with continuous operation.
- Well suited to traversing automated and robotic systems.
- Choice of electrode wire ($\varnothing$0.5–$\varnothing$1.5 mm) and shielding gas important cost considerations.
- Economical for low production runs. Can be used for one-offs.
- Tooling costs low to moderate.
- Equipment costs low to moderate, depending on degree of automation.
- Direct labor costs moderate to high. Skill level required is less than TIG.
- Finishing costs low generally. There is no slag produced at the weld area, however, some grinding back of the weld may be required.

Typical applications

- General fabrication
- Structural steelwork
- Automobile bodywork

Design aspects

- All levels of complexity possible.
- Typical joint designs possible using MIG: butt, lap, fillet and edge. MIG excellent for vertical and overhead welding (see Appendix B – Weld Joint Configurations).
- Design joints using minimum amount of weld, i.e. intermittent runs and simple or straight contours wherever possible.
- Welds should be balanced around the fabrication's neutral axis where possible.
- Design parts to give access to the joint area, for vision, electrodes, filler rods, cleaning, etc. MIG good for welds inaccessible by other methods.
- Sufficient edge distances should be designed for and avoid welds meeting at the end of runs.
- Provision for the escape of gases and vapors in the design important.
- Distortion can be reduced by designing symmetry in parts to be welded along weld lines.
- The fabrication sequence should be examined with respect to the above.
- Minimum sheet thickness = 0.5 mm (6 mm for cast iron).
- Maximum thickness, commonly:

  - Carbon, low alloy and stainless steels; cast iron, aluminum, magnesium, nickel, titanium alloys and copper = 80 mm
  - Refractory alloys = 6 mm.

- Multiple weld runs required on sheet thicknesses $\geq$5 mm.
- Unequal thicknesses possible.

Quality issues

- Clean, high quality welds with low distortion can be produced.
- Access for weld inspection important, e.g. Non-Destructive Testing (NDT).
- Joint edge and surface preparation important. Contaminates must be removed from the weld area to avoid porosity and inclusions.

- Shielding gas chosen to suit parent metal, i.e. it must not react when welding.
- Wire electrode must closely match the composition of the metals being welded.
- Slag created when using a flux-cored wire may aid the control of the weld profile and commonly used for site work (windy conditions where the shielding gas may be gusted or positional welding) and large fillet welds.
- A heat affected zone always present. Some stress relieving may be required for restoration of materials original physical properties.
- Cracking may be experienced when welding high alloy steels.
- Self-adjusting arc length reduces skill level required and increases weld uniformity.
- Backing strips can be used for avoiding excess penetration, but at added cost and increased setup times.
- Need for jigs and fixtures to keep joints rigid during welding and subsequent cooling to reduce distortion on large fabrications.
- Welding variables should be preset and controlled during production.
- Automation can limit the ability to weld mating parts with large size and shape variations, however, the use of dedicated tooling does reduce distortion, improve reproduction and produces fewer welding defects.
- 'Weldability' of the material important and combines many of the basic properties that govern the ease with which a material can be welded and the quality of the finished weld, i.e. porosity and cracking. Material composition (alloying elements, grain structure and impurities) and physical properties (thermal conductivity, specific heat and thermal expansion) are some important attributes which determine weldability.
- Surface finish of weld good.
- Fabrication tolerances typically ± 0.5 mm.

7.3 Manual Metal Arc Welding (MMA)

Process description

- An electric arc is created between a consumable electrode and the workpiece at the joint line. The parent metal is melted and the weld created with the manual feed of the electrode along the weld and downwards as the electrode is being consumed. Simultaneously, a flux on the outside of the electrode melts covering the weld pool and generates a gas shielding it from the atmosphere and preventing oxidation (see 7.3F).

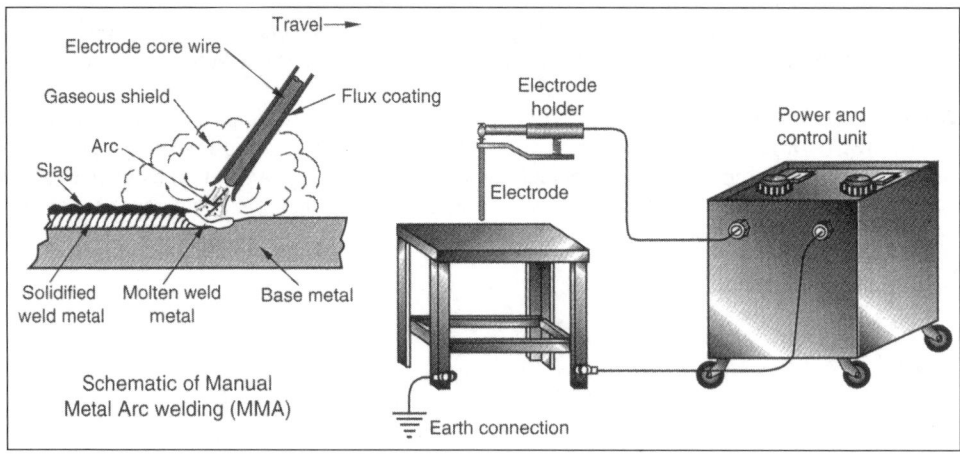

7.3F Manual metal arc welding process.

Materials

- Carbon, low alloy and stainless steels; nickel alloys and cast iron typically. Welding of non-ferrous metals is not recommended, but occasionally performed. Dissimilar metals are difficult to weld.

Process variations

- Manual d.c. and a.c. sets. Only a few fluxes give stable operation with a.c.
- Large selection of electrode materials with a variety of flux types for the welding of different metals and properties required. Core sizes are between $\varnothing 1.6$ and $\varnothing 9.5$ mm and the electrode length is usually 460 mm.
- Stud Arc Welding (SW): for welding pins and stud bolts to structures for subsequent fastening operations. Uses the pin or stud as a consumable electrode to join to the workpiece at one end. Portable semi-automatic or static automated equipment available.

Economic considerations

- Weld rates up to 0.2 m/min.
- Most flexible of all welding processes.

- Manually performed typically, although some automation possible.
- Can weld a variety of metals by simply changing the electrode.
- More power required for a.c. welding than d.c. welding.
- Suitable for site work. Welding can be performed up to 20 m away from power supply.
- Non-continuous process. Frequent changes of electrode are required.
- Economical for low production runs. Can be used for one-offs.
- Tooling costs low. Need for jigs and fixtures not as important as other methods and less accuracy required in setting up.
- Equipment costs low.
- Direct labor costs high. Skill level required is higher than MIG.
- Finishing costs high relative to other welding processes. Slag produced at the weld area, which must be removed during runs and some grinding back of the weld, may be required. Weld spatter often covers the surface which may need cleaning.

Typical applications

- Pressure vessels
- Structural steelwork
- Shipbuilding
- Pipework
- Machine frame fabrication
- Maintenance

Design aspects

- All levels of complexity possible.
- Typical joint designs possible using MMA: butt, lap, fillet and edge in heavier sections (see Appendix B – Weld Joint Configurations).
- Suitable for all welding positions.
- Design joints using minimum amount of weld, i.e. intermittent runs and simple or straight contours wherever possible.
- Balance the welds around the fabrication's neutral axis.
- Distortion can be reduced by designing symmetry in parts to be welded along weld lines.
- Design parts to give access to the joint area, for vision, electrodes, filler rods, cleaning, etc. MMA excellent for welds inaccessible by other methods.
- Sufficient edge distances should be designed for. Avoid welds meeting at end of runs.
- Provision for the escape of gases and vapors in the design important.
- The fabrication sequence should be examined with respect to the above.
- Minimum thickness = 1.5 mm (6 mm for cast iron).
- Maximum sheet thickness, commonly for carbon, low alloy and stainless steels, nickel alloys and cast iron = 200 mm.
- Multiple weld runs required on sheet thicknesses ≥10 mm.
- Unequal thicknesses difficult.

Quality issues

- Moderate to high quality welds with moderate, but acceptable levels of distortion can be produced.
- Quality and consistency of weld related to skill of welder to maintain correct arc length and burn-off rate.

- Access for weld inspection important, e.g. NDT.
- Joint edge and surface preparation important. Contaminates must be removed from the weld area to avoid porosity and inclusions after each pass.
- A heat affected zone always present. Some stress relieving may be required for restoration of materials original physical properties.
- Need for jigs and fixtures to keep joints rigid during welding and subsequent cooling to reduce distortion on large fabrications.
- Backing strips can be used for avoiding excess penetration, but at added cost and increased setup times.
- Can alter composition of weld by addition of alloying elements in the electrode. Addition of deoxidants in the flux minimizes carbon loss, which reduces weld strength.
- Electrodes must be dry and free from oil and grease to prevent weld contamination.
- Low hydrogen electrodes should be used when welding high carbon steels to reduce chance of hydrogen cracking.
- The protective slag can help the weld to keep its shape during positional welding.
- Weld ideally left to cool to room temperature before the slag removed.
- When the electrode's length reduced to approximately 50 mm it should be replaced.
- Welding current should be maintained during welding with a stable power supply.
- Arc deflection can sometimes occur with d.c. supplies, especially in magnetized metals. The workpiece may need demagnetizing or the return cable repositioned.
- Pre-heating of workpiece can reduce porosity and hydrogen cracking.
- 'Weldability' of the material important and combines many of the basic properties that govern the ease with which a material can be welded and the quality of the finished weld, i.e. porosity and cracking. Material composition (alloying elements, grain structure and impurities) and physical properties (thermal conductivity, specific heat and thermal expansion) are some important attributes which determine weldability.
- Surface finish of weld fair to good. Weld spatter often covers the surface.
- Fabrication tolerances typically ± 1 mm.

7.4 Submerged Arc Welding (SAW)

Process description

- A blanket of flux is fed from a hopper in advance of an electric arc created between a consumable electrode wire and the workpiece at the joint line. The arc melts the parent metal and the wire creates the weld as it is automatically fed downwards and traversed along the weld, or the work is moved under welding head. The flux shields the weld pool from the atmosphere preventing oxidation. Any flux that is not used is recycled (see 7.4F).

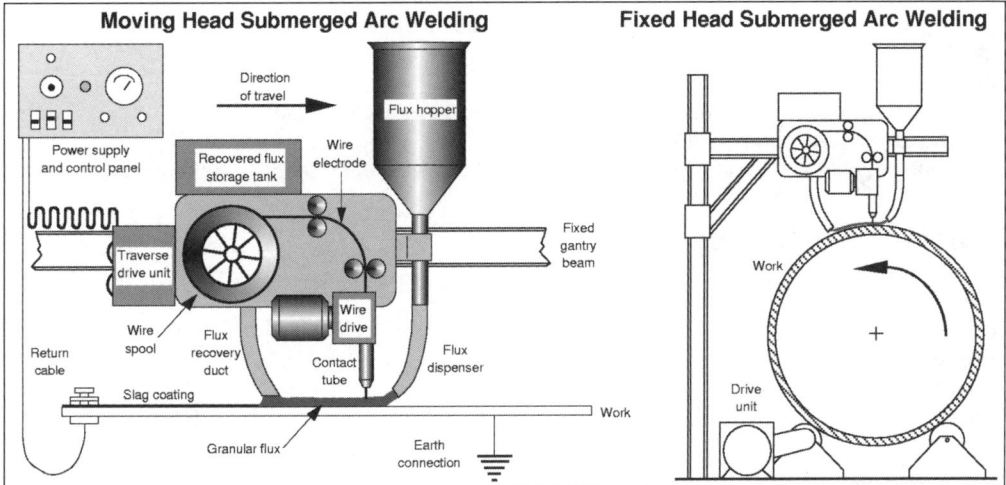

7.4F Submerged arc welding process.

Materials

- Carbon, low alloy and stainless steels, and some nickel alloys.
- Dissimilar metals are difficult to weld.

Process variations

- Self-contained, mainly automated a.c. or d.c. systems with up to three welding heads.
- Can have portable traversing welding unit using a wheeled buggy (for long welds on ship's deck plates for example), self-propelled traversing unit on a gantry or moving head type (for shorter weld lengths) and fixed head where the work rotates under the welding unit (for pressure vessels).
- Copper-coated electrode wire can be solid or tubular. Tubular is used to supply the weld with additional alloying elements. Wire sizes range from $\varnothing0.8$ to $\varnothing9.5$ mm.
- Can use a strip electrode for surfacing to improve corrosion resistance (pressure vessels) or for hardfacing parts subject to wear (bulk materials handling chute).

- Fluxes available in powdered or granulated form, either neutral or basic. Neutral fluxes used for low carbon steel and basic fluxes for higher carbon steels.
- Bulk welding: uses an iron powder placed in the joint gap in advance of the flux and electrode to increase deposition rates.
- For thin sections can use a flux-coated electrode wire.

Economic considerations

- Highest weld deposition rate of all arc welding processes.
- Speeds ranging 0.1 to 5 m/min.
- Economic for straight, continuous welds on thick plate using single or multiple runs.
- High power consumption offset by high productivity.
- Economical for low production runs. Can be used for one-offs.
- Tooling costs low to moderate. Need for jigs and fixtures important for accurate joint alignment.
- Equipment costs moderate to high.
- Direct labor costs low to moderate. Skill level required low to moderate.
- Flux handling costs can be high.
- Finishing costs moderate to high. Slag produced at the weld area needs to be removed.

Typical applications

- Ships
- Bridges
- Pressure vessels
- Structural steelwork
- Pipework

Design aspects

- Design complexity limited.
- Typical joint designs possible using SAW: butt and fillet in heavier sections (see Appendix B – Weld Joint Configurations).
- Suitable for horizontal welding, but can perform vertical welding with special copper side plates to retain flux and mold the weld pool.
- Welds should be designed with straight runs.
- Minimum sheet thickness $= 5$ mm (6 mm for nickel alloys).
- Maximum sheet thickness, commonly:

  - Carbon, low alloy and stainless steels $= 300$ mm
  - Nickel alloys $= 20$ mm

- Multiple weld runs required on sheet thicknesses ≥ 40 mm.
- Unequal thicknesses very difficult.

Quality issues

- High quality welds can be produced with low levels of distortion due to fast welding rates.
- Good weld uniformity and properties, although on large deposit welds a coarse grain structure is formed giving inferior weld toughness.

- Access for weld inspection important, e.g. NDT.
- Large weld beads can cause cracking. Weld penetration can be controlled by using a backing strip when using high currents.
- Joint edge and surface preparation important. Contaminates must be removed from the weld area to avoid porosity and inclusions on each pass.
- A heat affected zone always present. Some stress relieving may be required for restoration of materials original physical properties.
- Can alter composition of weld by addition of alloying elements in the electrode.
- Flux must be clean and free from moisture to prevent weld contamination.
- Weld ideally left to cool to room temperature to allow the slag to peel off.
- Welding variables automatically controlled. Monitoring of welding voltage is used to control arc length through varying the wire feed rate, and thereby improving weld quality.
- Pre-heating of workpiece can reduce porosity and hydrogen cracking, especially on high carbon steels.
- 'Weldability' of the material important and combines many of the basic properties that govern the ease with which a material can be welded and the quality of the finished weld, i.e. porosity and cracking. Material composition (alloying elements, grain structure and impurities) and physical properties (thermal conductivity, specific heat and thermal expansion) are some important attributes which determine weldability.
- Surface finish of weld good.
- Fabrication tolerances typically ± 2 mm.

7.5 Electron Beam Welding (EBW)

Process description

- A controlled high intensity beam of electrons ($\varnothing$0.5–$\varnothing$1 mm) is directed to the joint area of the work (anode) by an electron gun (cathode), where fusion of the base material takes place. The operation takes place in a vacuum, and the work is traversed under the electron beam typically (see 7.5F).

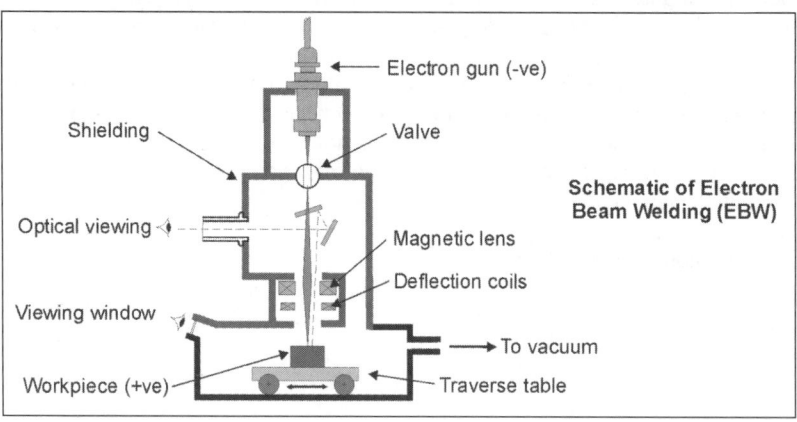

7.5F Electron beam welding process.

Materials

- Most metals and combination of metals weldable, including low to high carbon and alloy steels, aluminum, titanium, copper, refractory and precious metals.
- Copper alloys and stainless steel difficult to weld. Cast iron, lead or zinc alloys are not weldable.
- Metals that experience gas evolution or vaporization on welding difficult.

Process variations

- High-vacuum (most common), semi-vacuum and atmospheric (out-of-vacuum) equipment available, depending on type of work, size and location.
- Semi-vacuum setup used for transportable equipment. Only the area to be welded is surrounded by a vacuum using suction cups.
- Joint advanced under beam for high-vacuum EBW, but for short weld lengths, the beam can be moved along the joint using magnetic coils, rather than the work under the beam on a traversing system.
- EBM (see 5.3): an electron gun is used to generate heat and evaporating the workpiece surface for fusion.
- The electron beam process can also be used for cutting, profiling, slotting and surface hardening, using the same equipment by varying process parameters.

Economic considerations

- Weld rates ranging 0.2–2.5 m/min.
- Production rates range from 10–100/h using high-vacuum equipment.
- Lead times can be several weeks.
- Setup times can be short, but the time to create a vacuum in the chamber at each loading cycle an important consideration.
- High flexibility. Possible to perform many operations on same machine by varying process parameters.
- Full automation of process possible and gives best results.
- Economical for low to moderate production runs.
- Material utilization excellent.
- High power consumption.
- Tooling costs very high.
- Equipment costs very high.
- Direct labor varies depending on level of automation.
- No finishing needed typically.

Typical applications

- Aerospace assemblies (turbine vanes, filters, high pressure pump bodies)
- Automotive assemblies (crankshaft, gears, valves, bearings)
- Machine parts
- Instrumentation devices
- Pipes
- Reactor shells
- Hermetic sealing of assemblies
- Medical implants
- Bimetallic saw blades
- Repair work

Design aspects

- Typical joint designs possible using EBW: butt, fillet and lap (see Appendix B – Weld Joint Configurations). Horizontal welding position is the most suitable.
- Path to joint area from the electron beam gun must be a straight line.
- Beam and joint must be aligned precisely.
- Depth to width ratio can exceed 20:1.
- Balance the welds around the fabrication's neutral axis.
- Size limited by vacuum chamber dimensions unless semi-vacuum equipment used. Maximum height of work in a chamber is 1.2 m typically.
- Possible to weld thin and delicate sections due to no mechanical processing forces.
- Maximum thickness (dependent on vacuum integrity):

  - Aluminum and magnesium alloys = 450 mm
  - Carbon, low alloy and stainless steels = 300 mm
  - Copper alloys = 100 mm.

- Minimum thickness = 0.05 mm.
- Single pass maximum = 75 mm.
- Highly dissimilar thicknesses commonly welded.

Quality issues

- High quality welds possible with little or no distortion.
- No flux or filler used.
- Integrity of vacuum important. Beam dispersion occurs due to electron collision with air molecules.
- Out-of-vacuum systems must overcome atmospheric pressures at weld area.
- Beams can be generated up to 700 mm from workpiece surface for high-vacuum systems; can be reduced to less than 40 mm for out-of-vacuum.
- Precise alignment of work required and held using jigs and fixtures.
- Hazardous X-rays produced during processing which requires lead shielding.
- Vacuum removes gases from weld area, e.g. hydrogen to minimize hydrogen embrittlement in hardened steels.
- Localized thermal stresses leads to a very small heat affected zone. Distortion of thin parts may occur.
- Surface finish excellent.
- Fabrication tolerances a function of the accuracy of the component parts and the assembly/jigging method. Joints gaps less than 0.1 mm required. Therefore, abutment faces should be machined to close tolerances.

7.6   Laser Beam Welding (LBW)

Process description

- Heat for fusion is generated by the absorption of a high power density narrow beam of light, commonly known as a laser. Focusing of the laser is performed by mirrors or lenses (see 7.6F).

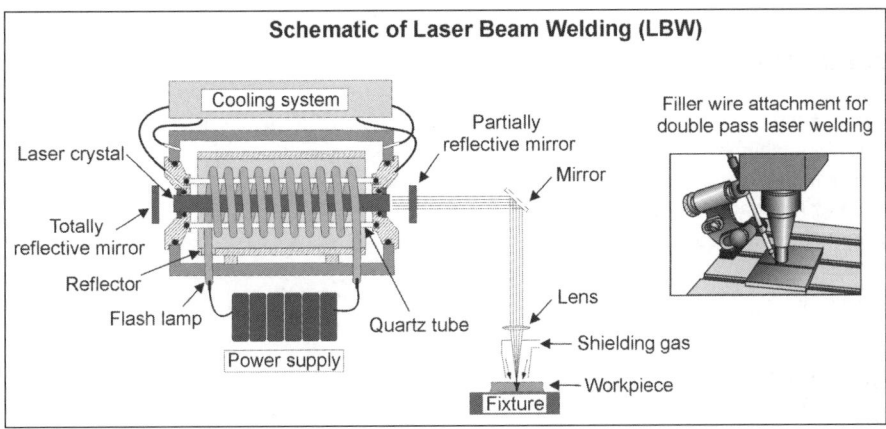

Schematic of Laser Beam Welding (LBW)

Cooling system

Laser crystal

Totally reflective mirror

Reflector

Flash lamp

Power supply

Partially reflective mirror

Mirror

Quartz tube

Lens

Shielding gas

Workpiece

Fixture

Filler wire attachment for double pass laser welding

7.6F Laser beam welding process.

Materials

- Dependent on thermal diffusivity and to a lesser extent the optical characteristics of material, rather than chemical composition, electrical conductivity or hardness.
- Stainless steel and carbon steels typically.
- Aluminum alloys and alloy steels difficult to weld. Not used for cast iron.

Process variations

- Many types of laser are available, used for different applications. Common laser types available are: CO_2, Nd:YAG, Nd:glass, ruby and excimer. Depending on economics of process, pulsed and continuous wave modes are used.
- Shielding gas such as argon sometimes employed to reduce oxidation.
- Laser beam machines can also be used for cutting, surface hardening, machining (LBM) (see 5.4), drilling, blanking, engraving and trimming, by varying the power density.
- Laser beam spot and seam welding can also be performed on same equipment.
- Laser soldering: provides very precise heat source for precision work.

Economic considerations

- Weld rates ranging 0.25–13 m/min for thin sheet.
- Production rates moderate.
- High power consumption.

- Lead times can be short, typically weeks.
- Setup times short.
- Material utilization excellent.
- High degree of automation possible.
- Possible to perform many operations on same machine by varying process parameters.
- Economical for low to moderate production runs.
- Tooling costs very high.
- Equipment costs high.
- Direct labor costs medium. Some skilled labor required depending on degree of automation.

Typical applications

- Structural sections
- Transmission casings
- Hermetic sealing (pressure vessels, pumps)
- Transformer lamination stacks
- Instrumentation devices
- Electronics fabrication
- Medical implants

Design aspects

- Laser can be directed, shaped and focused by reflective optics permitting high spatial freedom in 2-dimensions. Horizontal welding position is the most suitable.
- Typical joint designs using LBW: lap, butt and fillet (see Appendix B – Weld Joint Configurations).
- Mostly for horizontal welding.
- Balance the welds around the fabrication's neutral axis.
- Path to joint area from the laser must be a straight line. Laser beam and joint must be aligned precisely.
- Intimate contact of joint faces required.
- Filler rod rarely utilized, but for thick sheets or requiring multi-pass welds, a wire-feed filler attachment can be used.
- Minimal work holding fixtures required.
- Minimum thickness $= 0.1$ mm.
- Maximum thickness $= 20$ mm.
- Multiple weld runs required on sheet thickness ≥ 13 mm.
- Dissimilar thicknesses difficult.

Quality issues

- Difficulty of material processing dictated by how close the material's boiling and vaporization points are.
- Localized thermal stresses lead to a very small heat affected zone. Distortion of thin parts may occur.
- No cutting forces, so simple fixtures can be used.
- Inert gas shielding, argon commonly, employed to reduce oxidation.
- Control of the pulse duration important to minimize the heat affected zone, depth and size of molten metal pool surrounding the weld area.

- The reflectivity of the workpiece surface important. Dull and unpolished surfaces are preferred and cleaning prior to welding is recommended.
- Hole wall geometry can be irregular. Deep holes can cause beam divergence.
- Surface finish good.
- Fabrication tolerances a function of the accuracy of the component parts and the assembly/jigging method.

7.7 Plasma Arc Welding (PAW)

Process description

- A plasma column is created by constricting an ionized gas through a water-cooled nozzle reaching temperatures of around 20 000 °C. The plasma column flows around a non-consumable tungsten electrode, which provides the electrical current for the arc. The plasma provides the energy for melting and fusion of the base materials and filler rod (when used) (see 7.7F).

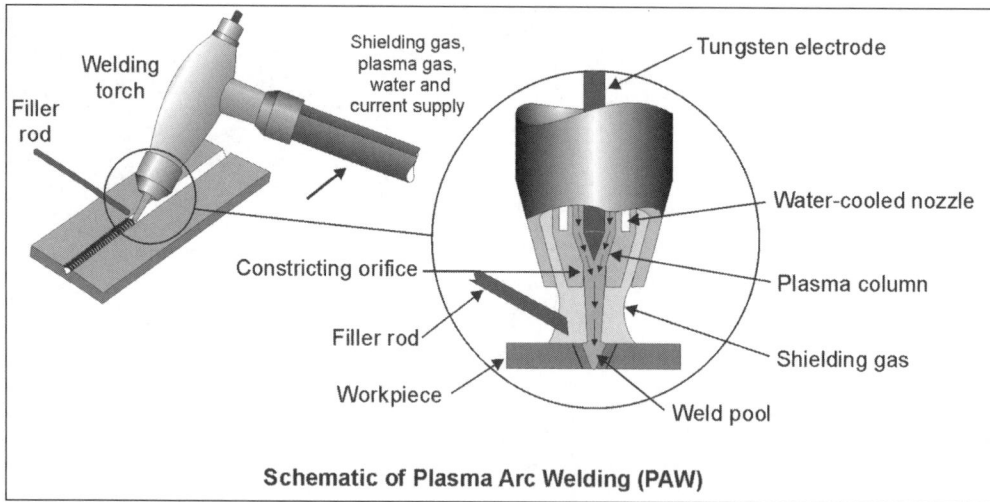

Schematic of Plasma Arc Welding (PAW)

7.7F Plasma arc welding process.

Materials

- Most electrically conductive materials.
- Commonly and stainless steels, aluminum, copper and nickel alloys, refractory and precious metals.
- Not cast iron, magnesium, lead or zinc alloys.

Process variations

- Portable manual or automated a.c. or d.c. systems: d.c. system most common.
- Two modes of operation used for welding:

  - Melt-in fusion for reduced distortion uses low currents
  - Key hole fusion at higher currents for full penetration on thick materials.

- Choice of gas and their proportions important for two modes of operation:

  - Plasma gas: argon or argon–hydrogen mix
  - Shielding gas: argon or argon–hydrogen mix. Also helium or helium–argon mix used.

- Plasma arc cutting: for cutting, slotting and profiling materials up to about 40 mm thickness using the key-holing mode of operation.

- Plasma arc spraying: melting of solid feedstock (e.g. powder, wire or rod) and propelling the molten material onto a substrate to alter its surface properties, such as wear resistance or oxidation protection.
- Filler rod sizes between $\emptyset$1.6 and $\emptyset$3.2 mm typically.

Economic considerations

- Weld rates vary from 0.4 m/min for manual welding to 3 m/min for automated systems.
- Alternative to TIG for high automation potential using key hole mode.
- Welding circuit and system more complex than TIG. Additional controls needed for plasma arc and filters and deionizers for cooling water mean more frequent maintenance and additional costs.
- Economical for low production runs. Can be used for one-offs.
- Tooling costs low to moderate.
- Equipment costs generally high.
- Direct labor costs moderate.
- Finishing costs low.

Typical applications

- Engine components
- Sheet-metal fabrication
- Domestic appliances
- Instrumentation devices
- Pipes

Design aspects

- Design complexity high.
- Typical joint designs possible using PAW: butt, lap, fillet and edge (see Appendix B – Weld Joint Configurations).
- Design joints using minimum amount of weld, i.e. intermittent runs and simple or straight contours wherever possible.
- Balance the welds around the fabrication's neutral axis.
- Distortion can be reduced by designing symmetry in parts to be welded along weld lines.
- The fabrication sequence should be examined with respect to the above.
- Design parts to give access to the joint area, for vision, filler rods, cleaning, etc.
- Sufficient edge distances should be designed for. Avoid welds meeting at end of runs.
- Mostly for horizontal welding, but can also perform vertical welding using higher shielding gas flow rates.
- Filler can be added to the leading edge of the weld pool using a rod, but not necessary for thin sections.
- Minimum sheet thickness $= 0.05$ mm.
- Maximum thickness, commonly:

  - Aluminum $= 3$ mm
  - Copper and refractory metals $= 6$ mm
  - Steels $= 10$ mm
  - Titanium alloys $= 13$ mm
  - Nickel $= 15$ mm.

- Multiple weld runs required on sheet thickness ≥ 10 mm.
- Unequal thicknesses difficult.

Quality issues

- High quality welds possible with little or no distortion.
- Provides good penetration control and arc stability.
- Access for weld inspection important, e.g. NDT.
- Tungsten inclusions from electrode not present in welds, unlike TIG.
- Joint edge and surface preparation important. Contaminates must be removed from the weld area to avoid porosity and inclusions.
- A heat affected zone always present. Some stress relieving may be required for restoration of materials original physical properties.
- Not recommended for site work in wind where the shielding gas may be gusted.
- Need for jigs and fixtures to keep joints rigid during welding and subsequent cooling to reduce distortion on large fabrications.
- Care needed to keep filler rod within the shielding gas to prevent oxidation.
- Tungsten inclusions can contaminate finished welds.
- Nozzle used to increase the temperature gradient in the arc, concentrating the heat and making the arc less sensitive to arc length changes in manual welding.
- Plasma arc very delicate and orifice alignment with tungsten electrode crucial for correct operation.
- Important process variables for consistency in manual welding: welding speed, plasma gas flow rate, current and torch angle.
- 'Weldability' of the material important and combines many of the basic properties that govern the ease with which a material can be welded and the quality of the finished weld, i.e. porosity and cracking. Material composition (alloying elements, grain structure and impurities) and physical properties (thermal conductivity, specific heat and thermal expansion) are some important attributes which determine weldability.
- Surface finish of weld excellent.
- Fabrication tolerances a function of the accuracy of the component parts and the assembly/jigging method, but typically ± 0.25 mm.

7.8 Resistance welding

Process description

- Covers a range of welding processes that use the resistance to electrical current between two materials to generate sufficient heat for fusion. A number of processes use a timed or continuous passage of electric current at the contacting surfaces of the two parts to be joined to generate heat locally, fusing them together and creating the weld with the addition of pressure, provided by current supplying electrodes or platens (see 7.8F).

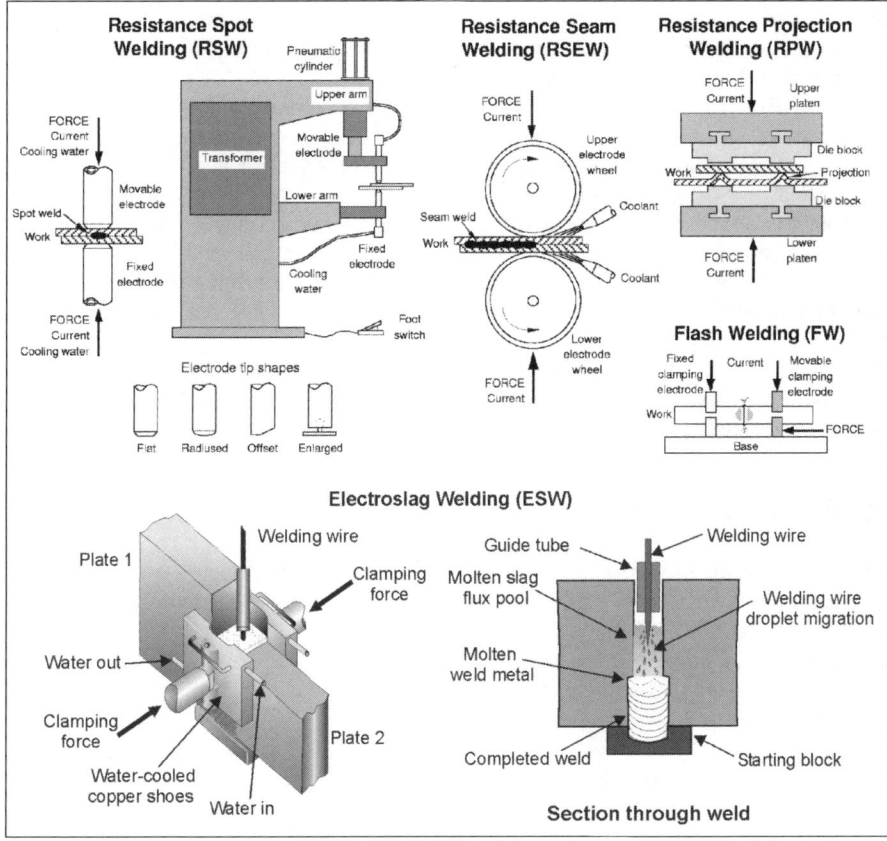

7.8F Resistance welding process.

Materials

- Low carbon steels commonly, however, almost any material combination can be welded using conventional resistance welding techniques. Not recommended for cast iron, low melting point metals and high carbon steels.
- Electroslag Welding (ESW) is used to weld carbon and low alloy steels typically. Nickel, copper and stainless steel less common.

Process variations

- Resistance Spot Welding (RSW): uses two water-cooled copper alloy electrodes of various shapes to form a joint on lapped sheet-metal. Can be manual portable (gun), single or multi-spot semi-automatic, automatic floor standing (rocker arm or press) or robot mounted as an end effector.
- Resistance Seam Welding (RSEW): uses two driven copper alloy wheels. Current is supplied in rapid pulses creating a series of overlapping spot welds which is pressure tight. Usually floor standing equipment, either circular, longitudinal or universal types.
- Resistance Projection Welding (RPW): a component and sheet-metal are clamped between current carrying platens. Localized welding takes place at the projections on the component(s) at the contact area. Usually floor standing equipment, either single or multi-projection press type.
- Upset resistance welding: electrical resistance between two abutting surfaces and additional pressure used to create butt welds on small pipe assemblies, rings and strips.
- Percussion resistance welding: rapid discharge of electrical current and then percussion pressure for welding rods or tubes to sheet-metal.
- Flash Welding (FW): parts are accurately aligned at their ends and clamped by the electrodes. The current is applied and the ends brought together removing the high spots at the contact area deoxidizing the joint (known as flashing). Second part is the application of pressure effectively forging the weld.
- ESW: the joint is effectively 'cast' between joint edges between a gap of about 20 to 50 mm. An electric arc is used initially to heat a flux within water-cooled copper molding shoes spanning the joint area. Resistance between the consumable electrode and the base material is then used to generate the heat for fusion. The weld pool is shielded by the molten flux as welding progresses up the joint.
- A variant of ESW is Electrogas Welding (EGW). However, the process doesn't use electrical resistance as a heat source, but a gas shielded arc, therefore the molten flux pool above the weld is not necessary. Used for thick sections of carbon steel.

Economic considerations

- Full automation and integration with component assembly relatively easy.
- High production rates possible due to short weld times, e.g. RSW = 20 spots/min, RSEW = 30 m/min, FW = 3 s/10 mm^2 area.
- Automation readily achievable using all processes.
- No filler metals or fluxes required (except ESW).
- Little or no post-welding heat treatment required.
- Minimal joint preparation needed.
- Economical for low production runs. Can be used for one-offs.
- Tooling costs low to moderate.
- Equipment costs low to moderate.
- Direct labor costs low. Skilled operators are not required.
- Finishing costs very low. Cleaning of welds is not necessary typically, except with Flash Welding (FW), which requires machining or grinding to remove excess material.
- High deposition rates for ESW, but can still be slow.

Typical applications

- RSW: car bodies, aircraft structures, light structural fabrications and domestic appliances
- RSEW: fuel tanks, cans and radiators

- RPW: reinforcing rings, captive nuts, pins and studs to sheet-metal, wire mesh
- FW: for joining parts of uniform cross section, such as bar, rods and tubes, and occasionally sheet-metal
- ESW: joining structural sections of buildings and bridges such as columns, machine frames and on-site fabrication

Design aspects

- Typical joint designs: lap (RSW and RSEW), edge (RSEW), butt (FW and ESW), attachments (PW).
- Access to joint area important.
- Can be used for joints inaccessible by other methods or where welded components are closely situated.
- Spot weld should have a diameter between four and eight times the material thickness.
- Can process some coated sheet-metals (except ESW).
- Same end cross sections are required for FW.
- For RSW, RSEW and PW:

  - Minimum sheet thickness $= 0.3$ mm
  - Maximum sheet thickness, commonly $= 6$ mm
  - Mild steel sheet up to 20 mm thick has been spot- and seam-welded, but requires high currents and expensive equipment.

- For FW, sizes ranging 0.2 mm thick sheet to sections up to 0.1 m^2 in area.
- Unequal thicknesses possible with RSW and RSEW (up to 3:1 thickness ratio).
- ESW applied to sheet thicknesses of same order from 25 up to 500 mm using several guide tubes and electrodes in one pass, but down to 75 mm for a single set. Vertical welds can restrict design freedom in ESW.

Quality issues

- Clean, high quality welds with very low distortion can be produced. Although a heat affected zone always created, can be small.
- Coarse grain structures may be created in ESW due to high heat input and slow cooling.
- Surface preparation important to remove any contaminates from the weld area such as oxide layers, paint and thick films of grease and oil. Resistance welding of aluminum requires special surface preparation.
- Welding variables for spot, seam and projection welding should be pre-set and controlled during production, these include: current, timing and pressure (where necessary).
- Electrodes or platens must efficiently transfer pressure to the weld, conduct and concentrate the current and remove heat away from the weld area, therefore, maintenance should be performed at regular intervals.
- Spot, seam and projection welds can act as corrosion traps.
- RSW, RSEW and PW welds can be difficult to inspect. Destructive testing should be intermittently performed to monitor weld quality.
- Depression left behind in RSW and RSEW serves to prevent cavities or cracks due to contraction of the cooling metal.
- Possibility of galvanic corrosion when resistance welding some dissimilar metals.
- High strength welds are produced by FW. Always leaves a ridge at the joint area which must be removed.

- 'Weldability' of the material important and combines many of the basic properties that govern the ease with which a material can be welded and the quality of the finished weld, i.e. porosity and cracking. Material composition (alloying elements, grain structure and impurities) and physical properties (thermal conductivity, specific heat and thermal expansion) are some important attributes which determine weldability.
- Surface finish of the welds fair to good for RSW, RSEW, FW and PW. Excellent for ESW.
- No weld spatter and no arc flash (except ESW initially).
- Alignment of parts to give good contact at the joint area important for consistent weld quality.
- Repeatability typically ± 0.5–± 1 mm for robot RSW.
- Axes alignment total tolerance for FW between 0.1 and 0.25 mm.

7.9   Solid state welding

Process description

- A range of methods utilizing heat, pressure and/or high energy to plastically deform the material at the joint area in order to create a solid phase mechanical bond (see 7.9F).

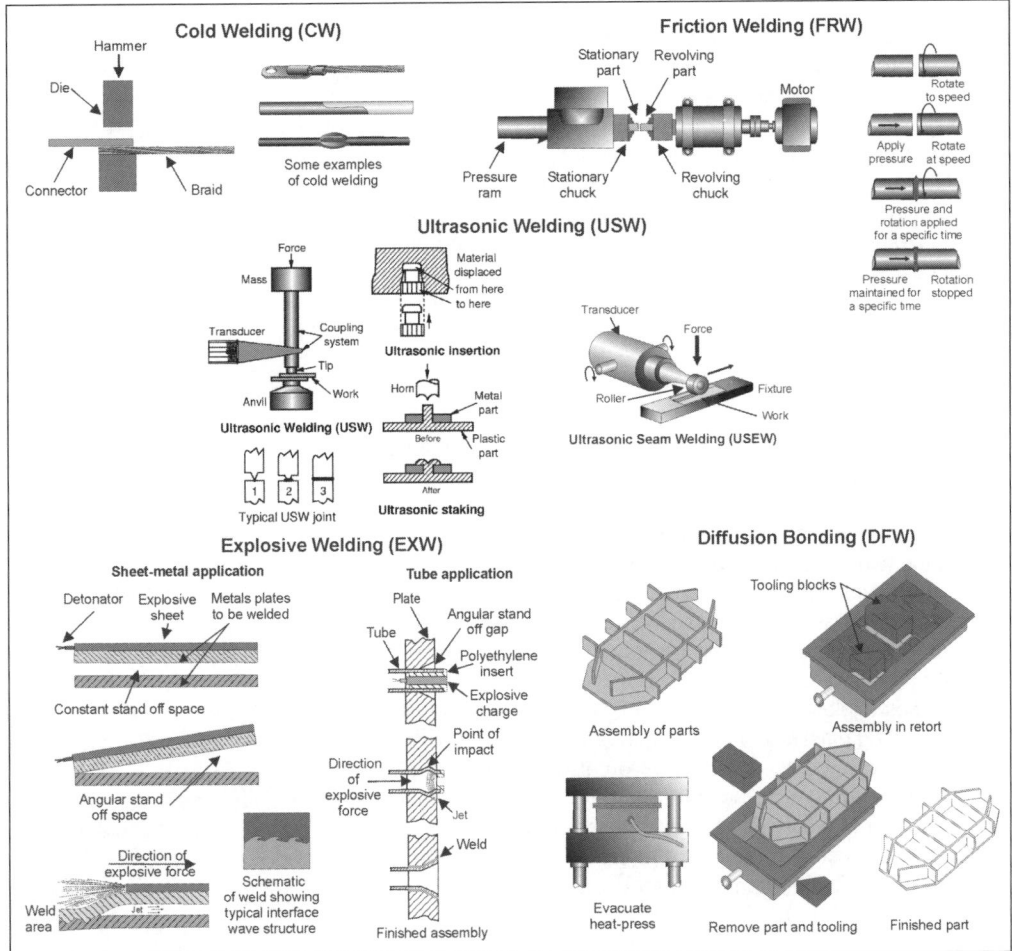

7.9F Solid state welding process.

Materials

- Cold Welding (CW): Ductile metals such as carbon steels, aluminum, copper and precious metals.
- Friction Welding (FRW): can weld many material types and dissimilar metals effectively, including aluminum to steel. Also thermoplastics and refractory metals.

- Ultrasonic Welding (USW): can be used for most ductile metals, such as aluminum and copper alloys, carbon steels and precious metals, and some thermoplastics. Can bond dissimilar materials readily.
- Explosive Welding (EXW): carbon steels, aluminum, copper and titanium alloys. Welds dissimilar metals effectively.
- Diffusion bonding (DFW): stainless steel, aluminum, low alloy steels, titanium and precious metals. Occasionally copper and magnesium alloys are bonded.

Process variations

- CW: process is performed at room temperature using high forces to create substantial deformation (up to 95 per cent) in the parts to be joined. Surfaces require degreasing and scratch-brushing for good bonding characteristics.
- Cold pressure spot welding: for sheet-metal fabrication using suitably shaped indenting tools.
- Forge welding: the material is heated in a forge or oxyacetylene ring burners. Hand tools and anvil used to hammer together the hot material to form a solid state weld. Commonly associated with the blacksmith's trade and used for decorative and architectural work.
- Thermocompression bonding: performed at low temperatures and pressures for bonding wires to electrical circuit boards.
- USW: hardened probe introduces a small static pressure and oscillating vibrations at the joint face disrupting surface oxides and raising the temperature through friction and pressure to create a bond. Can also perform spot welding using similar equipment.
- Ultrasonic Seam Welding (USEW): ultrasonic vibrations imparted through a roller traversing the joint line.
- Ultrasonic soldering: uses an ultrasonic probe to provide localized heating through high frequency oscillations. Eliminates the need for a flux, but requires pre-tinning of surfaces.
- Ultrasonic insertion: for introducing metal inserts into plastic parts for subsequent fastening operations.
- Ultrasonic staking: for light assembly work in plastics.
- FRW: the two parts to be welded, one stationary and one rotating at high speed (up to 3000 rpm), have their joint surfaces brought into contact. Axial pressure and frictional heat at the interface create a solid state weld on discontinuation of rotation and on cooling.
- Friction stir welding: uses the frictional heat to soften the material at the joint area using a wear resistant rotating tool.
- EXW: uses explosive charge to supply energy for a cladding sheet-metal to strike the base sheet-metal causing plastic flow and a solid state bond. Bond strength is obtained from the characteristic wavy interlocking at the joint face. Can also be used for tube applications.
- DFW: The surfaces of the parts to be joined are brought together under moderate loads and temperatures in a controlled inert atmosphere or vacuum. Localized plastic deformation and atomic interdiffusion occurs at the joint interface, creating the bond after a period of time.
- Superplastic diffusion bonding: can integrate DFW with superplastic forming to produce complex fabrications (see 3.7).

Economic considerations

- Production rates varying: high for CW and FW (30 s cycle time), moderate for USW and low for EXW and DFW.
- Lead times low typically.
- Material utilization excellent. No scrap generated.
- High degree of automation possible with many processes (except EXW).
- No filler materials needed.

- Economical for low production runs. Can be used for one-offs.
- Tooling costs low to moderate.
- Equipment costs low (CW, EXW) to high (USW, FRW, DFW).
- Direct labor costs low to moderate. Some skilled labor maybe required.
- Finishing costs low. Cleaning of welds not necessary typically, except with FRW, which requires machining or grinding to remove excess material.

Typical applications

- CW: welding caps to tubes, electrical terminations and cable joining
- USW: for sheet-metal fabrication, joining plastics, electrical equipment and light assembly work
- FRW: for welding hub-ends to axle casings, welding valve stems to heads and gear assemblies
- EXW: used mainly for cladding, or bonding one plate to another, to improve corrosion resistance in the process industry, for marine parts and joining large pipes in the petrochemical industry
- DFW: for joining high strength materials in the aerospace and nuclear industries, biomedical implants and metal laminates for electrical devices.

Design aspects

- Typical joint designs: lap (CW, USW, USEW, EXW, DFW), edge (USEW), butt (CW, FRW, ESW), T-joint (DFW), flange (EXW).
- Access to joint area important.
- Unequal thicknesses possible with CW, USW, EXW, DFW.
- CW: thicknesses ranging 5–20 mm.
- USW: thicknesses ranging 0.1–3 mm.
- EXW: thicknesses ranging 20–500 mm and maximum surface area $= 20\,m^2$.
- FRW: diameters ranging between $\varnothing2$ and $\varnothing150$ mm and maximum surface area $= 0.02\,m^2$. Parts must have rotational symmetry.
- DFW: thicknesses ranging 0.5–20 mm.

Quality issues

- Little or no deformation takes place (except EXW).
- No weld spatter and no arc flash.
- Alignment of parts crucial for consistent weld quality.
- Parts must be able to withstand high forces and torques to create bond over long period of times.
- Safety concerns for EXW include explosives handling, noise and provision for controlled explosion.
- Welds as strong as base material in many cases.
- Surface preparation important to remove any contaminates from the weld area such as oxide layers, paint and thick films of grease and oil.
- Possibility of galvanic corrosion when welding some material combinations.
- Surface finish of the welds good.
- Fabrication tolerances vary from close for DFW, moderate for FRW, CW, USW and low dimensional accuracy for EXW.

7.10   Thermit Welding (TW)

Process description

- A charge of iron oxide and aluminum powder is ignited in a crucible. The alumino-thermic reaction produces molten steel and alumina slag. On reaching the required temperature, a magnesite thimble melts and allows the molten steel to be tapped off to the mold surrounding the pre-heated joint area. On cooling, a cast joint is created (see 7.10F).

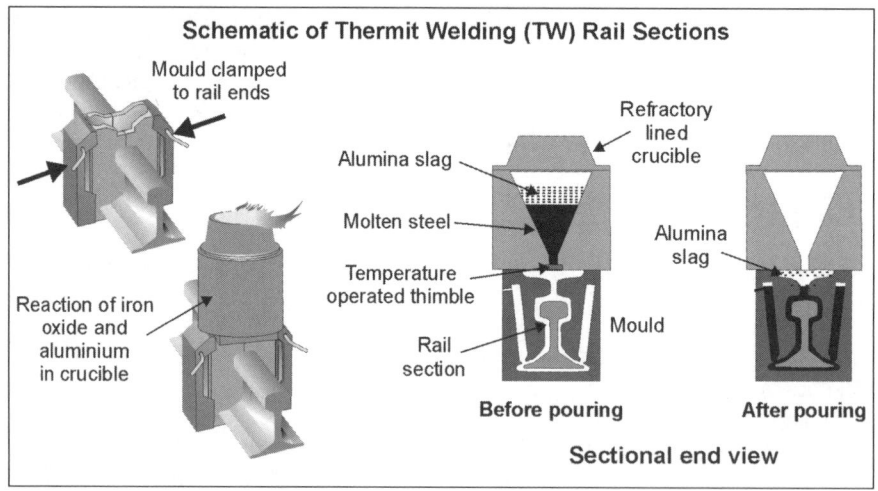

7.10F Thermit welding process.

Materials

- Carbon and low alloy steels, and cast iron only.

Process variations

- Molds can be refractory sand or carbon.
- Can be used to repair broken areas of structural sections using special molds.

Economic considerations

- Production rates very low. Cycle times typically 1 h.
- Lead time a few days.
- 20 per cent of welding metal lost in runners and risers.
- Scrap material cannot be recycled directly.
- Economical for low production runs. Can be used for one-offs.
- Manual operation only.
- Tooling costs low to moderate.
- Equipment costs low to moderate.

- Direct labor costs moderate to high. Some labor involved.
- Finishing costs moderate. Excess metal around joint not always removed, but gates and risers must be ground off.

Typical applications

- Site welding of rails to form continuous lengths
- Joining heavy structural sections and low-loaded structural joints
- Machine frame fabrication
- Shipbuilding
- Joining thick cables
- Concrete reinforcement steel bars
- Repair work

Design aspects

- The cross section of the parts to be joined can be complex, otherwise limited design freedom.
- Joint gaps typically 20–80 mm.
- Butt joint design possible only (see Appendix B – Weld Joint Configurations).
- Minimum sheet thickness $= 10$ mm.
- Maximum thickness $= 1000$ mm.

Quality issues

- Weld quality fair.
- The cast joint has inferior properties than that of the base material.
- Pre-heating times ranging 1–7 min depending on section thickness. Small section thicknesses may not require pre-heating.
- Joint area must be cleaned thoroughly.
- Joint edges must be aligned with a suitable gap dependent on section size.
- Alloying elements can be added to the charge to match physical properties of materials to be joined.
- Exothermic chemical reaction has safety concerns and proper precautions and ventilation necessary.
- Surface finish poor to fair.
- Fabrication tolerances a function of the accuracy of the component parts (hot-rolled sections usually which have poor dimensional accuracy) and the clamping/jigging method used, but typically ± 1.5 mm.

7.11 Gas Welding (GW)

Process description

- High pressure gaseous fuel and oxygen are supplied by a torch through a nozzle where combustion takes place, providing a controllable flame. The high temperature generated (greater than 3000 °C) is sufficient to melt the base metal at the joint area. Shielding from the atmosphere is performed by the outer flame. Filler metal can be supplied to the weld pool if needed (see 7.11F).

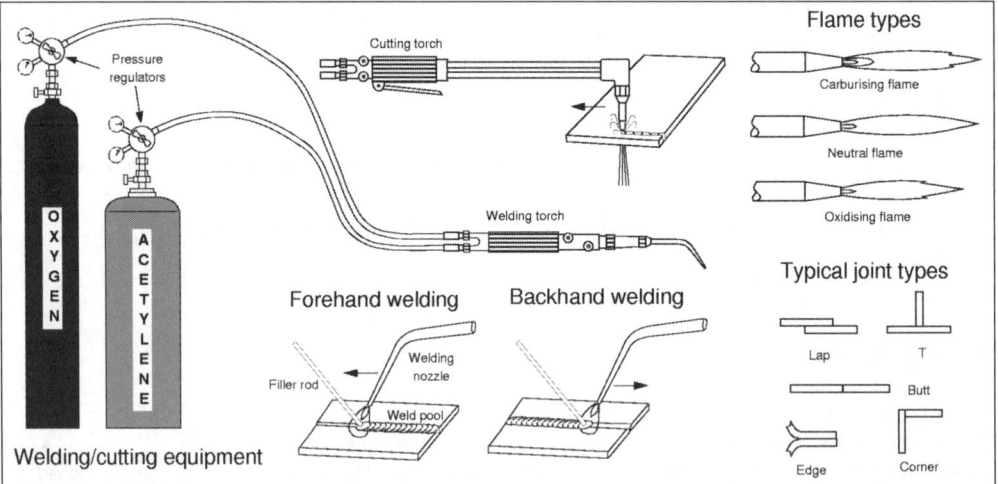

7.11F Gas welding process.

Materials

- Commonly ferrous alloys: low carbon, low alloy and stainless steels and cast iron.
- Also, nickel, copper and aluminum alloys, and some low melting point metals (zinc, lead and precious metals).
- Refractory metals cannot be welded.

Process variations

- Commonly manually operated, portable and self-contained welding sets.
- Can use forehand or backhand welding procedures.
- Gas fuel commonly used is acetylene for most welding applications and materials, known as oxyacetylene welding.
- Hydrogen, propane, butane and natural gas used for low temperature brazing and welding small and thin parts.
- Air can be used instead of oxygen for brazing, soldering and welding lead sheet.
- Flux may be necessary for welding metals other than ferrous alloys.

- By regulating the oxygen flow, three types of flame can be produced:

  - Carburizing: for flame hardening, brazing, welding nickel alloys and high carbon steels
  - Neutral: for most welding operations
  - Oxidizing: used for welding copper, brass and bronze.

- Braze welding: base metal is pre-heated with an oxyacetylene or oxypropane gas torch at the joint area. Brazing filler metal, usually supplied in rod form, and a flux is applied to joint area, where the filler becomes molten and fills the joint gap through capillary action. Although no fusion takes place, very high temperatures are required, typically 700 °C. Some finishing may be necessary to clean flux residue and excess braze.
- Pressure gas welding: heat from oxyacetylene burner is used to melt ends of the parts to be joined and then applied pressure creates the weld.
- Gas cutting: an oxyacetylene or oxypropane flame from a specially designed nozzle is used to preheat the parent metal and an additional high pressure oxygen supply effectively cuts the metal by oxidizing it. Can perform straight cuts or profiles (when automated) in plate over 500 mm thickness.

Economic considerations

- Weld rates very low, typically 0.1 m/min.
- Lead times very short.
- Very flexible process. Same equipment can be used for welding, cutting and several heat treatment processes.
- Economical for very low production runs. Can be used for one-offs.
- Automation not practical for most situations.
- Tooling costs low to moderate. Little tooling required and jigs and fixtures are simple for manual operation.
- Equipment costs low to moderate.
- Direct labor costs moderate. Skilled operators may be required.
- Finishing costs low to moderate. No slag produced, but cleaning may be required.

Typical applications

- Sheet-metal fabrication
- Ventilation ducts
- Small diameter pipe welding
- Repair work

Design aspects

- Moderate levels of complexity possible. Capability to weld parts with large size and shape variations.
- Typical joint designs possible using gas welding: butt, fillet, lap and edge, in thin sheet.
- All welding positions possible.
- Design joints using minimum amount of weld, i.e. intermittent runs and simple or straight contours wherever possible.
- Balance the welds around the fabrication's neutral axis.
- Distortion can be reduced by designing symmetry in parts to be welded along weld lines.
- The fabrication sequence should be examined with respect to the above.
- Sufficient edge distances should be designed for and avoid welds meeting at the end of runs.

- Minimum sheet thickness, commonly:

  - Carbon steel = 0.5 mm
  - Cast iron = 3 mm.

- Maximum sheet thickness, commonly:

  - Carbon steel and cast iron = 30 mm
  - Low alloy steel, stainless steel, nickel and aluminum alloys = 3 mm.

- Multiple weld runs required on sheet thicknesses ≥4 mm.
- Unequal thicknesses possible.

Quality issues

- Good quality welds with moderate but acceptable levels of distortion can be produced. Repeatability can be a problem.
- Access for weld inspection important.
- Attention to adequate jigs and fixtures when welding thin sheet recommended to avoid excessive distortion of parts by providing good fit-up and to take heat away from the surrounding metal.
- Heat affected zone always created. Some stress relieving may be required for restoration of materials original physical properties.
- Surface preparation important to remove any contaminates from the weld area such as oxide layers, paint and thick films of grease and oil.
- Gas flow rates should be pre-set and regulated during production. Even gas mix gives the neutral flame most commonly used for welding. Even heating of joint area required for consistent results.
- Shielding integrity at the weld area not as high as arc welding methods and some oxidation and atmospheric attack may occur.
- 'Weldability' of the material important and combines many of the basic properties that govern the ease with which a material can be welded and the quality of the finished weld, i.e. porosity and cracking. Material composition (alloying elements, grain structure and impurities) and physical properties (thermal conductivity, specific heat and thermal expansion) are some important attributes which determine weldability.
- Surface finish of weld fair to good.
- Fabrication tolerances typically ±1 mm.

7.12 Brazing

Process description

- Heat is applied to the parts to be joined which melts a manually fed or pre-placed filler braze metal (which has a melting temperature $\geq 450\,°C$) into the joint by capillary action. A flux is usually applied to facilitate 'wetting' of the joint, prevent oxidation, remove oxides and reduce fuming (see 7.12F).

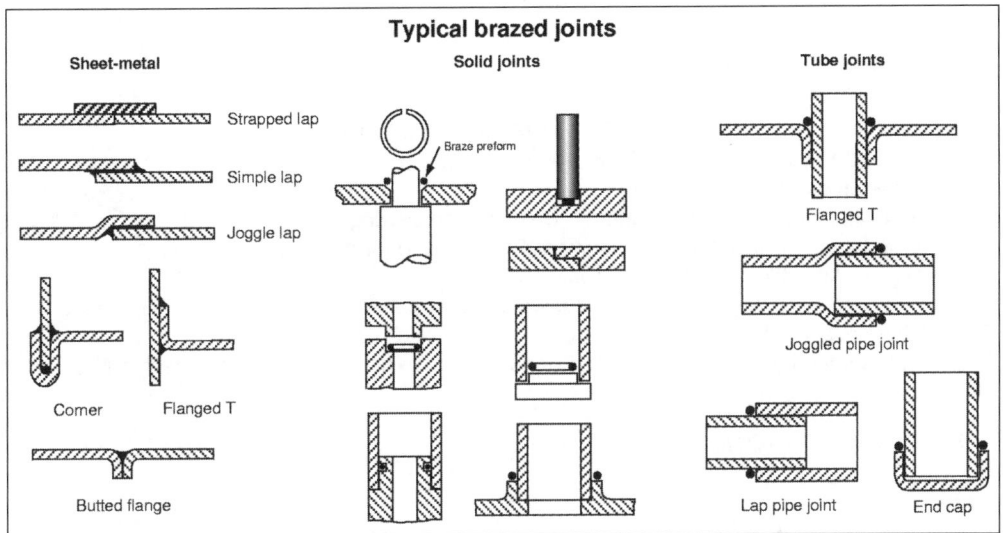

7.12F Brazing process.

Materials

- Almost any metal and combination of metals can be brazed. Aluminum difficult due to oxide layer.

Process variations

- Gas brazing: neutral or carburizing oxy-fuel flame is used to heat the parts. Can be manual Torch Brazing (TB) for small production runs or automated with a fixed burner (ATB).
- Induction Brazing (IB): components are placed in a magnetic field surrounding an inductor carrying a high-frequency current giving uniform heating.
- Resistance Brazing (RB): high electric resistance at joint surfaces causes heating for brazing. Not recommended for brazing dissimilar metals.
- Dip Brazing (DB): parts immersed to a certain depth in a bath of molten chemical or brazing alloy covered with molten flux. Commonly used for brazing aluminum.
- Furnace Brazing (FB): heating takes place in carburizing/inert atmosphere or a vacuum. The filler metal is preplaced at the joint and no additional flux is needed. Large batches of parts of varying sizes and joint types can be brazed simultaneously. Good for parts that may distort using localized heating methods and dissimilar metals.

- Infrared Brazing (IRB): uses quartz-iodine incandescent lamps as heat energy. For joining pipes typically.
- Diffusion Brazing (DFB): braze filler actually diffuses into the base metal creating a new alloy at the joint interface. Gives a strong bond of equal strength to that of the base metal.
- Braze welding: base metal is pre-heated with an oxyacetylene or oxypropane gas torch at the joint area. Brazing filler metal, usually supplied in rod form, and a flux is applied to joint area where the filler becomes molten and fills the joint gap through capillary action (see 7.11).
- Filler metal can be in preforms, wire, foil, coatings, slugs and pastes in a variety of metal alloys, commonly the alloys are based on: copper, silver, nickel and aluminum.
- Flux types: borax, borates, fluoroborates, alkali-fluorides and alkali-chlorides (for brazing aluminum and its alloys) in powder, pastes or liquid form.

Economic considerations

- High production rates possible using FB and IB, but low with TB.
- Cycle times vary. Long for FB and DFB, short for TB.
- Very flexible process.
- Large fabrications may be better suited to welding than brazing.
- Economical for very low production volumes. Can be used for one-offs.
- Tooling costs low. Little tooling required.
- Equipment costs vary depending on process and degree of automation. Low for TB, high for FB.
- Direct labor costs low to moderate. Cost of joint preparation can be high.
- Finishing costs moderate. Cleaning of the parts to remove corrosive flux residues is critical.

Typical applications

- Machine parts
- Pipework
- Bicycle frames
- Repair work
- Cutting tool inserts

Design aspects

- All levels of complexity.
- Joints should be designed to operate in shear or compression, not tension.
- Typical joint designs using brazing: lap and scarf in thin joints with large contact areas or a combination of lap and fillet. Fillets can help to distribute stresses at the joint. Butt joints are possible but can cause stress concentrators in bending.
- Lap joints should have a length to thickness ratio of between three and four times that of the thinnest part for optimum strength.
- Joints should be designed to give a clearance between the mating parts of typically, 0.02–0.2 mm depending on the process to be used and the material to be joined (can be zero for some process/material combinations). The clearance directly affects joint strength. If the clearance is too great the joint will loose a considerable amount of strength.
- Tolerances on mating parts should maintain the joint clearances recommended.
- Parts in the assembly should be arranged to promote capillary action by gravity.
- Machine marks should be in line with the flow of solder.
- Joint strength between that of the base and filler metals in a well-designed joint.

- Vertical brazing should integrate chamfers on parts to create reservoirs.
- Jigs and fixtures should be used only on parts where self-locating mechanisms (staking, press fits, knurls and spot welds) not practical. If jigs and fixtures are used they should support the joint as far from the joint area as possible, have minimum contact and have low thermal mass.
- Provision for the escape of gases and vapors in the joint design important.
- Metals with a melting temperature less than 650 °C cannot be brazed.
- Minimum sheet thickness = 0.1 mm.
- Maximum thickness = 50 mm.
- Unequal thicknesses possible, but sudden changes in section can create stress concentrators.
- Dissimilar metals can cause thermal stresses on cooling.

Quality issues

- Good quality joints with very low distortion produced.
- Virtually a stress free joint created with proper control of cooling.
- Choice of filler metal important in order to avoid joint embrittlement. Possibility of galvanic corrosion.
- A limited amount of inter-alloying takes place between the filler metal and the part metal, however, excessive alloying can reduce joint strength. Control of the time and temperature of the applied heat important with respect to this.
- Subsequent heating of assembly after brazing could melt the filler metal again.
- Filler metal selection based upon the metals to be brazed, process to be used and its economics, and the operating temperature of the finished assembly.
- Surface preparation important to remove any contaminates from the joint area such as oxide layers, paint and thick films of grease and oil and promote wetting. Pickling and degreasing commonly performed before brazing of parts.
- Smooth surfaces preferred to rough ones. Sand blasted surfaces not recommended as they tend to reduce joint strength. Abrading the joint area using emery cloth acceptable.
- Correct clearance, temperature gradients and use of effective use of gravity promote flow of braze filler through capillary action.
- Flux residues after the joint has been made must be removed to avoid corrosion.
- Surface finish of brazed joints good.
- Fabrication tolerances a function of the accuracy of the component parts and the assembly/jigging method.

7.13 Soldering

Process description

- Heat is applied to the parts to be joined which melts a manually fed or pre-placed filler solder metal (which has a melting temperature $< 450\,°C$) into the joint by capillary action. A flux is usually applied to facilitate 'wetting' of the joint, prevent oxidation, remove oxides and reduce fuming (see 7.13F).

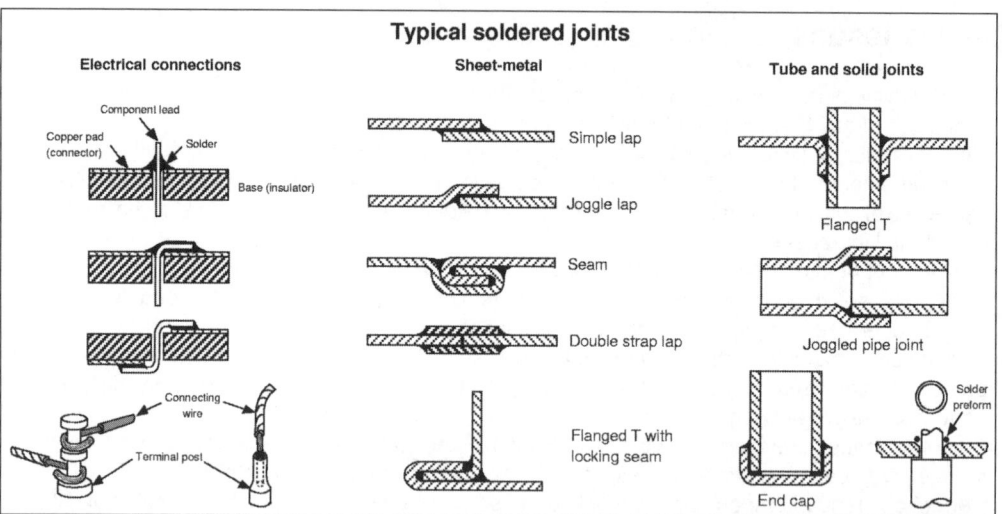

7.13F Soldering process.

Materials

- Most metals and combination of metals can be soldered with the correct selection of filler metal, heating process and flux. Commonly, copper, tin, mild and low alloy steels, nickel and precious metals are soldered. Some ceramics can be soldered.
- Magnesium, titanium, cast iron and high carbon or alloy steels are not recommended.

Process variations

- Gas soldering: air-fuel flame is used to heat the parts. Can be manually performed with a torch (TS) for small production runs or automated (ATS) with a fixed burner for greater economy.
- Furnace Soldering (FS): uniform heating takes place in an inert atmosphere or vacuum.
- Induction Soldering (IS): components are placed in a magnetic field surrounding an inductor carrying a high frequency current giving uniform heating.
- Resistance Soldering (RS): high electric resistance at joint surfaces causes heating for brazing. Not recommended for brazing dissimilar metals.
- Dip Soldering (DS): assemblies immersed to a certain depth in bath of molten solder. Can require extensive jigging and fixtures.

- Wave Soldering (WS): similar to dip soldering, but the solder is raised to the joint area on a wave. Used extensively for soldering electronic components to printed circuit boards.
- Contact or iron soldering (INS): uses an electrically heated iron or hot plate. Most common soldering process used for general electrical and sheet-steel work.
- Infra Red Soldering (IRS): heat application through directed spot of infrared radiation. Used or small precision work and difficult to reach joints.
- Laser beam soldering: provides very precise heat source for precision work, but at high cost.
- Ultrasonic soldering: uses an ultrasonic probe to provide localized heating through high-frequency oscillations. Eliminates the need for a flux, but requires pre-tinning of surfaces.
- Filler metal (solder) can be in preforms, wire, foil, coatings, slugs and pastes in a variety of metal alloys, commonly: tin-lead, tin-zinc, lead-silver, zinc-aluminum and cadmium-silver. The selection is based upon the metals to be soldered.
- Flux types: either corrosive (rosin, muriatic acid, metal chlorides) or non-corrosive (aniline phosphate), in powder, pastes or liquid form.

Economic considerations

- High production rates possible for WS.
- Very flexible process.
- Economical for very low production runs. Can be used for one-offs.
- Tooling costs low. Little tooling required.
- Equipment costs vary depending on degree of automation.
- Direct labor costs low to moderate. Cost of joint preparation can be high.
- Finishing costs moderate. Cleaning of the parts to remove corrosive flux residues is critical.

Typical applications

- Electrical connections
- Printed-circuit boards
- Light sheet-metal fabrication
- Pipes and plumbing
- Automobile radiators
- Precision joining
- Jewelery
- Food handling equipment

Design aspects

- Design complexity high, but low load capacity joints.
- Most common joint the lap with large contact areas or a combination of lap and fillet. Fillet joints predominantly used in electrical connections.
- Can be used to provide electrical or thermal conductivity or provide pressure tight joints.
- Joints should be designed to operate in shear and not tension. Additional mechanical fastening is recommended on highly stressed joints.
- Joints should be designed to give a clearance between the mating parts of 0.08–0.15 mm.
- Joint strength directly affected by clearance. If the clearance is too great the joint will loose a considerable amount of strength.
- Tolerances on mating parts should maintain the joint clearances recommended.
- On lap joints the length of lap should be between three and four times that of the thinnest part for optimum strength.

- Parts in the assembly should be arranged to promote capillary action by gravity.
- Machine marks should be in line with the flow of solder.
- Design joints using minimum amount of solder.
- Jigs and fixtures should be used only on parts where self-locating mechanisms, i.e. seaming, staking, knurls, bending or punch marks not practical.
- If jigs and fixtures used they should support the joint as far from the joint as possible, have minimum contact with the parts to be soldered and have low thermal mass.
- Soldered joints in electronic printed circuit boards should be spaced more than 0.8 mm apart.
- Provision for the escape of gases and vapors in the design important with vent-holes.
- Minimum sheet thickness $= 0.1$ mm.
- Maximum thickness, commonly $= 6$ mm.
- Unequal thicknesses possible but may create unequal joint expansion.
- Dissimilar metals can cause thermal stresses at the joint on cooling due to different expansion coefficients.

Quality issues

- Virtually stress and distortion free joints can be produced.
- Solderability improved by coating metals with tin.
- Coatings should be used on parts to protect the parent metal prior to soldering, classed as: protective, fusible, soluble, non-soluble and stop-off coatings.
- Control of the time and temperature of the applied heat important.
- Contamination free environment important for electronics soldering.
- Subsequent operations should have a lower processing temperature than the solder melting temperature.
- Heat sinks should be used when soldering heat-sensitive components, especially in electronics manufacture.
- Jigs and fixtures should be used to maintain joint location during solder cooling for delicate assemblies.
- Choice of solder important in order to avoid possibility of galvanic corrosion.
- Surface preparation important to remove any contaminates from the joint area such as oxide layers, paint and thick films of grease and oil and promote wetting. Degreasing and pickling of the parts to be soldered is recommended.
- Smooth surfaces preferred to rough ones. Abrading the joint area using emery cloth is acceptable.
- Correct clearance, temperature gradients and use of effective use of gravity promote flow of solder metal through capillary action.
- Flux residues after the joint has been made must be removed to avoid corrosion.
- Surface finish of soldered joints excellent.
- Fabrication tolerances a function of the accuracy of the component parts and the assembly/jigging method.

7.14   Thermoplastic welding

Process description

- Joint edges are heated using hot gas from a hand held torch causing the thermoplastic material to soften. A consumable thermoplastic filler rod of the same composition as the base material is used to fill the joint and create the bond with additional pressure from the filler rod at the joint area (see 7.14F).

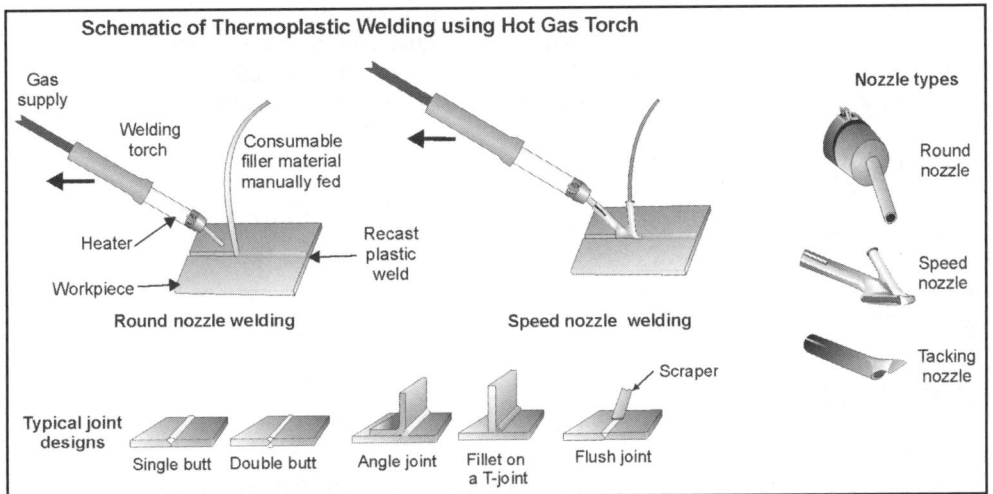

7.14F Thermoplastic welding process.

Materials

- Only thermoplastic materials.

Process variations

- Hot gas can be either nitrogen or air, depending on thermoplastic to be joined. Nitrogen minimizes oxidation of some thermoplastic materials.
- Various nozzle types for normal welding, speed welding and tacking.
- Other thermoplastic welding techniques available:

  - Spin welding: similar to Friction Welding (FRW), where the two parts to be joined, one stationary and one rotating at speed, have their joint surfaces brought into contact. Axial pressure and frictional heat at the interface create a solid state weld on discontinuation of rotation and on cooling (see 7.9).
  - Ultrasonic Welding (USW): hardened probe introduces a small static pressure and oscillating vibrations at the joint face disrupting surface oxides and raising the temperature through friction and pressure to create a bond. Can also perform spot welding using similar equipment (see 7.9).
  - Hot plate welding: electrically heated platens are used to soften base material at the joint and a bond is created with additional pressure giving good joint strength.

Economic considerations

- Production rates very low.
- Weld rates typically less than 1.5 m/min.
- Lead time typically hours.
- Manual operation typically using transportable equipment.
- Automation possible using a trolley system traversing over joint.
- Economical for low production runs. Can be used for one-offs.
- Tooling costs low.
- Equipment costs generally low.
- Direct labor costs moderate to high. Some skill needed by operator.
- Finishing costs low. Scraping the joint flush may be required for aesthetic reasons.
- Other thermoplastic welding techniques have a moderate to high production rate, are applicable to large volumes, have a moderate to high equipment cost and are more readily automated.

Typical applications

- Joining plastic pipes
- Ducts
- Containers
- Repair work

Design aspects

- Moderate levels of complexity possible.
- Typical joint designs possible using hot gas welding: butt, lap and fillet, in thin sheet.
- Horizontal welding position only.
- Parts to be joined must be in contact.
- Minimum overlap for lap joints = 13 mm.
- Minimum sheet thickness = 2 mm.
- Maximum sheet thickness = 8 mm.
- Multiple weld runs required on sheet thicknesses $\geq$5 mm.

Quality issues

- Filler rods must be same thermoplastic as base material.
- The force from the filler rod is applied to encourage mixing of softened material and must be consistent through the operation.
- Joints are weakened by incomplete softening, oxidation and thermal degradation of plastic material.
- Process variables are hot gas temperature, pressure (either from filler rod or fixtures) and speed of welding.
- Hot gas needs excess moisture and contaminants removed using filters.
- Weld strength is between 50 and 100 per cent of base material.
- Recast plastic filler at the joint can be made flush with base material using a scraper.
- Tack welding of parts to be joined should be performed before welding commences.
- Use of additional fixtures is advised for large parts, also to provide additional pressure to aid joint formation.
- Surface finish of weld is fair to good.
- Fabrication tolerances are typically $\pm$0.5 mm.

7.15 Adhesive bonding

Process description

- Joining of similar or dissimilar materials (adherent) by the application of a natural or synthetic substance (adhesive) to their mating surfaces which subsequently cures to form a bond (see 7.15F).

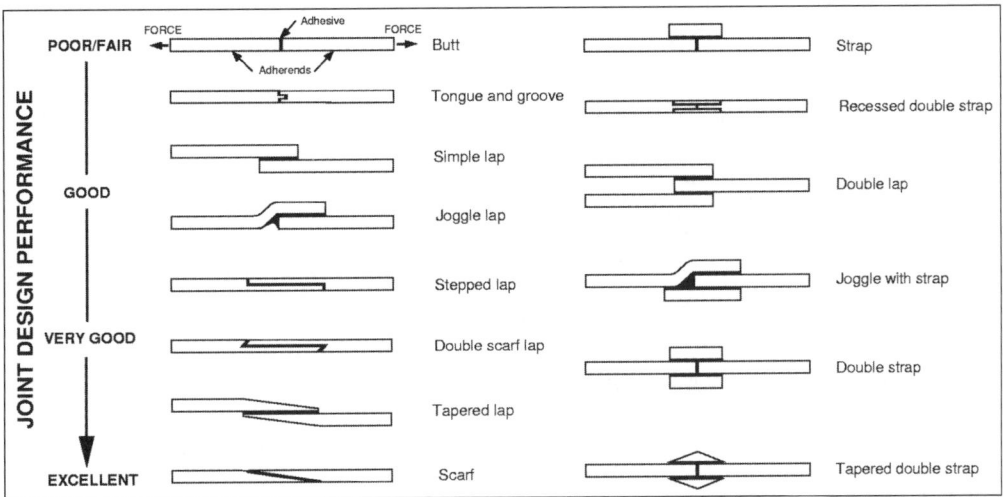

7.15F Adhesive bonding process.

Materials

- Most materials can be bonded with the correct selection of adhesive, surface preparation and joint design. Metals, plastics, composites, wood, glass, paper, leather and ceramics are bonded commonly.
- Can join dissimilar materials readily with proper adhesive selection, even materials with marked differences in coefficient of linear expansion, strength and thickness.

Process variations

- Adhesives available in many forms: liquids, emulsions, gels, pastes, films, tapes, powder, rods and granules.
- Curing mechanisms: heat, pressure, time, chemical catalyst, UV light, vulcanization or reactivation, or a combination of these.
- Various additives: catalysts, hardeners, accelerators and inhibitors to alter curing characteristics, silver metal flakes for electrical conduction and aluminum oxide to improve thermal conduction.
- Adhesives can be applied manually or automatically by: brushing, spreading, spraying, roll coating, placed using a backing strip or dispensed from a nozzle.

- Many types of adhesive are available:

  - Natural animal (beeswax, casein), vegetable (gum, wax, dextrin, starch) and mineral- (amber, paraffin, asphalt) based glues. Commonly low strength applications such as paper, cardboard (packaging) and wood.
  - Epoxy resins: typically uses a two-part resin and hardener or single part cured by heat for large structural applications.
  - Anaerobics: set in the absence of atmospheric oxygen. Commonly known as thread locking compounds and used for locating and sealing closely mated machined parts such as bearings and threads.
  - Cyanoacrylates: better known as super glues and use the presence of surface moisture as the hardening catalyst. Creates good bonds when using assembling small plastic, rubber and most metal parts.
  - Hot melts: thermoplastic resin bonds as it cools. Used for low load situations.
  - Phenolics: based on phenol formaldehyde thermosetting resins, two-part cold or heat and pressure cured. More expensive than most adhesives, but gives strong bonds for structural applications and good environmental resistance.
  - Plastisols: based on Polyvinyl Chloride (PVC) and uses heat to cure. For larger parts such as furniture and automotive panels.
  - Polyurethanes: similar to epoxies. Fast acting adhesive for low temperature applications and low loads. Footwear commonly uses this type of adhesive.
  - Solvent-borne rubber adhesives: rubber compounds in a solvent which evaporates to cure for minimal load applications.
  - Toughened adhesives: acrylic or epoxy-based adhesives cured by a number of methods and can withstand high shock loads and high loads in large structures.
  - Tapes: pressure sensitive adhesives on a backing strip for light loading applications such as packaging, automotive trim, cable secure and craft work.
  - Emulsions: based on Polyvinyl Acetate (PVA), highly versatile suitable for cold bonding of plastic laminates, wood, plywood, paper, cardboard, cork and concrete.
  - Polyimides: requires very high curing temperatures and pressures. Used in electronics and aerospace industries. High temperature capability.

Economic considerations

- High production rates possible.
- Lead time hours typically, but weeks if automated.
- Time for curing heavily dictates achievable production rate: tapes are instant, cyanoacrylates take several seconds, anaerobics can take 15–30 min, epoxy resins may take 2–24 h, although this can be reduced using catalysts.
- The viscosity of the adhesive must be suitable for the mixing and dispersion method chosen in production.
- Very flexible process.
- Simplifies the assembly process and therefore can reduce costs.
- Can replace or complement conventional joining methods such as welding and mechanical fasteners.
- Very little waste produced. Liquid adhesives require accurate metering to avoid excess.
- Economical for low production runs. Can be used for one-offs.
- Tooling costs low to medium. Jigs and fixtures recommended during curing procedure to maintain position of assembled parts can be costly.
- Equipment costs generally low.

- Direct labor costs low to moderate. Cost of joint preparation can be high.
- Finishing costs low. Little or no finishing required except removal of excess adhesive in some situations.

Typical applications

- Building and structural applications
- Electrical, electronic, automotive, marine and aerospace assemblies
- Packaging and stationery
- Furniture and footwear
- Craft and decorative work

Design aspects

- All levels of complexity.
- Can be used where other forms of joining not possible or practical.
- Joints should be designed to operate in shear, not tension or compression.
- Adhesives have relatively low strength and additional mechanical fixing recommended on highly stressed joints to avoid peeling.
- Most common joint is the lap or variations on the lap, for example, the tapered lap and scarf (preferred). Can also incorporate straps and self-locating mechanisms. Butt joints are not recommended on thin sections.
- A loaded lap joint tends to produce high stresses at the ends of the joints due to the slight eccentricity of the force line. Excessive joint overlap also increases the stress concentrations at the joint ends.
- For lap joints, the length of lap should be approximately 2.5 times that of the thinnest part for optimum strength. Increasing the width of the lap, adhesive thickness or increasing the stiffness of the parts to be joined can improve joint strength.
- Adhesive selection should also be based on: joint type and loading, curing mechanism and operating conditions.
- Can aid weight minimization in critical applications or where other joining methods are not suitable or where access to joint area limited.
- Inherent fluid sealing and insulation capabilities (electricity, heat and sound).
- Life prediction at operating temperature and should be assessed.
- Adequate space should be provided for the adhesive at the joint (~0.05 mm optimum clearance).
- Adhesives can be used to provide electrical, sound and heat insulation.
- Can provide a barrier to prevent galvanic corrosion between dissimilar metals or to create a pressure tight seal.
- Design joints using minimum amount of adhesive and provide for uniform thin layers.
- Jigs and fixtures should be used to maintain joint location during adhesive curing.
- Provision for the escape of gases and vapors in the design important.
- Minimum sheet thickness = 0.05 mm.
- Maximum sheet thickness, commonly = 50 mm.
- Unequal thicknesses commonly bonded.

Quality issues

- Excellent quality joints with little or no distortion.
- Residual stresses may be problematic with long curing time adhesives in combination with poor surface condition of base material, but otherwise not problematic.

- Dissimilar materials can cause residual stresses on cooling due to different expansion coefficients especially if heat is used in the curing process.
- Problems encountered with materials which are prone to solvent attack, stress cracking, water migration or low surface energy.
- Problems may be encountered in bonding materials which have surface oxides, loose surface layers or which are plated or painted (de-lamination may occur from the base material).
- Stress distribution over the joint area more uniform than other joining techniques.
- Joint fatigue resistance improved due to inherent damping properties of adhesives to absorb shocks and vibrations.
- Heat sensitive materials can be joined without any change of base material properties.
- Adhesives generally have a short shelf life.
- Optimum joint strength may not be immediate following assembly.
- Various adhesives can operate in temperatures up to approximately 250 °C.
- Control of surface preparation, adhesive preparation, assembly environment and curing procedure important for consistent joint quality.
- In surface preparation important to remove any contaminates from the joint area such as oxide layers, paint and thick films of grease and oil to aid 'wetting' of the joint. Mechanical abrasion (grit blasting, abrasive cloth), solvent degreasing, chemical etching, anodizing or surface primers may be necessary depending on the base materials to be joined.
- Adhesive almost invisible after assembly. Joint surface free of irregular shapes and contours as produced by mechanical fastening techniques and welding.
- Joint inspection difficult after assembly and NDT techniques currently inadequate. Quality control should include intermittent testing of joint strength from samples taken from the production line.
- Quality control of adhesive mix also important.
- Consideration of joint permanence important for maintenance purposes. Bonded structures are not easily dismantled.
- Joint strength may deteriorate with time, and severe environmental conditions (UV, radiation, chemicals, humidity and water) can greatly reduce joint integrity.
- Flammability and toxicity of adhesives can present problems to the operator. Fume extraction facilities may be required and safety procedures for chemical spillage need to be observed.
- Rough surfaces preferred to smooth ones to provide surface locking mechanisms.
- Fabrication tolerances a function of the accuracy of the component parts and the assembly/jigging method during curing time.

7.16 Mechanical fastening

Process description

- A mechanical fastening system is a separate device or integral component feature that will position and hold two or more components in a desired relationship to each other. The joining of parts by mechanical fastening systems can be generally classified as:

  - Permanent: can only be separated by causing irreparable damage to the base material, functional element or characteristic of the components joined, for example, surface integrity. A permanent joint is intended for a situation where it is unlikely that a joint will be dismantled under any servicing situation.
  - Semi-permanent: can be dismantled on a limited number of occasions, but may result in loss or damage to the fastening system and/or base material. Separation may require an additional

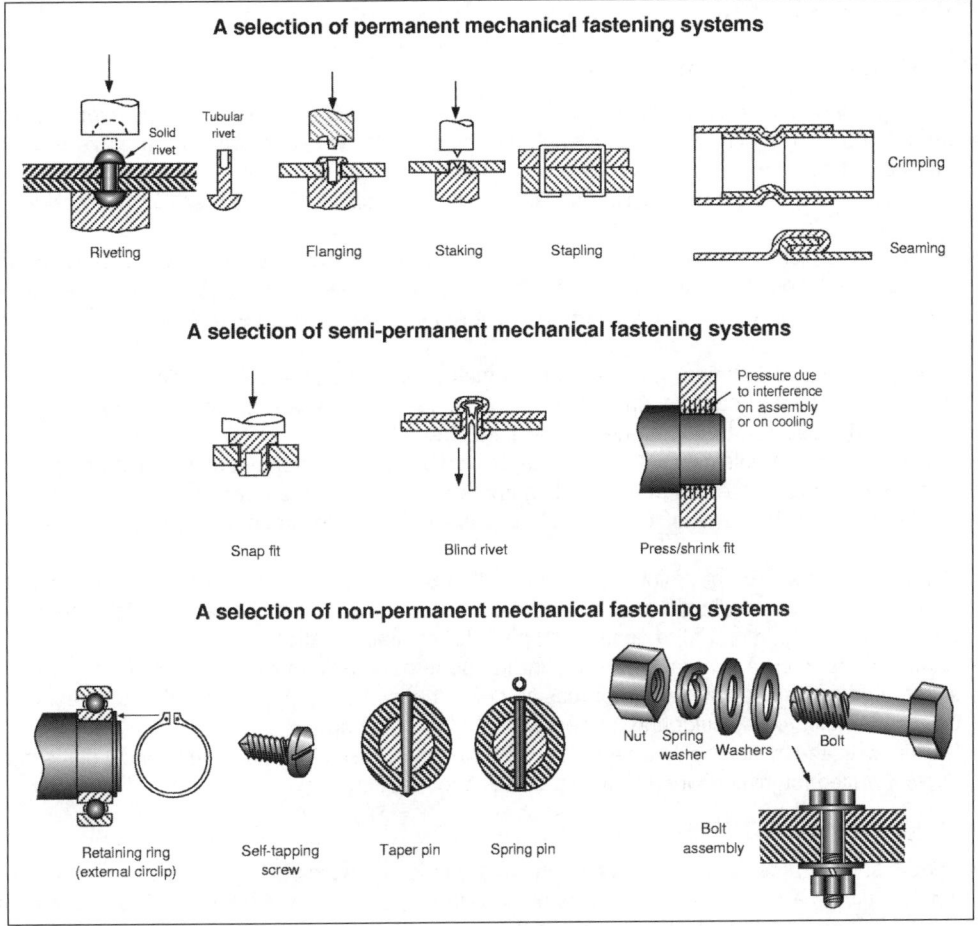

7.16F Mechanical fastening process.

process, for example, plastic deformation. A semi-permanent joint can be used when disassembly is not performed as part of regular servicing, but for some other need.
- Non-permanent: can be separated without special measures or damage to the fastening system and/or base material. A non-permanent joint is suited to situations where regular dismantling is required, for example, at scheduled maintenance intervals (see 7.16F).

Materials

- Can join most materials and combinations of materials using various processes. Metals, plastics, ceramics and wood are commonly joined.
- Fastening elements made from most metal alloys such as ferrous (steel most common), copper, nickel, aluminum and titanium, depending on strength of joint and environmental requirements. Use of plastics for fastening methods common for low loading conditions.
- Variety of coatings available for metal fasteners to improve corrosion resistance, commonly: zinc (electroplated and hot-dip), cadmium, chromate, phosphate and bluing.

Process variations

- Permanent fastening systems:

  - Riveting: used to create a closed mechanical element spanning an assembly. The rivet is located through a previously created hole through the materials to be joined and then the rivet shank is plastically deformed (either hot or cold) on one side typically. Used for joining sheet materials of varying type and thickness by solid, tubular (both semi-tubular and eyelet), split, compression and explosive types.
  - Flanging: the plastic deformation of an amount of excess material exposed on one component to locate and hold it to an adjacent face of another component. Readily lends itself to full automation. Deformation can be performed through direct pressure, rotary or vibratory tool movement.
  - Staking: similar to flanging, but plastic deformation is localized to where the components are closely assembled through a punch mark in the center of a protrusion. Location of the parts is by friction and pressure at their interface. Low joint strengths.
  - Stapling: joins materials using U-shaped staples fed on strips to the head of a semi-automatic tool. Can join dissimilar materials of thin section and no hole prior to the operation is needed.
  - Stitching: similar to stapling, but the stitching is made by the machine itself into a U-shaped form.
  - Crimping: a pressure tight joint is created on thin section assembled components by localized plastic deformation at dimple points, by swaging or shrinkage. Also notching which shears and bends the same portion of the assembled parts to maintain location.
  - Seaming: creation of a pressure tight joint in sheet-metal assemblies by hooking together two sheets through multiple bends and pressing down the joint area. Joint strength and integrity can be further improved by soldering, adhesive bonding or brazing.
  - Nailing: uses the friction between a nail and the pierced materials to maintain location of the parts. Typically used for joining wood to wood, or wood to masonry.

- Semi-permanent fastening systems:

  - Snap fits: integral features of the components to be joined typically hooked tabs which lock into notches on the adjacent part to be assembled with the application of a modest force. Commonly used for large volume production of plastic assemblies. Require special design attention to determine deflections and dimensional clearances.

- Press fits: use of the negative difference in dimensions (or interference) on the components to impart an interface pressure through the force for assembly.
- Shrink fits: use of the negative difference in component dimensions to impart an interface pressure on assembly by heating one component (usually the external) causing expansion and then allowing it to cool and contract in situ.
- Blind rivets: located into a previously created hole in the assembly from a single direction using a special tool. The tool retracts a headed pin from the rivet body deforming it enough to hold the components. The head is left inside the rivet body on joint completion. Used for thin sheet material fabrication.

- Non-permanent fastening systems:

  - Retaining rings: provide a removable shoulder within a groove of a bore or on the surface of a shaft to locate and lock components assembled to it. Presented either axially, radially or pushed into the groove using special tools. Self-locking, circlip, E-clip and wireformed types available for various applications. Made from spring steel typically.
  - Self-tapping screws: for assembling thin sheet material by passing a large pitch screw through previously created holes in the parts. Also self-drilling and thread forming types for soft materials.
  - Quick release mechanisms: for rapid securing and release of parts, e.g. doors, access panels, tooling jigs and fixtures. Various types available, such as clips, locks, latches, cams, clamps and quarter turn fastening systems.
  - Pins: for locating and retaining collars, hubs, gears and wheels on shafts, or to act as pivots in machinery or stops. Various types available, such as taper, spring, grooved, split and cotter.
  - Tapered and gib-head keys: for locating and holding gears, wheels and hubs on shafts through friction.
  - Magnetic devices: for locating or holding items such as doors and work holders for machine tools. Can be permanent type, mechanically or electrically actuated. Parts must be ferrous, nickel or cobalt based if direct magnetic attraction is required.
  - Threaded fastening systems: includes a number of standard thread forms and pitches. Variety of drive types (hexagonal head, socket head, slotted head), washers (plain, spring, double coil, toothed locking, crinkle, tab), nuts (plain, thin, nyloc, castle nut), locking mechanisms (split pin, lock plate, wiring), and bolt, screw, stud and set screw configurations.
  - Anchor and rag bolts: used for fixing structural sections and fabrications to concrete.
  - Threaded inserts: for use in brittle and flexible materials such as ceramics and plastics. Can be molded or cast in situ or inserted in previously threaded holes. Also Helicoil wire thread inserts for protecting and strengthening previously tapped threads.
  - Collets: for locating gears, hubs and wheels on shafts through friction mechanisms. Various types, such as expanding, taper and Morse.
  - Zips, studs, buttons, plastic tie-wraps, wire and Velcro are all very useful non-permanent fastening systems which have from time to time been used in engineering assemblies, particularly the last three.

- All mechanical fastening systems can be manually or semi-automatically performed during assembly or installation, however, not all fastening systems readily lend themselves to full automation.

Economic considerations

- High production rates possible depending on the fastening system and degree of automation. Also dependent on time to 'open' and 'close' fastening system.
- Economical for very low production runs.
- All production quantities viable.

- Regular use of same fastening system type on an assembly more cost effective than the use of many different types.
- A smaller number of large fasteners may be more economical than many small ones.
- Consideration must be given to fastener replacement costs for maintenance or service requirements.
- Tooling costs low to moderate depending on degree of automation.
- Equipment costs low.
- Direct labor costs low to moderate.
- Cost and skill of joint preparation can be high.
- Finishing costs very low. Usually no finishing is required.
- Little or no scrap, except where hole generation concerned.

Typical applications

- Structures for buildings and bridges
- Automotive, aerospace, electrical and marine assemblies
- Domestic and office appliances
- Machine tools
- Pipework and ducting
- Furniture
- Clothing

Design aspects

- Applicable to all levels of design complexity.
- Identification of possible failure modes (tension, shear, bearing, fatigue) and calculation of stresses in the fastener at the design stage recommended in joints subjected to high static, impact and/or fluctuating loads.
- Examination of the stresses in the joint area under the fastener important to determine the load bearing capability and stiffness of the parts to be joined.
- Use of recommended torque values for bolted connections critical for obtaining correct preloads and should be indicated on assembly drawings.
- Differentials in thermal expansion must be taken into consideration when using a fastener of different material to that of the base material.
- Provision for anti-vibration mechanisms in the fastening system where necessary, e.g. Nyloc, lock nuts in combination with split pins, spring washers.
- The damping characteristics of the assembled product must be considered when using a specific fastening system with fluctuating loads.
- Can incorporate pressure tight seals with most bolted joints, e.g. gaskets.
- Try to use standard fastener sizes, lengths and common fastening systems for a product.
- Keep the number of fasteners to a minimum for economic reasons.
- Design for the easy disassembly and maintenance of non-permanent fasteners, i.e. provide enough space for spanners, sockets and screwdrivers.
- Placing fasteners too close to the edge of parts or too close to each other avoided because of assembly difficulty and reduced strength capacity, i.e. pull out and rupture.
- Maximum operating temperatures of mechanical fastenings approximately 700 °C using nickel-chromium steel bolts.
- When joining plastics it is good practice to use metal threaded inserts or plastic fasteners.
- Minimum section thickness $= 0.25$ mm.
- Maximum section thickness, typically $= 200$ mm.
- Unequal section thicknesses commonly joined.

Quality issues

- Galvanic corrosion between dissimilar metals requires careful consideration, e.g. aluminum and steel.
- There is a risk of damage to joined parts or fasteners when using permanent systems or non-permanent fasteners that have been disassembled many times.
- Stress relaxation can cause the joint to loosen over time (especially in high temperature operating conditions over long periods). Subsequent re-torquing is recommended at regular intervals. This should be written into the service requirements for critical applications.
- High temperature applications in combination with harsh environments accelerate creep and fatigue failure.
- Rolled threads on bolts and screws are preferred over machined threads due to improved strength and surface integrity.
- Variations in flatness and squareness of abutment faces in assemblies can affect joint rigidity, corrosion resistance and sealing integrity.
- Variations in tolerances and accumulations of tolerances can result in mismatched parts and cause high assembly stresses. Dissimilar materials will also cause additional stresses, if reactions to the assembly environment result in unequal size changes.
- Variation in bolt preload is dependent on degree of automation of torquing method and frictional conditions at the component interfaces. Both should be controlled wherever possible.
- Lubricants and plate finishes on fasteners can help reduce torque required and improve corrosion resistance.
- Hydrogen embrittlement in electroplated steel fasteners can be problematic and accelerates failure.
- Stress concentrations in fastener and joint designs should be minimized by incorporating radii, gradual section changes and recesses.
- Hole size and preparation (where required) is important. Holes can act as stress concentrations. Fatigue life can be improved by inducing compressive residual stresses in the hole, e.g. by caulking.
- Reliability of joint and consistency of operation are improved with automation generally. Can be highly reliant on operator skill where automation not feasible.
- Fabrication tolerances are a function of the accuracy of the component parts and the fastening system used.

2.5 Combining the use of the selection strategies and PRIMAs

2.5.1 Manufacturing processes

Consider the problem of specifying a manufacturing process for a chemical tank made from thermoplastic with major dimensions – 1 m length, and 0.5 m in depth and width. A uniform thickness of 2 mm is considered initially with the requirement of a thicker section if needed. The likely annual requirement is 5000 units, but this may increase over time. The manufacturing process PRIMA selection matrix in Figure 2.2 shows that there are four possible processes considered economically viable for a thermoplastic material with a production volume of 1000–10 000. These are:

- Compression molding   (2.3)
- Vacuum forming   (2.5)
- Blow molding   (2.6)
- Rotational molding   (2.7).

Next we proceed to compare relatively the data in each PRIMA for the candidate processes against product requirements. Figure 2.8 provides a summary of the key data for each process upon which a decision for final selection should be based. An '✗' next to certain process data indicates that they should be eliminated as candidates. Vacuum forming is found to be the prime candidate as it is suitable for the manufacture of tub-shaped parts of uniform thickness within the size range required. Vacuum forming is also relatively inexpensive compared to the other processes and has low to moderate tooling, equipment and labor costs, with a reasonably high production rate achievable. Production volumes over 10 000 make it a very competitive process.

With reference to the manufacturing process PRIMA selection matrix in Figure 2.2, it can be seen that the requirement to process carbon steel in low to medium volumes (1000–10 000) returns thirteen candidate processes. This is a large number of processes from which to select a frontrunner. However, some processes can be eliminated very quickly, for example, those that are on the border of economic viability for the production volume requested. The process of elimination is also aided by the consideration of several of the key process selection drivers (as shown in Figure 1.11) in parallel. For example:

- For the required major or critical dimension does the tolerance capability of the process achieve specification and avoid secondary processing?
- What is the labor intensity and skill level required to operate the process, and will labor costs be high as dictated by geographical location?
- Is the initial material costly and can any waste produced be easily recycled?
- Is the lead time high together with initial equipment investment indicating a long time before a return on expenditure?

In this manner, a process of elimination can be observed which gives full justification to the decisions made. An overriding requirement is of course component cost, and the methodology provided in Part III of this book may be used in conjunction with the selection process when deciding the most suitable process from just several candidates. However, not all processes are included in the component-costing analysis and in this case it must be left to the designer to gather all the detailed requirements for the product and relate these to the data in the relevant PRIMAs.

2.3 COMPRESSION MOULDING	2.5 VACUUM FORMING
Economic Considerations • Production rates from 20 to 140/h. • The greater the thickness of the part, the longer the curing time. • Lead times may be several weeks according to die complexity. • Material utilisation is high. No sprues or runners. • Tooling costs are moderate to high. ✗ • Equipment costs are moderate. • Direct labour costs are low to moderate. • Finishing costs are generally low. Flash removal required. **Typical Applications** • Dishes. **Design Aspects** • Shape complexity is limited to relatively simple forms. Moulding in one plane only. • Threads, ribs, inserts, lettering, holes and bosses possible. • Thin walled parts with minimum warping and dimensional deviation may be moulded. • Maximum section, typically = 13mm. • Minimum section = 0.8mm. • Maximum dimension, typically = 450mm. ✗ • Maximum area = 1.5m². **Quality Issues** • Variation in raw material charge weight results in variation of part thickness and scrap. • Dimensions in the direction of the mould opening and the product density will tend to vary more than those perpendicular to the mould opening.	**Economic Considerations** • Production rates from 60 to 360/h commonly. • Lead times of a few days typically. • Material utilisation is moderate to low. Unformed parts of the sheet are lost and cannot be directly recycled. • Set-up times and change-over times are low. • Sheet material much more expensive than raw pellet material. • Tooling costs are low to moderate, depending on complexity. • Equipment costs are low to moderate, but can be high if automated. • Labour costs are low to moderate. • Finishing costs are low. Some trimming of unformed material after moulding. **Typical Applications** • Open plastic containers and panels. • Bath tubs, sink units and shower panels. **Design Aspects** • Shape complexity limited to mouldings in one plane. • Open forms of constant thickness. • Undercuts possible with a split mould. • Cannot produce parts with large surface areas. • Bosses, ribs and lettering possible, but at large added cost. • Maximum section = 3mm. • Minimum section = 0.05mm to 0.5mm, depending on material used. • Sizes range from 25mm² to 7.5m x 2.5m in area. **Quality Issues** • Thermoplastic material must possess a high uniform elongation otherwise tearing at critical points in the mould may occur. • Sheet material will have a plastic memory and so at high temperatures the formed part will revert back to original sheet profile. Operating temperature therefore important.
2.6 BLOW MOULDING	2.7 ROTATIONAL MOULDING
Economic Considerations • Production rates between 100 and 2,500/h, depending on size. • Lead times are a few days. • There is generally little material waste, but can increase with some complex geometries using extrusion blow moulding. • Set-up times and change-over times are relatively short. • Tooling costs are moderate to high. • Equipment costs are moderate to high, especially for full automation. • Direct labour costs are low. One operator can mange several machines. • Finishing costs are low. Some trimming required. **Typical Applications** • Hollow plastic parts with relatively thin walls. ✗ **Design Aspects** • Complexity limited to hollow, well rounded, thin walled parts with low degree of asymmetry. ✗ • Undercuts, bosses, ribs, lettering, inserts and threads possible. • Maximum section = 6mm. • Minimum section = 0.25mm. • Sizes range from 12mm in length to volumes up to 3m³. **Quality Issues** • Poor control of wall thickness, typically ±50% of nominal. ✗ • Good surface detail and finish possible.	**Economic Considerations** • Production rates of 3 to 50/h, but dependent on size. • Lead time several days. • Material utilisation very high. Little waste material. • Tooling costs are low. • Equipment costs are low to moderate. • Labour costs are moderate. • Finishing costs are low. Little finishing required. **Typical Applications** • Water tanks. • Buckets. • Drums. **Design Aspects** • Complexity limited to large, hollow parts of uniform wall thickness. ✗ • Large flat surfaces should be avoided due to distortion and difficulty to form. Use stiffening ribs. • Maximum section = 13mm. • Minimum section is typically 2mm, but can be as low as 0.5mm for certain applications. ✗ • Sizes up to 4m³. **Quality Issues** • Control of inside surface finish is not possible. ✗ • Wall thickness is determined by the close control of the amount of raw material used. • Wall thickness tolerances are generally between ±5 to ±20% of the nominal.

Fig. 2.8 Comparison of Key PRIMA data for the candidate processes.

2.5.2 Assembly systems

The case studies that follow describe where an automation technology has been successfully implemented as an economic and high quality alternative to manual assembly. The intention is to illustrate the application of the selection criteria and mapping given in Section 2.3.2, and also to indicate some of the opportunities for businesses associated with the implementation of assembly automation in industry. In the design of assembly systems, machine manufacturers have tended to adopt, where at all possible, a modular philosophy, coupled with the application of a well-trusted technology. This enables the suppliers to create systems for their customers that can be realistically priced, be effective and highly reliable. The case studies used illustrate what might be considered to be applications of automation, but with differing forms and degree of flexibility. The cases are discussed under the headings of products and customer requirements, the assembly process and machine design and selection considerations. The case studies are all in the public domain and for more information on the studies the reader is directed to reference (2.17).

Case study 1 – Assembly of medical non-return valves

The product and customer requirements

The product to be assembled was a non-return check-valve used in medical equipment including catheters and tracheotomy tubes. The requirement was for a highly process capable system with a defect rate (valve failure rate) of less than one part per million. Therefore, there was a requirement for checks to be built-in to the assembly system to reject any part that does not conform to the process capability standard. The valve comprises six very small components and was configured in four different versions. The variants result from the requirement for the use of different material types and differences in the diameter of the caps that seal the valves. The demand for the product necessitated a production rate of 200 items/min, and cleanliness was a critical requirement for the assembly process.

Assembly process and machine design

To achieve the level of reliability needed at the required production rate, a linear assembly system was specially developed to assemble the six components of the valve. The cell was equipped with six vibratory bowl feeders of different sizes to feed and orient the valve's components onto pallets containing four sets of nests. The assembly system was designed with 21 stations and to enable the operator to select random samples for inspection from each of four nests. The system was configured with an operating speed of 50 cycles/min to realize the required overall production output of 200 items/min, and the flexible cell was capable of producing the four different versions of the product. Despite this high rate of production, the valves produced were of the required quality, and displayed no surface faults (damage to the plastic components) that could have led to rejects. To meet the cleanliness requirements, the parts of the assembly system that come into contact with the valve's components were made from stainless steel, and the machine was carefully designed to operate without traces of dust or particulates. In addition, precise component fitting operations were required by the product design, with some of the items having to be inserted into the body of the valve within a tolerance of 0.05 mm.

Selection considerations

Factors driving the selection of the assembly technology adopted for the application could be considered to include:

- High production volumes and continuous demand
- Four different product variants
- Very high levels of process capability (< 1 ppm)
- Clean assembly process environment, free from contamination.

The product volume, number of variants and process capability requirements support the application of flexible assembly system for the product.

Case study 2 – Assembly and test of diesel injector units

Product and customer requirements

The requirement was for a flexible system to assemble a family of diesel unit injectors that could yield economic operation at fluctuating demand volumes. To realize the demanding tolerances necessitated by the product technology, the injector unit makes use of precision shims to compensate for machining variation and the inevitable variation in the characteristics of the spring embedded within the injector body. By choosing a shim of the correct characteristic thickness and capability, the business can vary the opening pressure of the valve to achieve an injector unit assembly that operates correctly first time. The customers' 'Lean Manufacturing' philosophy required that automation should only be introduced where there is a clear quality and economic case to do so. The automation project had to respect the customer's principle of balancing the relative benefits of automation against that of well-known manual assembly processes.

Assembly process and machine design

The system created by the assembly machine supplier operated on the 'Negari' principle which readily allows production volumes to be varied depending on the number of operators allocated to the system at any one time. The machine was designed such that a single operator could operate all machine stations in sequence; however, up to four operators could work on the same machine system to create a proportionate increase in production rates. The system was designed to enable assembled injectors to be 'wet tested' to verify the functional performance of the unit. The system provides the business with a means of directly responding to fluctuations in demand for the product. The system was also designed so that when the product is eventually withdrawn from service, the Negari facility will be able to provide 'service' components to reflect demand with the minimum of downtime.

Selection considerations

Considerations driving the selection of the assembly technology adopted include:

- Medium/high production volumes
- Fluctuating demand patterns
- Very high levels of process capability
- Integrated product testing

In order to meet the requirement for volume flexibility, the assembly system needs flexibilities in areas including: parts handling and fitting processes, machine capacity and processing routes. Adopting the Negari machine layout with multi-stations and manual handling and loading of parts provides a natural way of dealing with this problem.

Case study 3 – Accelerator pedal sensor assembly

Product and customer requirements

The electronic pedal sensor provides a means of throttle control that is more accurate and more reliable than cables, and provides a product that is essentially maintenance free. The sensor design is supplied to a leading (tier 1) manufacturer whose generic throttle pedal design places it well to meet the requirements of many major Original Equipment Manufacturers (OEM). Given the safety-critical nature of accelerator pedal sensor it is essential to electronically test each completed assembly to make sure it works correctly. The product comprised eight components and was to be assembled on a 9 s cycle-time.

Process and assembly machine design

The process is essentially automatic, but requires two operators to load critical components. Each operation is checked to make sure that it took place correctly (any incorrect assemblies are flagged on the pallet and pass through without further work). Laser trimming calibrates the resistance of the unit, and an electronic test also ensures that each completed assembly works properly. A modular approach was adopted for the design of the machine. The operator first loads a housing and rotor onto a flagged pallet. The first automatic station then loads the substrate, ensures that it is laid flat, and heat stakes it into position. The system checks only that the substrate is present, not that it performs correctly. Electronic testing of the final unit is carried out later in the process. Further operations load the spring, rotor and cover, which are heat staked into position to complete the assembly.

Three of the assembly operations are particularly technically demanding: wire bonding, spring contact assembly and laser trimming. A proprietary wire bonding station is used to weld the thin wire contacts into position to link the substrate with the electrical contacts molded in the sensor body. The spring contact assembly positions three small twin-spoked contacts and heat stakes them in position. The contacts must be secured without deformation and a force gauge is used to measure the pressure exerted by every spoke of the contact on the substrate track to ensure proper connections are made. Laser trimming of the substrate track calibrates the final assembly to ensure it has the correct resistance at a reference position. The system checks the resistance before and after the trimming process. This is a critical operation that ensures the correct operation of the sensor.

Selection considerations

The assembly technology adopted for the application could be considered as driven by factors including:

- High production volumes and continuous demand
- Very high levels of process capability
- Complex assembly processes
- Integrated testing processes.

The product volume and the safety critical nature of the process, coupled with complex assembly processes point to the need for a special purpose automatic machine with operator loading of critical components.

2.5.3 Joining processes

In order to illustrate the selection methodology, two sample case studies are presented. The case studies show just how many different joining processes can be used on essentially the same design and how this affects part-count, assemblability and functional performance in support of DFA.

Case study 1 – Rear windscreen wiper motor

The first case study shows important DFA measures and highlights where joining methods have had a detrimental effect on the design. The joining process selection methodology has been applied and the suggested joining processes compared to those used in the DFA redesigns. The architecture of the original design is shown in Figure 2.9. The DFA evaluation shows six functional parts and 23 non-functional parts, giving a DFA design efficiency of 17.8 per cent using the Lucas DFA methodology (1.36). Twelve of the non-functional components are only present for joining and to support the joining method, two bolts and two nuts to attach the housing and four rivets (and associated spacers) to join the brush plate to the retaining plate. The motor as intended is a throwaway module, that is, if a failure occurred during operation, the motor would be replaced, not repaired. Based on this information, all joints can be stated as permanent.

The redesign based on the DFA analysis suggestions is shown in Figure 2.10. The design proposed has six functional components, and no non-functional components, giving a DFA design efficiency of 100 per cent. The redesign eliminates all twelve components used for joining. The rivets and spacers have been removed, as the components they join are not in the redesign. Integrated snap fit fasteners have replaced the nut and bolt assemblies for fastening the housing.

The first step in selecting a joining process from the matrix is to determine the joint's requirements. The joint parameters for the housing are high volume (100 000+), permanent joint, thermoset material and thin ($\leq$3 mm) material thickness. Based on these constraints, the selection matrix shows the only suitable process to be a snap fit fastener. However, the quantity column must also be evaluated for all quantities. This search identifies tubular rivets, split rivets, compression rivets, nailing, cyanoacrylate adhesives, epoxy resin adhesives, polyurethane adhesives and solvent-borne rubber adhesives as alternatives. In this case study, the geometry and material are unsuitable for riveting and nailing. A comparison of adhesives and snap fit fasteners indicates that adhesives require more time for application, including a setting phase, and additional alignment features would need to be built into the components. Therefore, it is clear that the snap fit fasteners are the most appropriate joining method.

Although the rivets have been removed along with the components they joined, they formed part of the assembly that held the bearing in place. Consequently, the joint between the bearing and housing needs to be considered. The joint parameters for the bearing to housing

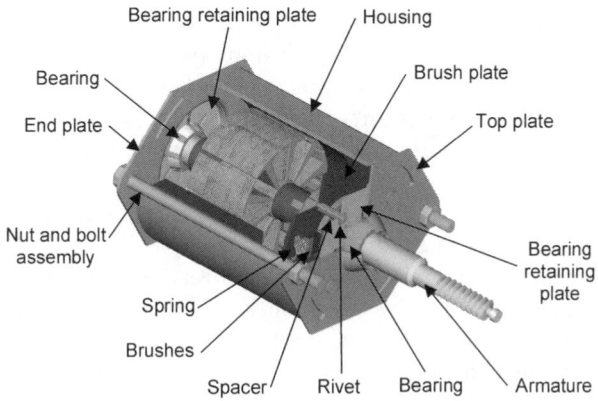

Fig. 2.9 Motor original design.

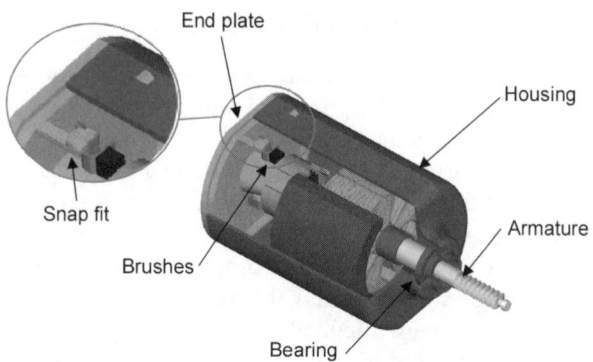

Fig. 2.10 Motor redesign.

are high volume, permanent joint, thermoset and steel material, and thin and medium thickness materials respectively. For this evaluation, the joining processes must match both material requirements. The search indicated two adhesive types: cyanoacrylate and epoxy resin as candidates. A search based on the same parameters for all quantities indicates toughened adhesives as a third candidate. As all the candidate joining processes are similar, the final decision would be based on process, detailed design requirements and economic factors, such as cost and availability as provided in the PRIMAs. The proposed redesign suggested adhesive bonding for fixing the bearing into the housing.

Case study 2 – Gas meter diaphragm assembly

This case study details a sample set of designs from a case study involving 12 designs from different manufacturers. Here three designs from different manufactures are considered. The designs incorporate different joining processes for the same problem. Essentially all the designs are the same with moderately different geometry as shown in Figure 2.11. In each case there is a top-plate, base-plate, supports for the flow measurement arm and a rubber/fabric diaphragm. The diaphragm is sandwiched between the base-plate and top-plate with the flow measurement arm support on top. The joining process used fixes all components together.

The consequences of the joining process selection are highlighted by the influence on part-count and DFA design efficiency. The design processes can now be compared with the results from the joining process selection matrix. The joint parameters and results are shown in Figure 2.12. It must be noted that in cases where two thicknesses are used, a match must be found for both. Also, although the quantity is high, the 'ALL QUANTITIES' column must be considered. While a permanent joint is required, as the joining strategy is 'through hole', it is also necessary to consider non-permanent solutions.

The matrix results show both riveting and retaining rings (including clips), along with a number of additional processes as candidates. This example clearly identifies the importance of selecting the most appropriate joining process. It shows that considering the impact on part-count and manufacturing processes helps to optimize the fastening of a joint. For example, both designs B and C use clips, although design C needs two extra pins to form the joint. A possible redesign would be to combine ideas from designs B and C, by integrating the top plate with the flow measurement arm support component (incorporating fastening pins as in design B) molded from a polymer. This would eliminate the separate support component, remove the need for separate fastening pins and provide location features as part of a functional part.

DESIGN	DESCRIPTION	DFA MERITS*
A	The top-plate and base-plate are steel pressings and the diaphragm is rubber. The flow measurement arm support component is a thermoplastic moulding. All four components are joined with two rivets.	**Part-count = 6** **Design Efficiency = 50%** **No. Joining Parts = 2**
B	The top-plate and base-plate are steel pressings and the diaphragm is rubber. The flow measurement arm support component is a thermoplastic moulding. Two pins integrated into the support component, which are fastened with two push-on retaining rings to fasten all the parts.	Uses identical top-plates and base-plates reducing the number of unique components. Integrated pins aid with alignment during assembly. **Part-count = 6** **Design Efficiency = 50%** **No. Joining Parts = 2**
C	The top-plate and base-plate are steel pressing and the diaphragm is fabric. The support features are pressed into the top-plate. Two pins pushed through the assembly fastened by two push-on retaining rings.	Uses identical top-plates and base-plates reducing the number of unique components and the measurement arm support component is integrated with the plates reducing the initial part-count. **Part-count = 7** **Design Efficiency = 42.8%** **No. Joining Parts = 4**

* Part-count, design efficiency and number of joining components used to relate to this sub-assembly only and do not necessarily reflect the overall design. Design Efficiency has been determined from the Lucas DFA methodology (1.36).

Fig. 2.11 Diaphragm assembly designs.

The case studies show that selecting an inappropriate joining process can have a large detrimental effect on a design. It could be argued that a DFA analysis would highlight poor fastening methods and suggest the need for redesign. This point is demonstrated by the examples shown above; however, a DFA analysis requires a completed design, and while highlighting the need for redesign, DFA offers no support for generating redesign solutions. If a proactive DFA approach is to be realized, it is essential that joining process selection be performed. Applying the joining process selection methodology and supporting data during product development allows the geometry of components to be tailored to the selected joining process, eliminating the need for redesign.

DESIGN	MATERIAL	QUANTITY	THICK-NESS	PERMANENCE	RESULTS
A	Steel and Thermoplastic	High & All quantities	Thin-Medium	Permanent	Ultrasonic welding Tubular rivet Staking Nailing Solid rivet Cyanoacrylate Epoxy resin Solvent-borne rubber Toughened adhesives
				Non-permanent	Retaining ring Quick release devices Threaded fasteners
B	Steel and Thermoplastic	High & All quantities	Thin-Medium	Permanent	Ultrasonic welding Tubular rivet Staking Nailing Solid rivet Cyanoacrylate Epoxy resin Solvent-borne rubber Toughened adhesives
				Non-permanent	Retaining ring Quick release devices Threaded fasteners
C	Steel	High & All quantities	Thin	Permanent	Ultrasonic welding Spot welding Plasma arc welding Laser beam welding Electron beam welding Furnace brazing Diffusion brazing Anaerobic Cyanoacrylate Epoxy resin Hot melt Solvent-borne rubber Toughened adhesives Solid rivet Flanging Stapling / Stitching Seaming Staking Crimping Nailing
				Non-permanent	Retaining ring Quick release devices Threaded fasteners

Fig. 2.12 Diaphragm assembly joint parameters and results.

Part-count optimization is one of the main aims of DFA, significantly influencing economic feasibility and often the technical performance of a design. Joining has been proved to have a large influence on part-count. In many designs, a significant proportion of the components are only present to support the joining process. Consequently, it can be concluded that a joining selection methodology is an important aspect of DFA. The case studies presented highlight the importance of joining process selection and its effect on the assemblability of a design. It can be seen that selecting an appropriate joining process at early stages of the design process encourages a right-first-time design philosophy, reducing the need for costly redesign work.

Part III

Costing designs

Procedures to enable the exploration of design and process combinations for manufacturing and assembly cost.

3.1 Introduction

For financial control and successful marketing it is necessary to have cost targets and realizations throughout the product introduction process. Product cost is virtually always a prime element in decision making, in manufacturing industry. The main problem in product introduction is the provision of reliable cost information in the early stages of the design process, for the comparison of alternative conceptual designs and assessment of the myriad of ways in which a product may be structured during concept development.

Cost estimates are needed to determine the viability of projects and to minimize project and product costs. The inadequate nature of the historical standard costing methods and cost estimating practices found in most companies has been highlighted by researchers over a number of years (3.1–3.5). One signal that emerges from all workers is that it is crucial to reject uneconomic designs early, for it is not often possible to reduce costs productively once production has commenced, largely due to the high cost of change at this stage in the product life cycle. Hence, cost analysis is best utilized at the stage in the design process when rough designs for a component have been prepared.

The aim of the component costing analysis presented here (3.6)(3.7) is to highlight expensive and difficult to manufacture designs, thus indicating areas that will benefit from further attention, before the design has been completed. Benefits of the methodology include:

- Lower component costs
- Systematic component costing
- Identification of feasible manufacturing processes
- Rapid comparison of alternative designs and competitor products
- Reduced engineering change
- Shorter development time and reduced time-to-market
- Education and training.

The methodology described is ideally applicable to team-based applications, both manually and in the form of computer software. The initial work was primarily designed to cater for components found in the light engineering, aerospace and automotive business sectors.

The section on assembly costing is intended to support the process of assembly-orientated design through the provision of assembly performance metrics. As with conventional DFA approaches, the methodology allows the user to match designed features with typical assembly situations (and associated penalties) on charts for each aspect of the assembly analysis. In this way ambiguity is reduced, and the user may identify features that are of high penalty and redesign these where necessary. The assembly cost measure should not strictly be taken as an absolute value. In practice, assembly costs are difficult to quantify and measure, and correlation requires testing a large number of industrial case studies. Nevertheless, the analysis results are useful when used in a relative mode of application.

◼ 3.2 Component costing

In order to produce a practical and widely applicable tool for designers with the capability to provide feedback on the technological and economic consequences of component design decisions, it was considered useful to develop a sample model that is widely applicable to a number of different manufacturing processes. In addition, the model was designed such that appropriate manufacturing processes and equipment requirements can be specified early in the product introduction process. Recognizing the problem, that the relationship between a design and its manufacturing feasibility and cost, is not easily amenable to precise scientific formulation; the model has come out of knowledge-engineering work in a number of user companies and those specializing in particular manufacturing techniques.

3.2.1 Development of the model

The model is logically based on material volume and processing considerations. The process cost is determined using a basic processing cost (the cost of producing an ideal design for that process) and design-dependent relative cost coefficients (which enable any component design to be compared with the ideal). Material costs are calculated taking into account the transformation of material to yield the final form.

Thus a single process model for manufacturing cost, M_i, can be formulated as:

$$M_i = VC_{mt} + R_c P_c \qquad [3.1]$$

where V is the volume of material required in order to produce the component, C_{mt} is the cost of the material per unit volume in the required form, P_c is the basic processing cost for an ideal design of component by a specific process and R_c is the relative cost coefficient assigned to a component design (taking account of shape complexity, suitability of material for processing, section dimensions, tolerances and surface finish).

The initial hypothesis can be expanded to allow for secondary processing, and thus the model can take the general form:

$$M_i = VC_{mt} + (R_{c_1} P_{c_1} + R_{c_2} P_{c_2} + \cdots + R_{c_n} P_{c_n})$$

$$M_i = VC_{mt} + \sum_{i=1}^{n} (R_{c_i} P_{c_i}) \qquad [3.2]$$

where n is the number of operations required to achieve the finished component.

In order for such a formulation to be used in practice it is necessary to define relationships enabling the determination of the quantities P_c and R_c for design-process combinations. In practice, it has been found that Equation (3.1) is the form preferred by industry. This is based on the need to work in the early stages of the design process with incomplete component data and without the necessity for detailing the sequence of manufacturing operations. The approach has been to build the secondary processing requirements into the relative cost coefficient. More will be said about this in Section 3.2.3.

3.2.2 Basic processing cost (P_c)

In order to represent the basic processing cost of an ideal design for a particular process, it is first necessary to identify the factors on which it is dependent. These factors include:

- Equipment costs including installation
- Operating costs (labor, number of shifts worked, supervision and overheads, etc.)
- Processing times
- Tooling costs
- Component demand

The above variables are taken account of in the calculation of P_c using the simple equation:

$$P_c = \alpha T + \beta/N \qquad [3.3]$$

where α is the cost of setting up and operating a specific process, including plant, labor, supervision and overheads, per second, β is the process specific total tooling cost for an ideal design, T is the process time in seconds for processing an ideal design of component by a specific process and N is the total production quantity *per annum*.

Values for α and β are based on expertise from companies specializing in producing components in specific technological areas. Using these process specific values in Equation (3.3), it is possible to produce comparative cost curves for any process.

Data for P_c against annual production quantity, N, is illustrated in Figures 3.1–3.5 for several main process groups (casting and molding, forming, machining, continuous extrusion and chemical milling) covering 20 individual manufacturing processes. While the data presented might be adequate in most cases, the methodology was devised with the idea that users would develop their own data for the process they would wish to consider. Such an approach has many benefits to a business, including ownership of the data and a confidence in the results produced. The values of P_c represent the minimum likely costs associated with a particular manufacturing process at a given annual production quantity. In this way, it is possible to indicate the lowest likely cost for a component associated with a particular manufacturing process route assuming an ideal design for the process, one-shift working and a two-year payback on investment.

A process key for the figures is provided below:

AM	Automatic Machining
CCEM	Cold Continuous Extrusion (Metals)
CDF	Closed Die Forging
CEP	Continuous Extrusion (Plastics)
CF	Cold Forming
CH	Cold Heading
CM2.5	Chemical Milling (2.5 mm depth)
CM5	Chemical Milling (5 mm depth)
CMC	Ceramic Mold Casting
CNC	Computer Numerical Controlled Machining

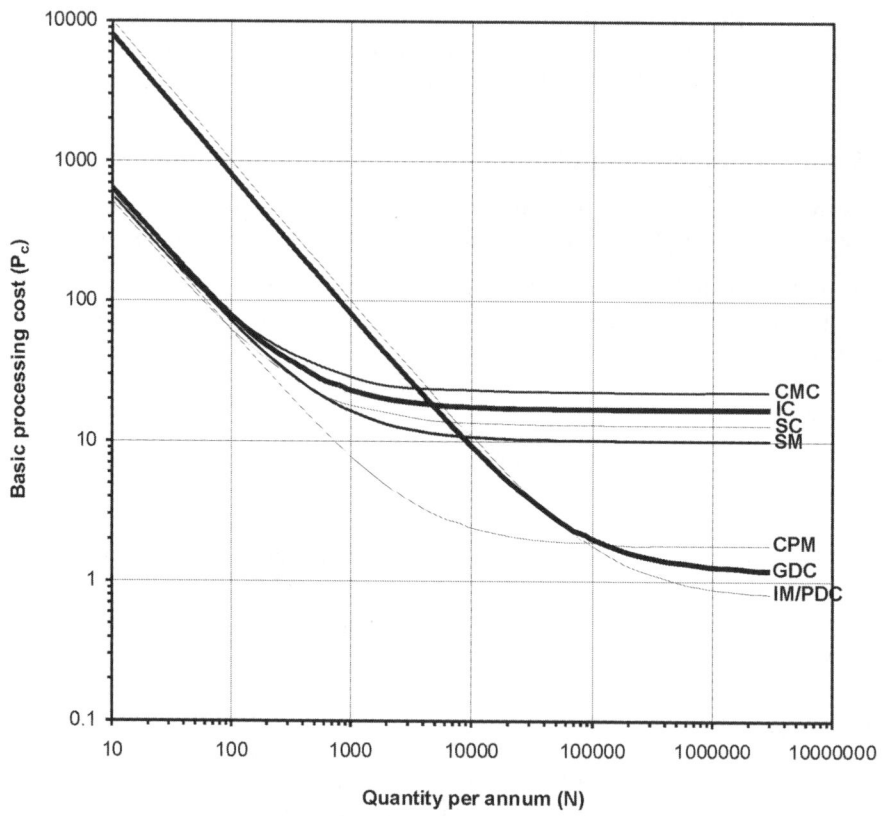

Fig. 3.1 Basic processing cost (P_c) against annual production quantity (N) for casting and molding processes.

CPM	Compression Molding
GDC	Gravity Die Casting
HCEM	Hot Continuous Extrusion (Metals)
IC	Investment Casting
IM	Injection Molding
MM	Manual Machining
PDC	Pressure Die Casting
PM	Powder Metallurgy
SM	Shell Molding
SC	Sand Casting
SMW	Sheet Metal Work
VF	Vacuum Forming

Having defined P_c, it is necessary to determine the design-dependent factors. The variables, shape complexity, tolerances, etc. modify the relationship between the curves. The relative cost coefficient R_c in Equation (3.1) is one way in which these variables can be expressed.

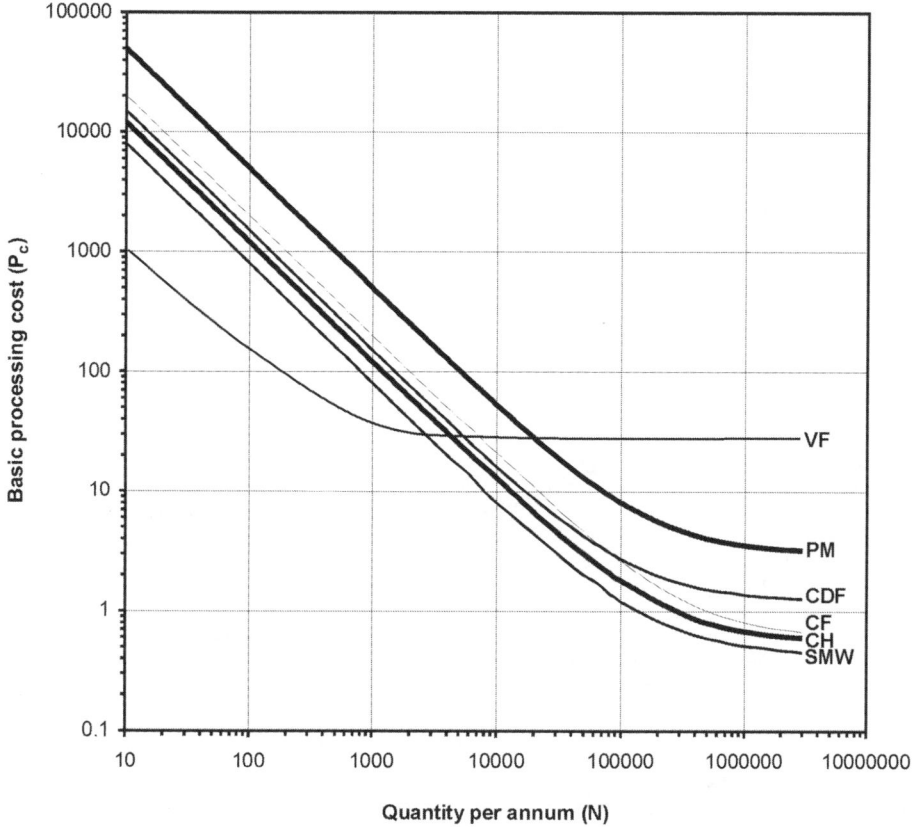

Fig. 3.2 Basic processing cost (P_c) against annual production quantity (N) for forming processes.

3.2.3 Relative cost coefficient (R_c)

This coefficient will determine how much more expensive it will be to produce a component with more demanding features than the 'ideal design'. The characteristics which we have assumed to influence the relative cost coefficient, R_c, are given below:

$$R_c = \phi(C_{mp}, C_c, C_s, C_t, C_f)$$

where C_{mp} is the relative cost associated with material-process suitability, C_c is the relative cost associated with producing components of different geometrical complexity, C_s is the relative cost associated with size considerations and achieving component section reductions/thickness, C_t is the relative cost associated with obtaining a specified tolerance and C_f is the relative cost associated with obtaining a specified surface finish.

Analysis of the influence of the above quantities and discussions with experts led to the idea that these could be combined as shown below:

$$R_c = C_{mp}^a C_c^b C_s^c C_t^d C_f^e \qquad [3.4]$$

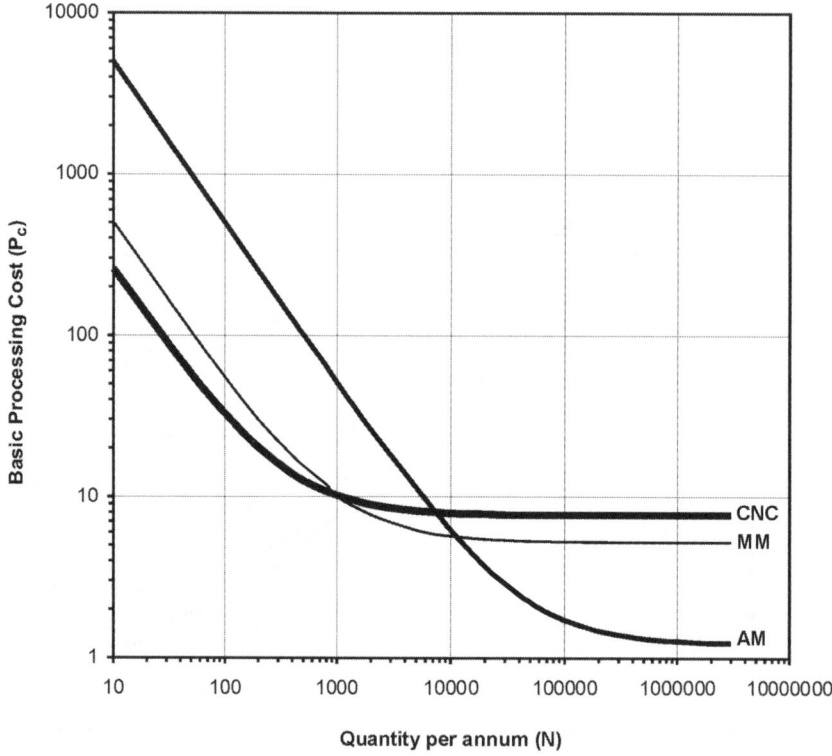

Fig. 3.3 Basic processing cost (P_c) against annual production quantity (N) for machining processes.

where a, b, c, d and e are weighting exponents. However, it was found that the knowledge could be structured to enable each of the exponents to be assigned the value of unity. Therefore, the relative cost coefficient can be represented by the formula:

$$R_c = C_{mp} C_c C_s C_{ft} \qquad [3.5]$$

where C_{ft} is the higher of C_f and C_t, but not both.

This was refined on the basis that when a fine surface finish is being produced, fine tolerances could be attained at the same time, and thus it would be somewhat dubious to compound both relative cost coefficients.

Knowledge engineering indicated that Equation (3.5) was the most appropriate combination of coefficients at the present stage of development. The method of comparison and accumulation of costs was shown to be analogous to those methodologies employed by experts in the field of cost engineering/estimating. For the ideal design, each of these coefficients is unity, but as the component design moves away from this state, then one or more of the coefficients may increase in magnitude, thus changing the manufacturing cost term, M_i, in Equation (3.1).

Figure 3.6 shows how the cost curves for P_c are modified according to the model proposed, as a component design shifts away from the ideal for that process. As the design becomes more difficult to process, because of material types or geometrical features for example, its cost curve progresses up the cost axis as illustrated, moving from Design A to B in the figure.

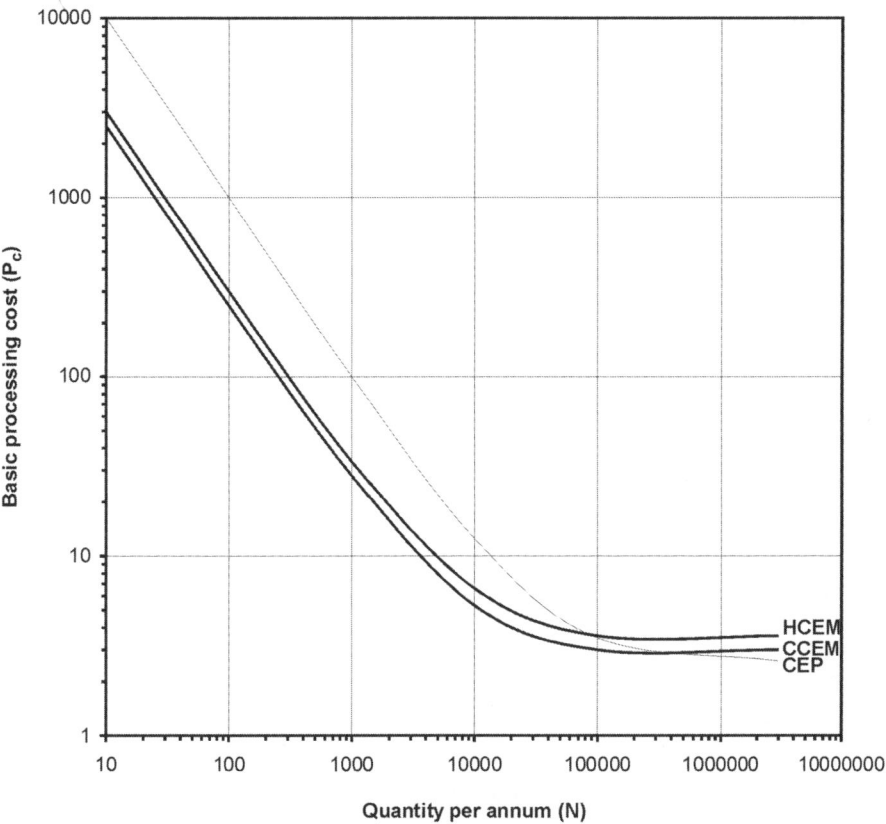

Fig. 3.4 Basic processing cost (P_c) against annual production quantity (N) for continuous extrusion processes.

Sets of data for the above relative cost coefficients C_{mp}, C_c, C_s, C_t and C_f can be found in Figures 3.8, 3.9, 3.10, 3.11 and 3.12 respectively.

Material to process suitability (C_{mp})

In Figure 3.7, the C_{mp} data indicates the suitability of using various materials with different processes. Clearly some combinations are inappropriate, and C_{mp} values only appear at nodes currently considered to be technologically and economically feasible.

Shape complexity (C_c)

Figures 3.8 and 3.9 present a system to identify the shape classification used, in order to determine C_c. The first step is to read the supporting notes and complexity definitions provided in Figure 3.8. There are three basic shape categories: solid of revolution, prismatic solid and flat or thin wall section components. These three fundamental shape categories can be sub-divided into five bands of complexity as shown pictorially in Figure 3.9. The classifica-

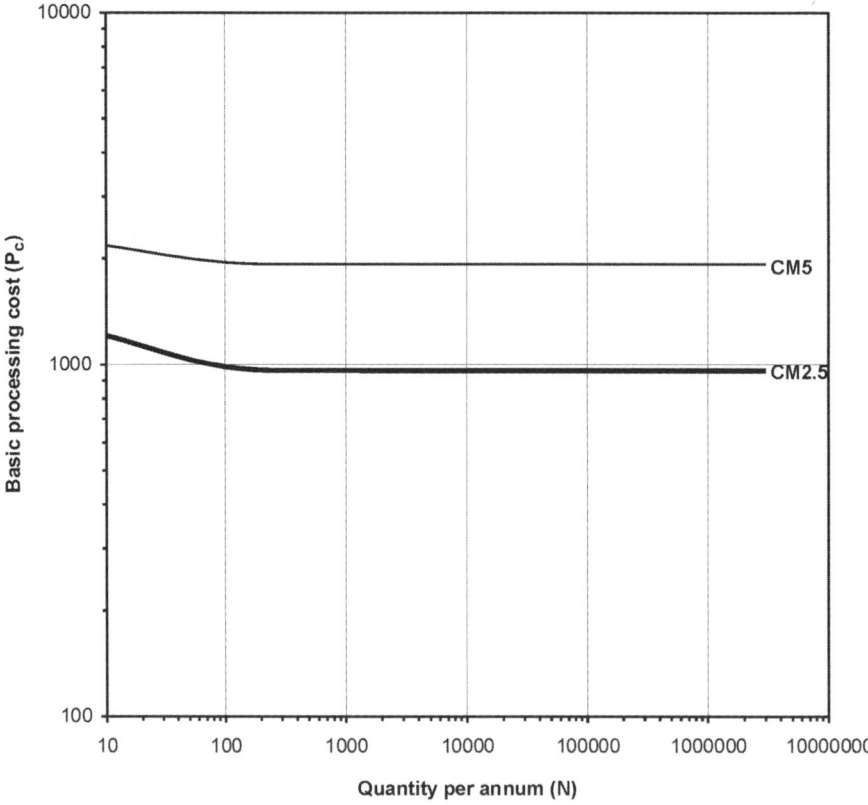

Fig. 3.5 Basic processing cost (P_c) against annual production quantity (N) for chemical milling.

tion process must reflect the finished form of the component, and the features listed in the tables should be used as an aid to the selection of the appropriate value of C_c from Figures 3.10, 3.11 or 3.12 for classification categories 'A', 'B' or 'C' respectively. Determination of the shape complexity is important. Failure to classify the geometry properly may affect the final component cost result, M_i, quite significantly, and studying the shape complexity definitions is crucial in this connection.

Section coefficient (C_s)

Figures 3.13, 3.14 and 3.15 show the relative cost consequences of producing specific section/wall thicknesses for the sample set of processes. Data required are the maximum dimension where the section acts, and the specified section size. Using a process outside its normal size domain can result in an additional cost. In the situation where a process is being considered for a component that is longer or smaller than the size range given in the macro (length, area, volume or weight), the estimate of cost produced is likely to be a lower bound value. Note that for Chemical Milling (CM2.5 and CM5), $C_s = 1$, as this penalty is taken account of in the formulation of the basic processing cost, P_c.

Tolerance (C_t) and surface finish (C_f) coefficients

The sample data on the effects of tolerance (C_t) and surface finish (C_f) can be found in Figures 3.16–3.18 and Figures 3.19–3.21 respectively. These indicate the relative cost consequences of achieving specific tolerance and surface finish levels for the various manufacturing processes. The process of analysis is:

1 Determine the most important tolerance values.
2 Identify the tolerance band on the C_t table.
3 Count the number of tolerances in the same band.
4 Identify the number of planes on which the critical values lie.
5 Select the appropriate C_t index from the table.

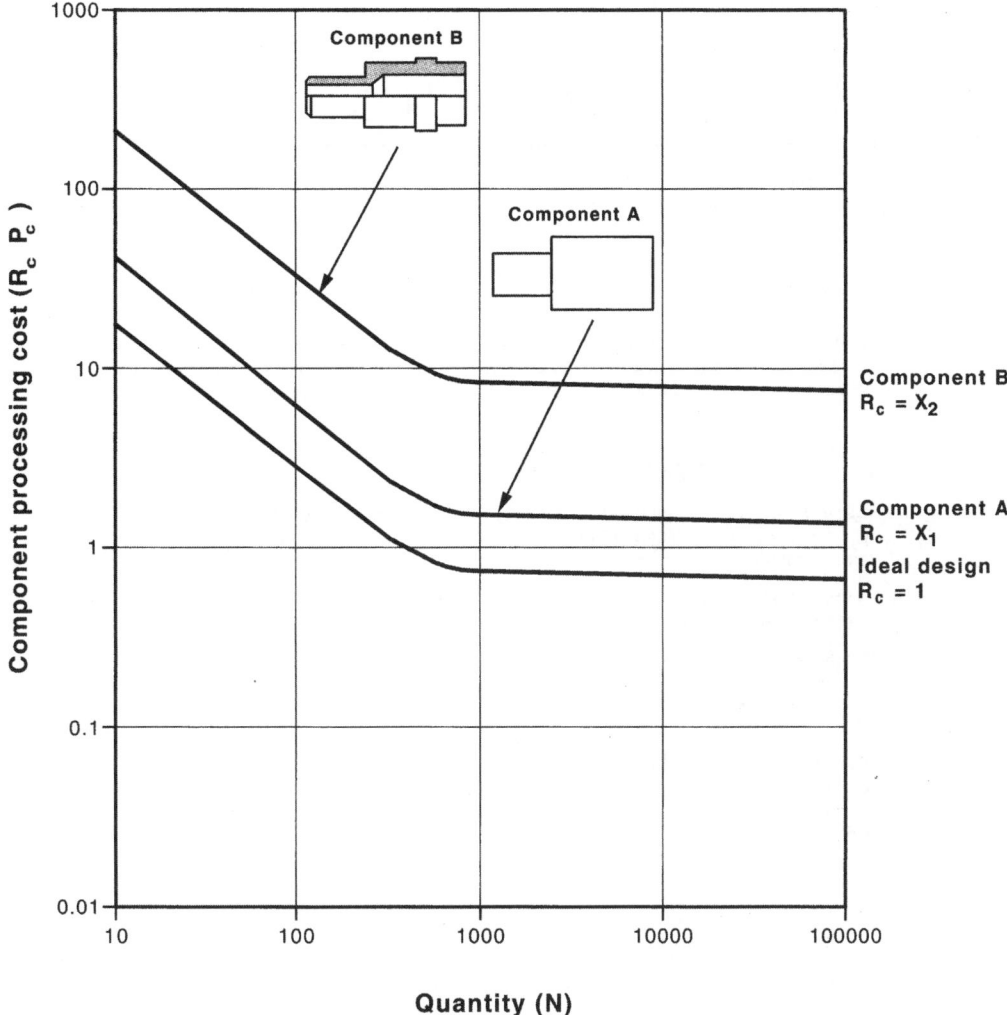

Fig. 3.6 Relative manufacturing cost versus quantity curves.

Process / Material	AM	CCEM	CDF	CEP	CF	CH	CM2.5	CM5	CMC	CNC	CPM	GDC	HCEM	IC	IM	MM	PDC	PM	SM	SC	SMW	VF
Cast Iron	1.2					1	1	1	1	1.2				1		1.2		1.6	1	1		
Low Carbon Steel	1.4	1.3	1		1.3	1.3	1	1	1.2	1.4			1.3	1		1.4		1.2	1.2	1.2	1.2	
Alloy Steel	2.5	2	2		2	2	1	1	1.3	2.5			2	1		2.5		1.1	1.3	1.3	1.5	
Stainless Steel	4	2	2		2	2	1	1	1.5	4			2	1		4		1.1	1.5	1.5	1.5	
Copper Alloy	1.1	1.1	1		1	1			1	1.1			1	1		1.1	3	1	1	1	1	
Aluminium Alloy	1	1.1	1		1	1			1	1		1.5	1.1	1		1	1.5	1	1	1	1	
Zinc Alloy	1.1	1	1		1	1			1	1.1		1.2	1	1		1.1	1.2	1	1	1	1	
Thermoplastic	1.1			1						1.1	1.2				1	1.1						1
Thermoset	1.2			1.2						1.2	1				1	1.2						
Elastomer	1.1			1.5						1.1	1.5				1.5	1.1						

Fig. 3.7 Relative cost data for material processing suitability (C_{mp}).

Notes - Geometrical Considerations

The shape complexity index is obtained by using a feature based classification system which enables the important design/manufacturing issues to be taken into account. Firstly determine the shape category:

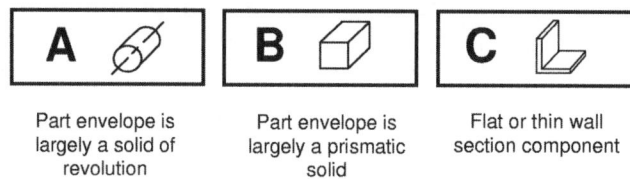

A	B	C
Part envelope is largely a solid of revolution	Part envelope is largely a prismatic solid	Flat or thin wall section component

Within the above classes, components are divided into five bands of complexity.

Note that the classification process should reflect the finished form of the component and the features listed in the tables should be used to aid the selection of the appropriate band. Always determine the classification by working from the left hand side of the table.

Notes - Shape Complexity Definitions

- **Basic Features** - Straightforward processing where the operation can be carried out without a change of setting or the need of complex tooling. Parts are usually uniform in cross section.

- **Secondary Features** - As above, but where additional processing is necessary or more complex tooling is required.

- **Multi-axis Features** - Parts require to be processed in more than a single axis/set-up.

- **Non-uniform Features** - Parts require the development of more complex processing techniques/set-up.

- **Complex Forms** - Parts need dedicated tooling and the development of specialised processing techniques.

- **Single Axis** - This is usually the axis along the components largest dimension, however, in the case of cylindrical or disc shaped components, it is more convenient to consider the axis of revolution as the primary axis.

- **Through Features** - Features which run along, across or through a component from one end or side to the other.

- **Important** - If the component falls into more than one category, always choose the one that gives the highest value of C_c.

Fig. 3.8 Notes on shape classification used in the determination of C_c.

 A   **Part Envelope is Largely a Solid of Revolution**

Single/Primary Axis		Secondary Axes: Straight line features parallel and/or perpendicular to primary axis		Complex Forms
Basic rotational features only	Regular secondary/ repetitive features	Internal	Internal and/or external features	Irregular and/or complex forms
A 1	**A 2**	**A 3**	**A 4**	**A 5**
Category Includes: Rotationally symmetrical/ grooves, undercuts, steps, chamfers, tapers and holes along primary axis/centre line.	Internal/external threads, knurling and simple contours through flats/splines/keyways on/around the primary axis/centre line.	Holes/threads/ counterbores and other internal features not on the primary axis .	Projections, complex features, blind flats, splines, keyways on secondary axes.	Complex contoured surfaces,and /or series of features which are not represented in previous categories.

B   **Part Envelope is Largely a Rectangular or Cubic Prism**

Single Axis/Plane		Multiple Axes		Complex Forms
Basic features only	Regular secondary/ repetitive features	Orthogonal/straight line based features	Simple curved features on a single plane	Irregular and/or contoured forms
B 1	**B 2**	**B 3**	**B 4**	**B 5**
Category Includes: Through steps, chamfers and grooves/channels/slots and holes/threads on a single axis.	Regular through features, T-slots and racks/plain gear sections etc. Repetitive holes/threads/counter bores on a single plane.	Regular orthogonal/straight line based pockets and/or projections on one or more axis. Angled holes/threads/ counter bores.	Curves on internal and/or external surfaces.	Complex 3-D contoured surfaces/geometries which cannot be assigned to previous categories.

C   **Flat Or Thin Wall Section Components**

Single Axis	Secondary/Repetitive Regular Features		Regular Forms	Complex Forms
Basic features only	Uniform section/ wall thickness	Non-uniform section/ wall thickness	Cup, cone and box-type parts	Non-uniform and/or contoured forms
C 1	**C 2**	**C 3**	**C 4**	**C 5**
Category Includes: Blanks, washers, simple bends, forms and through features on or parallel to primary axis.	Plain cogs/gears, multiple or continuous bends and forms.	Component section changes not made up of multiple bends or forms. Steps, tapers and blind features.	Components may involve changes in section thickness.	Complex or irregular features or series of features which are not represented in previous categories.

Fig. 3.9 Shape classification categories used in the determination of C_c.

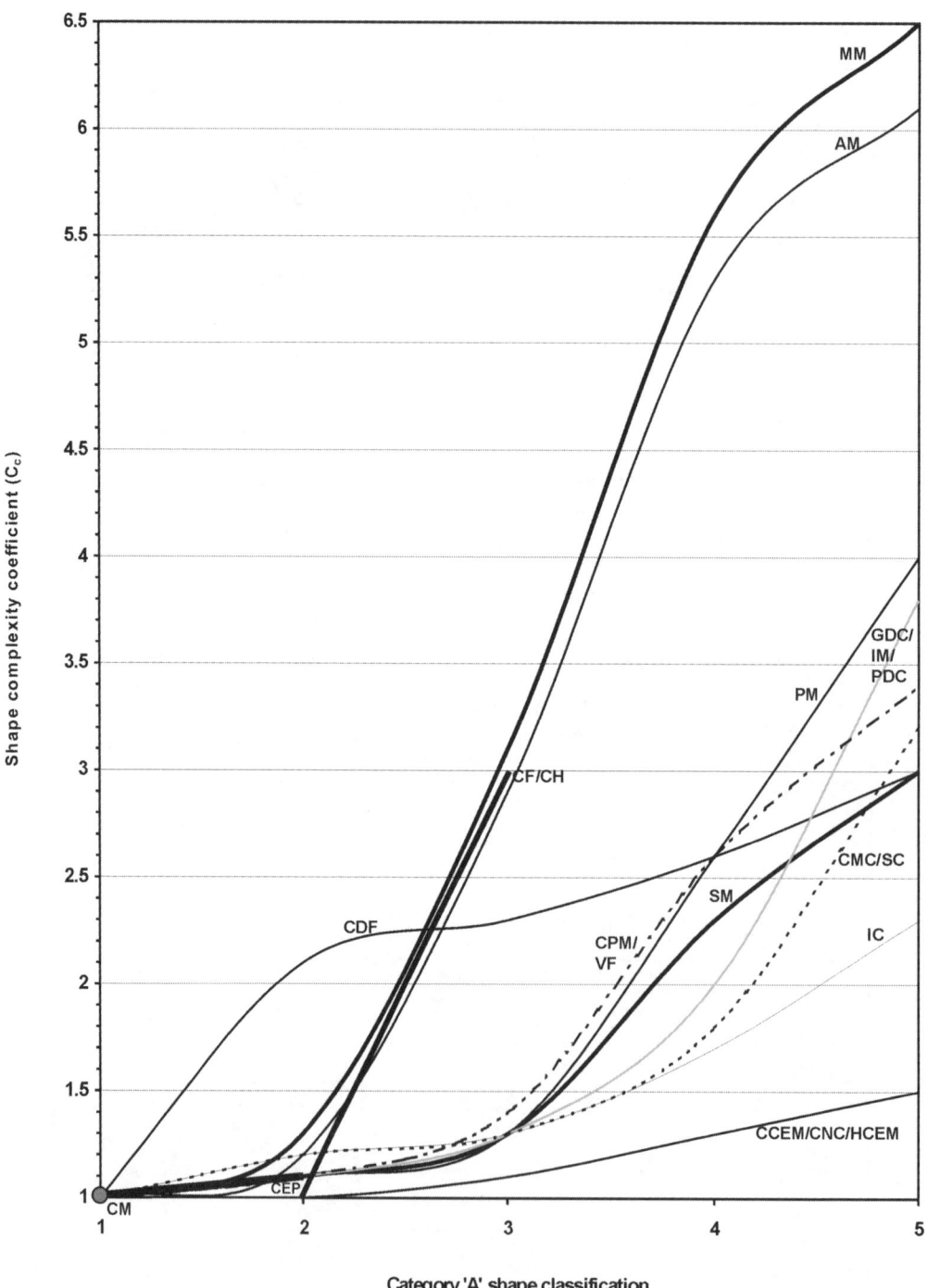

Fig. 3.10 Determination of shape complexity coefficient (C_c) – category 'A' shape classification.

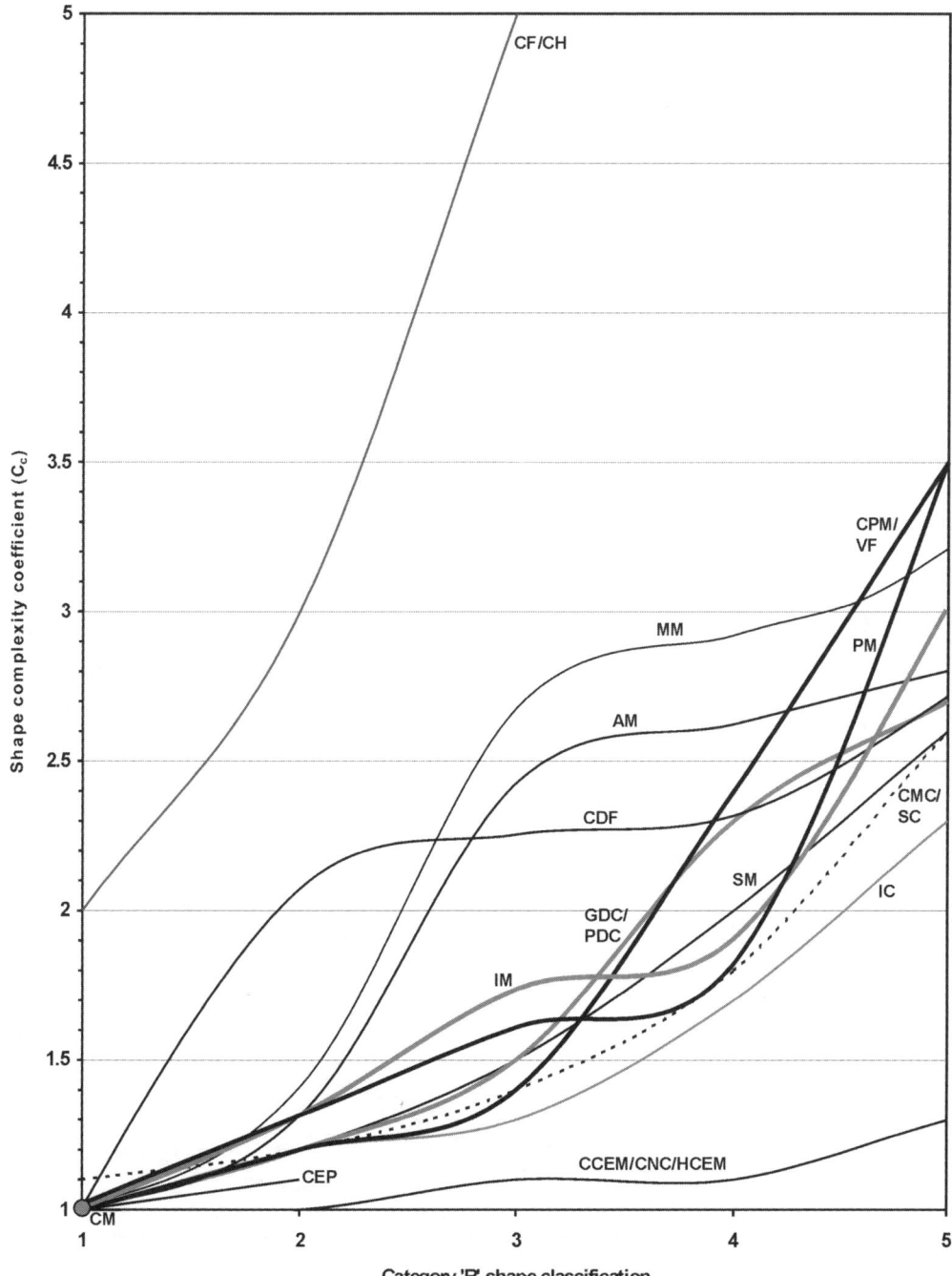

Fig. 3.11 Determination of shape complexity coefficient (C_c) – category 'B' shape classification.

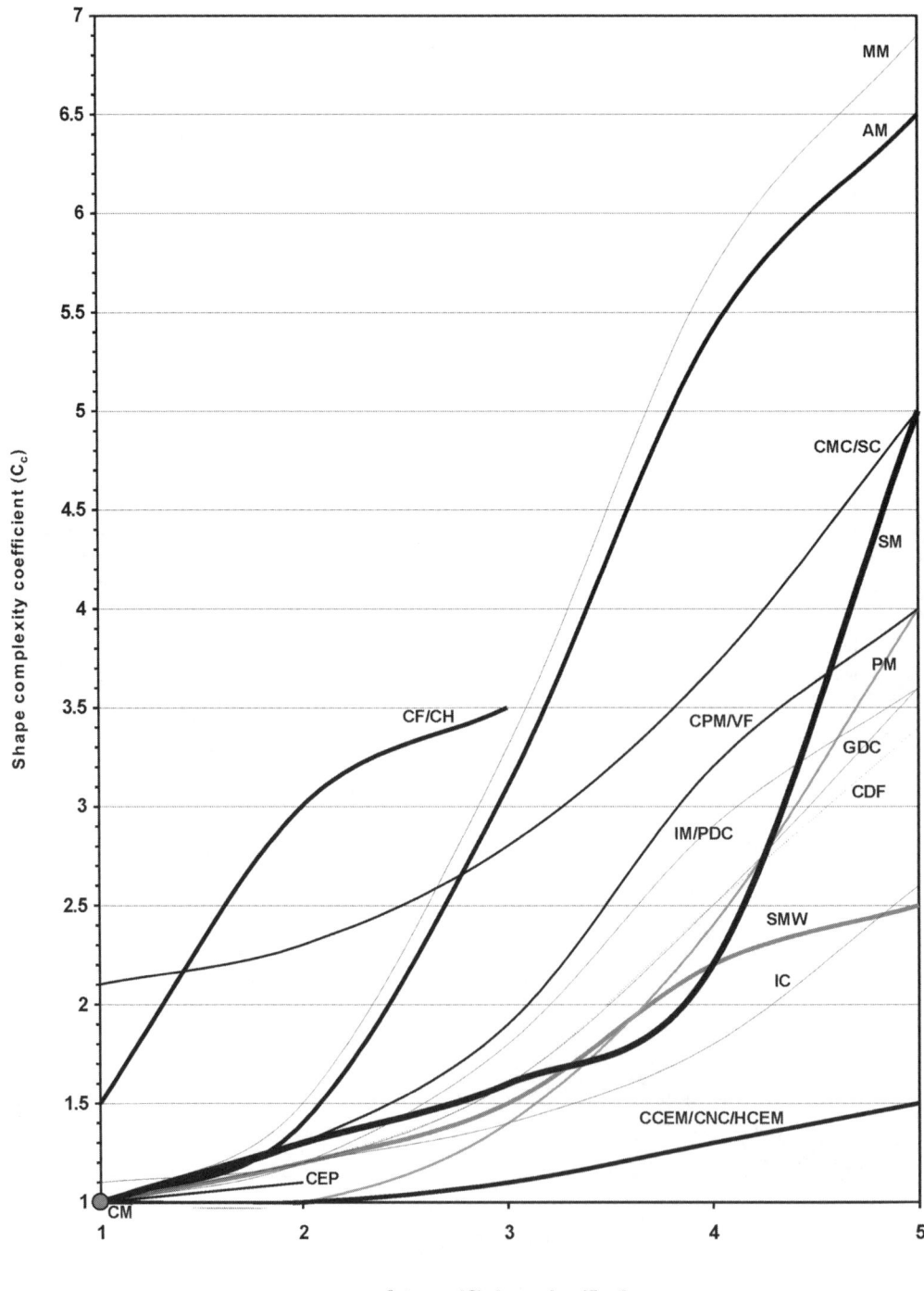

Category 'C' shape classification

Fig. 3.12 Determination of shape complexity coefficient (C_c) – category 'C' shape classification.

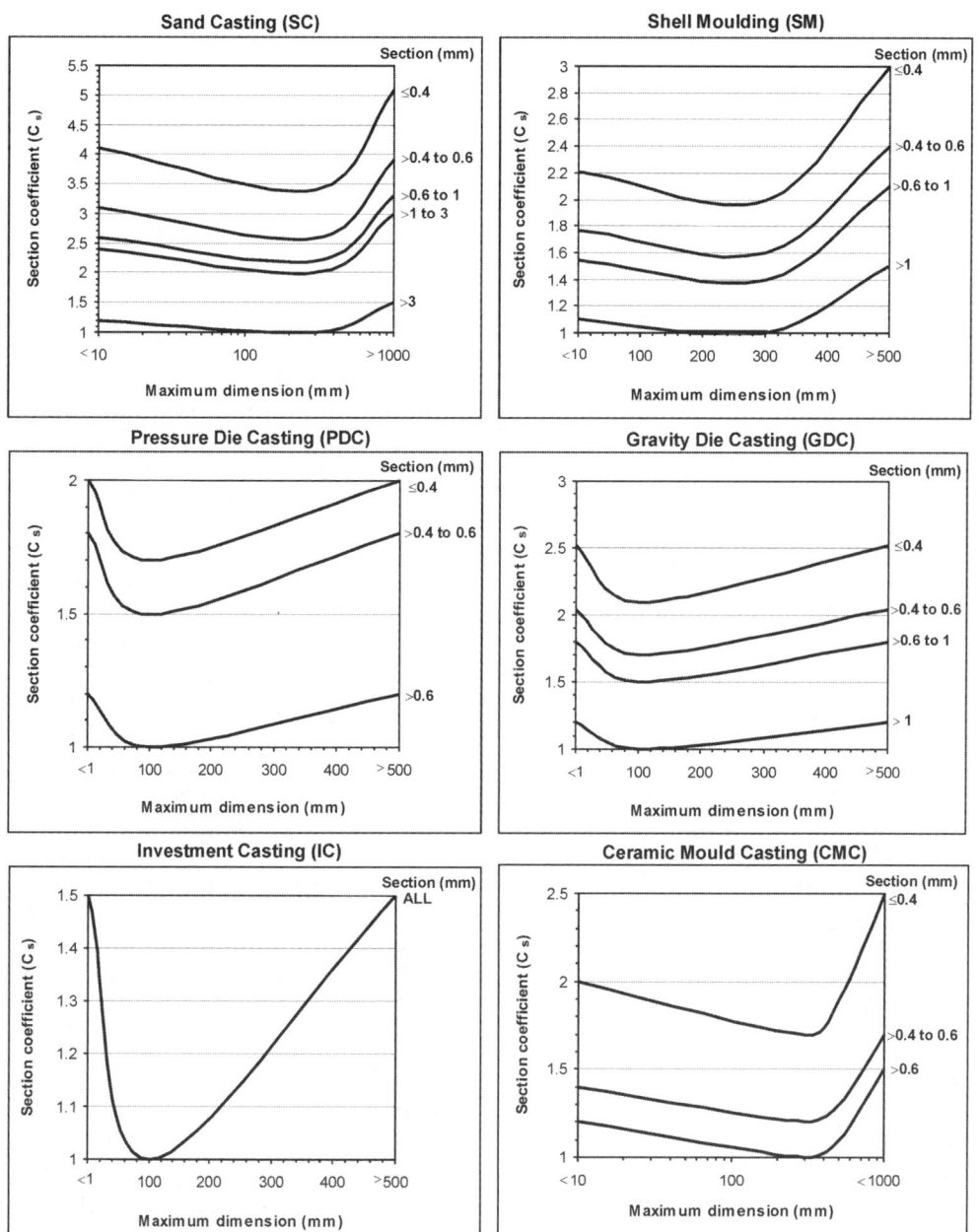

Fig. 3.13 Chart used for the determination of the section coefficient (C_s) for casting processes.

If the tolerance falls to the left of the thick gray line, a final machining, lapping, honing, polishing or grinding process is necessary to achieve the tolerance. This is already taken into account in the indices shown. Only the tightest tolerance required should be used, even if it only occurs on one plane. Included in the graph are separate lines for the number of

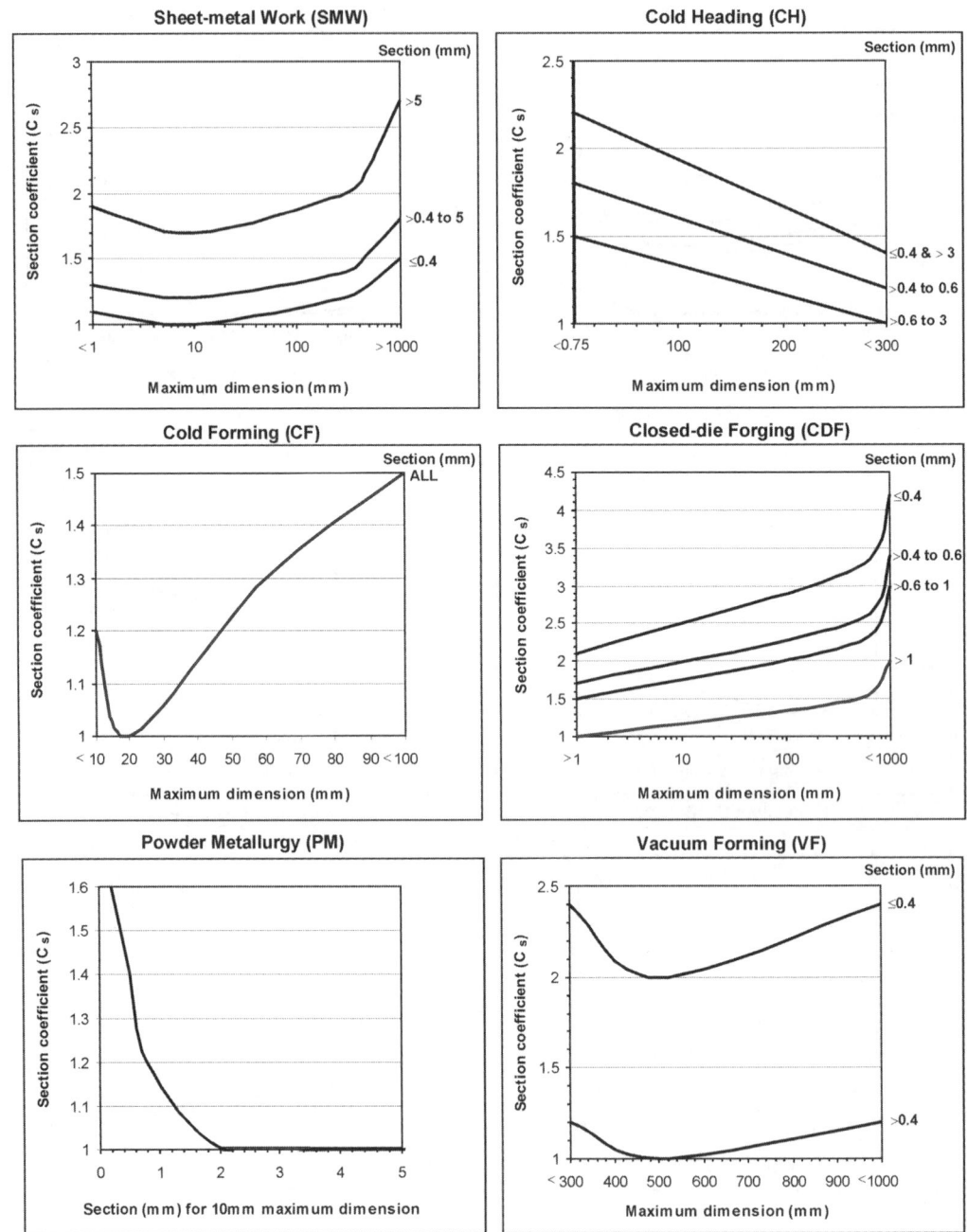

Fig. 3.14 Chart used for the determination of the section coefficient (C_s) for forming processes.

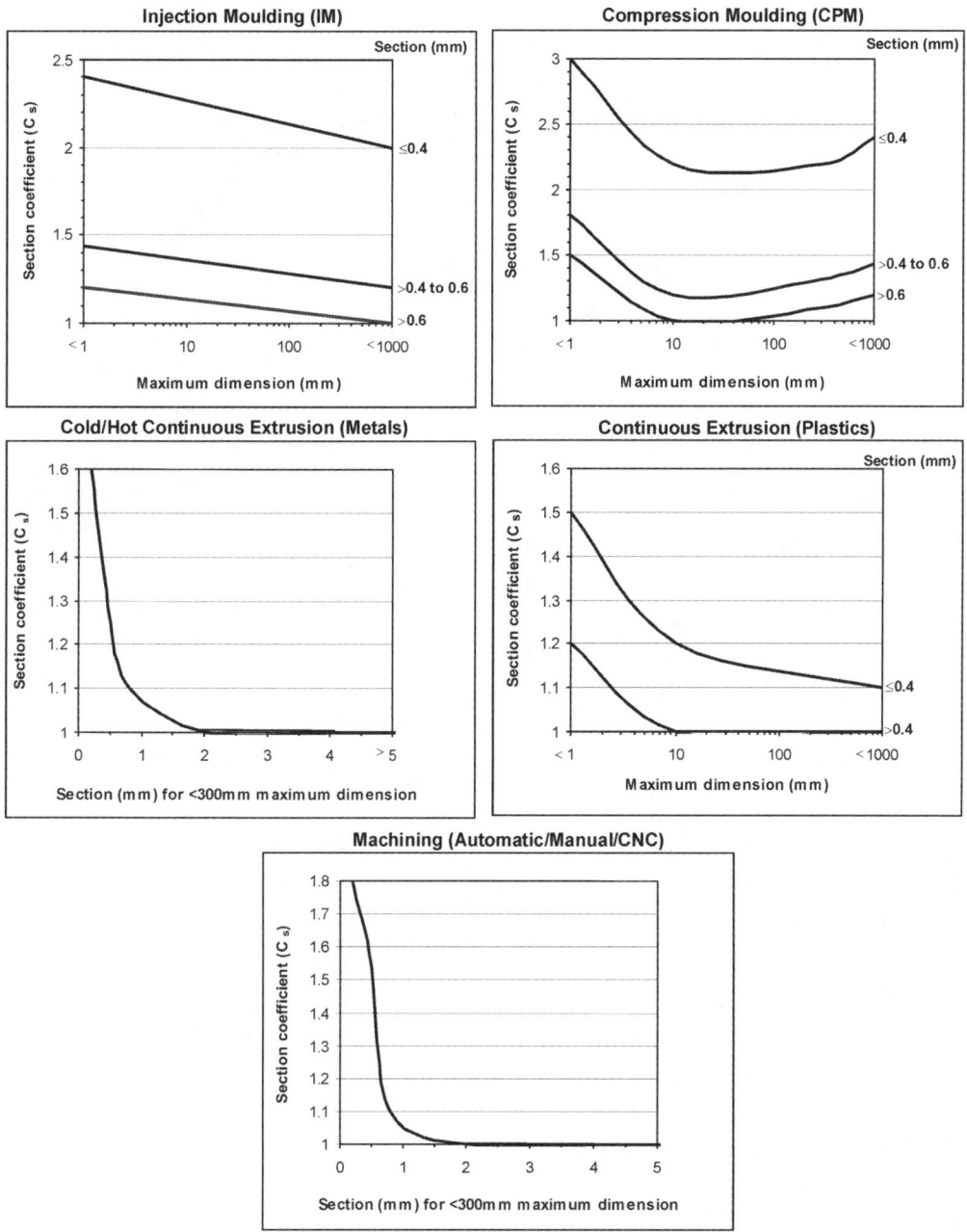

Fig. 3.15 Chart used for the determination of the section coefficient (C_s) for plastic molding, continuous extrusion and machining processes.

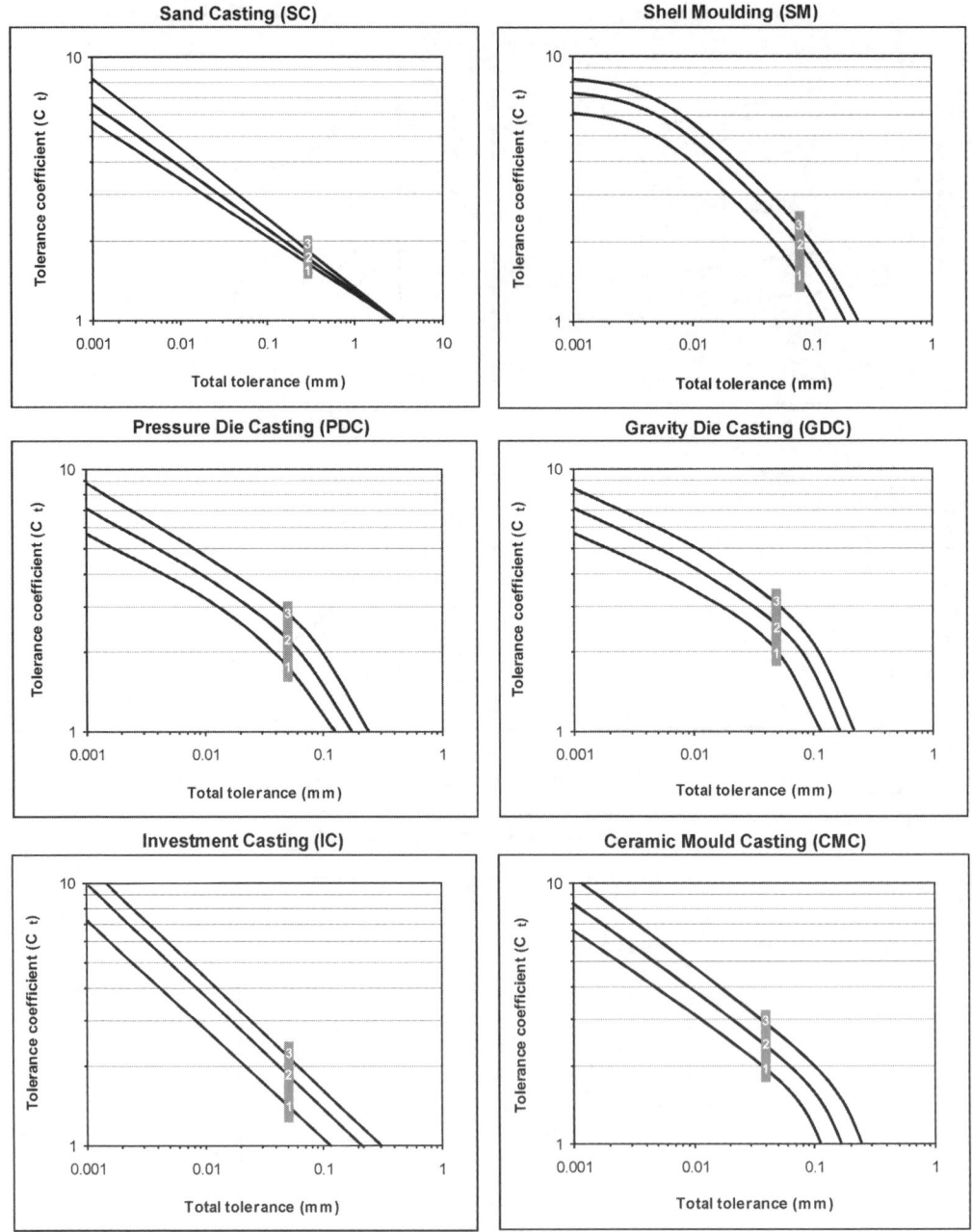

Fig. 3.16 Chart used for the determination of the tolerance coefficient (C_t) for casting processes.

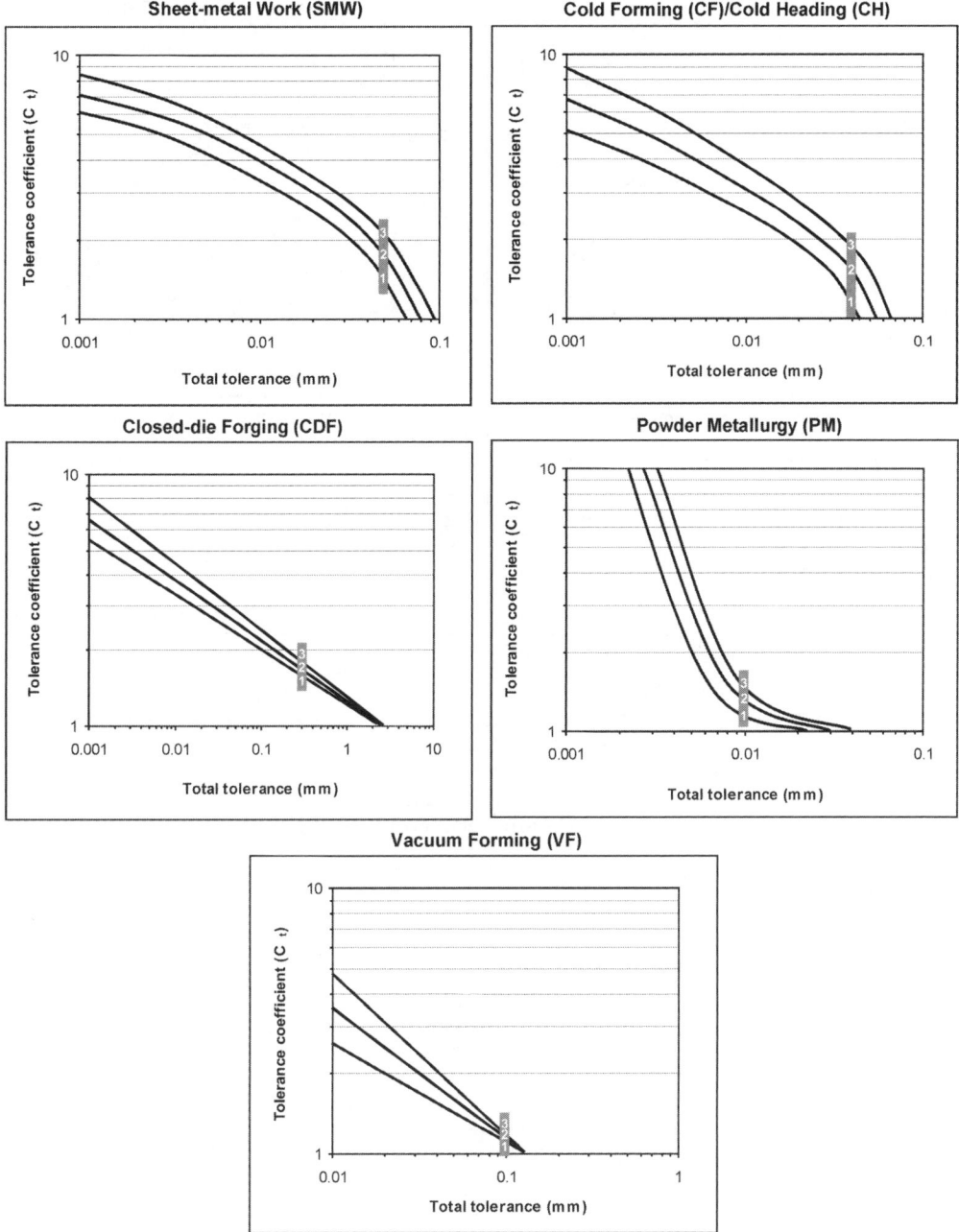

Fig. 3.17 Chart used for the determination of the tolerance coefficient (C_t) for forming processes.

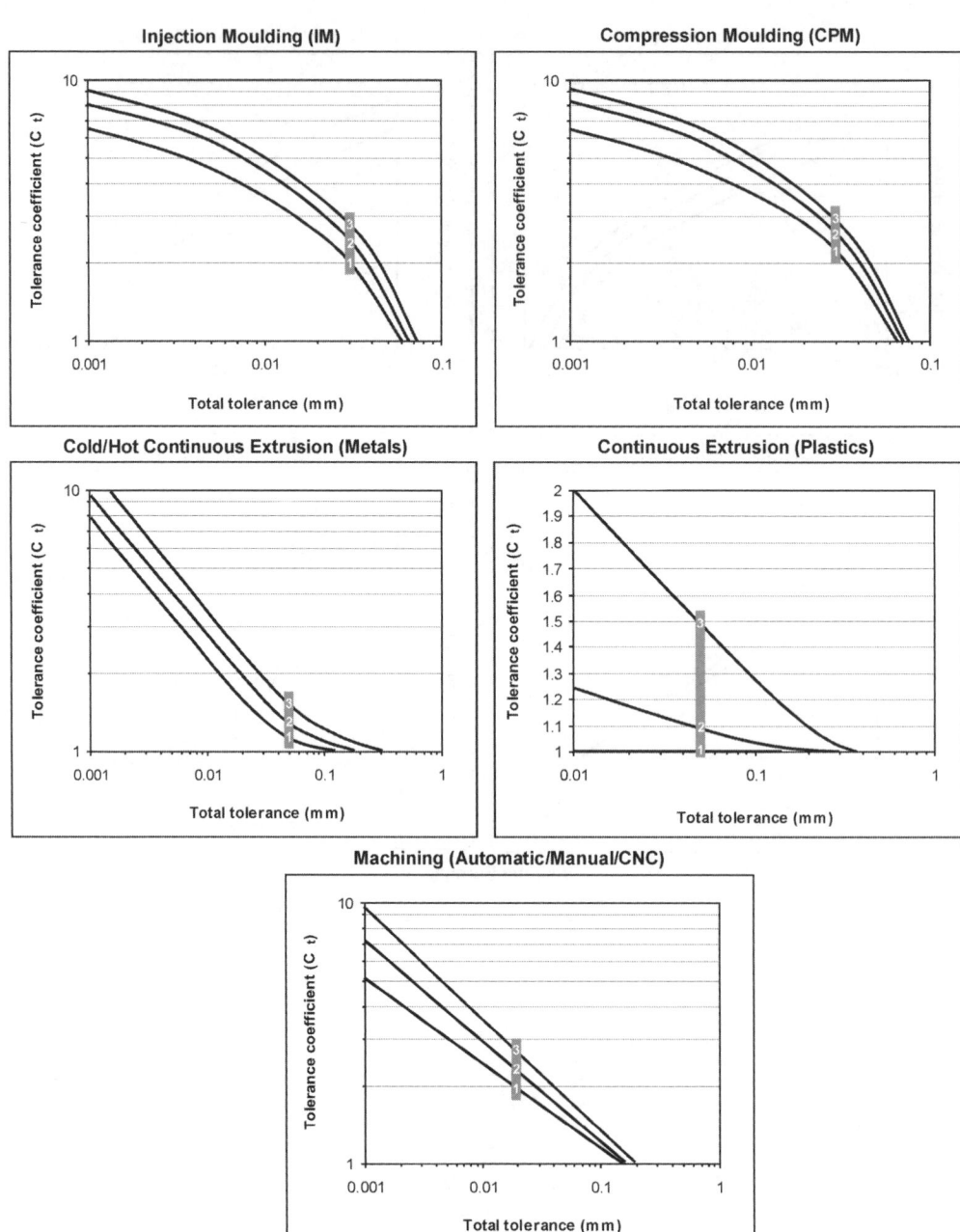

Fig. 3.18 Chart used for the determination of the tolerance coefficient (C_t) for plastic molding, continuous extrusion and machining processes.

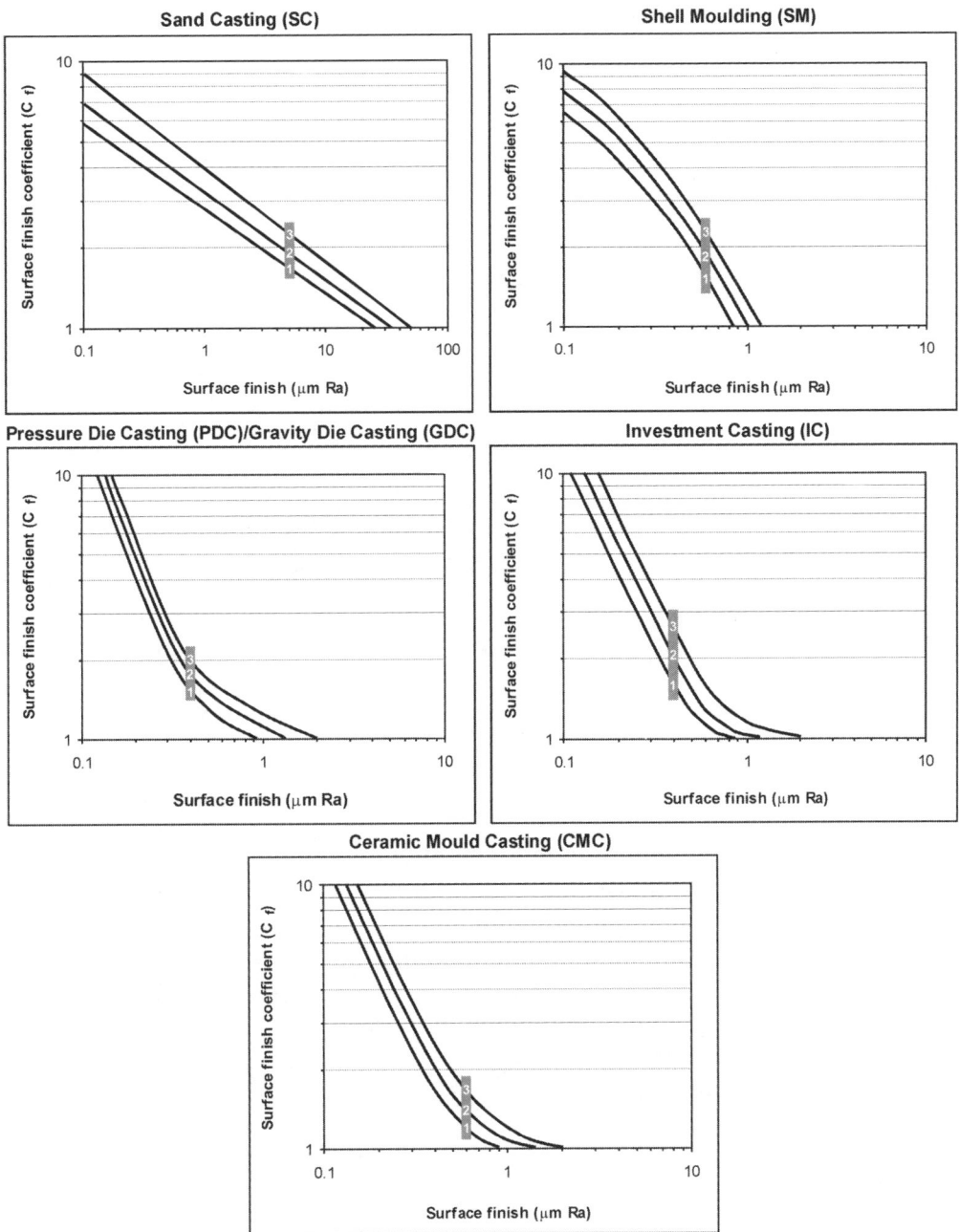

Fig. 3.19 Chart used for the determination of the surface finish coefficient (C_f) for casting processes.

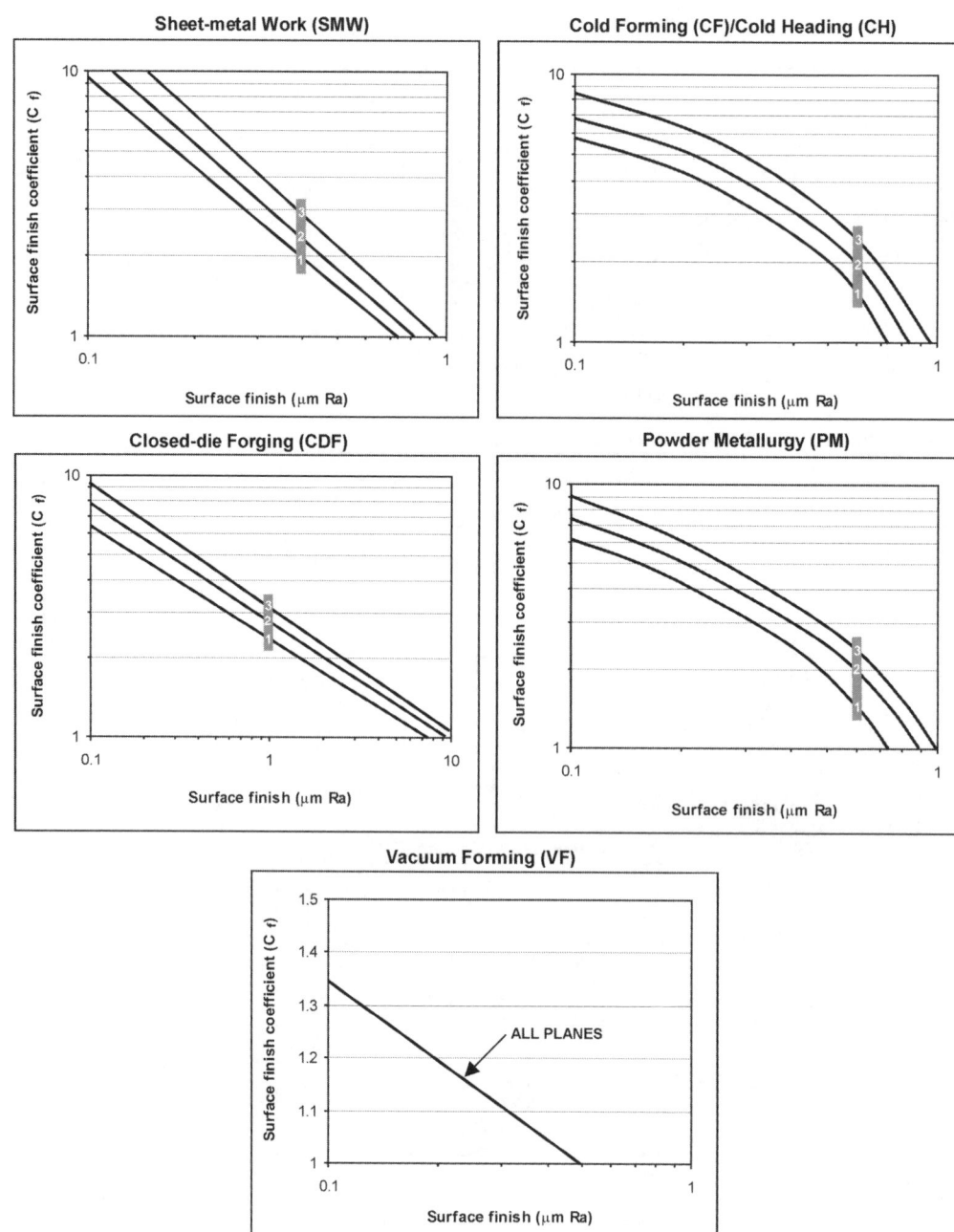

Fig. 3.20 Chart used for the determination of the surface finish coefficient (C_f) for forming processes.

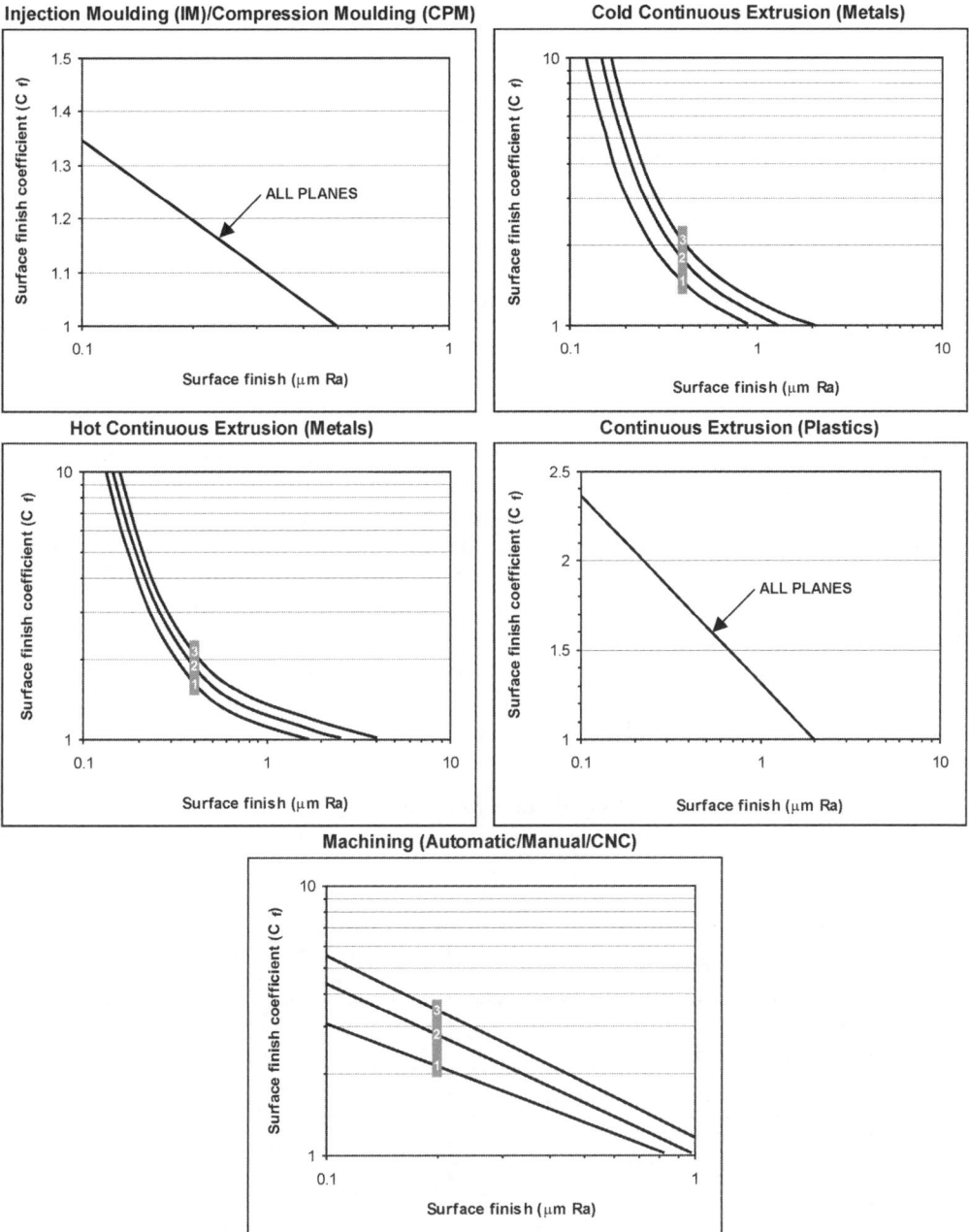

Fig. 3.21 Chart used for the determination of the surface finish coefficient (C_f) for plastic molding, continuous extrusion and machining processes.

orthogonal axes or planes (either 1, 2 or 3+), on which the critical tolerances lie, and which cannot be achieved from a single direction using the manufacturing process. Repeat the above process exactly for C_f using the graphs in Figures 3.19–3.21.

$C_{ft} = C_t$ or C_f, whichever gives the highest value.

Note that for Chemical Milling (CM2.5 and CM5), $C_{ft} = 1$, as the penalty is taken account of in the formulation of the basic processing cost, P_c.

3.2.4 Material cost (M_c)

The material cost, M_c, was defined in Equation (3.1) as the volume of raw material required to process the component multiplied by the cost of the material per unit volume in the required form, C_{mt}:

$$M_c = VC_{mt} \qquad [3.6]$$

Sample average values for C_{mt} for commonly used material classes can be found in Figure 3.22. Company specific data should be used wherever possible. In many situations the material cost can form a large proportion of the total component cost, therefore a consistent approach should be taken in the volume calculation if valid comparisons are to be produced. Note that the volume, V, in Equation (3.6) must be worked out in cubic millimeters (mm^3). Reference (1.39) has relative cost data for a number of material classes that can be used where specific data is not available.

Component manufacture may involve surface coating and/or heat treatments, and have some effect on manufacturing cost. Development of models for this aspect of component manufacturing cost can be found in reference (3.8).

The volume may be calculated in one of two ways:

1 Using the total volume – If the total volume of material required to produce the component is known (i.e. the volume including any processing waste), then this value is used for 'V' and the waste coefficient, W_c, is ignored.

Material	Material Cost (C_{mt}) (pence/mm^3)*
Cast Iron	0.00048
Low Carbon Steel	0.00041
Alloy Steel	0.00157
Stainless Steel	0.00206
Copper Alloy	0.00259
Aluminium Alloy	0.00083
Zinc Alloy	0.00124
Thermoplastic - nylon, PMMA	0.00065
- other (PVC, PE, PS)	0.00018
Thermoset	0.00035
Elastomer	0.00018

*Average cost per unit volume sourced in the UK, 2002.

Fig. 3.22 Sample material cost values per unit volume (C_{mt}) for commonly used material classes.

Shape Classification / Process	AM	CCEM	CDF	CEP	CF	CH	CMC	CNC	CPM	GDC	HCEM	IC	IM	MM	PDC	PM	SM	SC	SMW	VF
A1	1.6	1	1.1	1	1	1	1.1	1.6	1	1	1	1	1.1	1.6	1	1	1	1.1		1
A2	2	1	1.1	1.1	1	1	1.1	2	1.1	1.1	1	1	1.1	2	1.1	1	1.1	1.1		1.1
A3	2.5	1.5	1.2		1	1	1.2	2.5	1.1	1.1	1.5	1.1	1.1	2.5	1.1	1	1.1	1.2		1.1
A4	3	2	1.2				1.3	3	1.2	1.2	2	1.1	1.1	3	1.2	1	1.2	1.3		1.2
A5	4	3	1.3				1.4	4	1.3	1.3	3	1.2	1.2	4	1.3	1.2	1.3	1.4		1.3
B1	1.7	1	1.1	1	1	1	1.1	1.7	1	1	1	1	1.1	1.7	1	1	1	1.1		1
B2	2.2	1	1.1	1.1	1	1	1.1	2.2	1.1	1.1	1	1	1.1	2.2	1.1	1	1.1	1.1		1.1
B3	2.8	1.5	1.2		1	1	1.2	2.8	1.1	1.1	1.5	1.1	1.1	2.8	1.1	1	1.1	1.2		1.1
B4	4	2	1.2				1.3	4	1.1	1.2	2	1.1	1.1	4	1.2	1	1.2	1.3		1.1
B5	6	3	1.3				1.4	6	1.2	1.3	3	1.2	1.2	6	1.3	1.2	1.3	1.4		1.2
C1	1.8	1	1.1	1	1	1	1.1	1.8	1	1	1	1	1.1	1.8	1.1	1	1.1	1.1	1.2	1
C2	2.4	1	1.1	1.1	1	1	1.2	2.4	1.1	1.1	1	1	1.1	2.4	1.1	1	1.1	1.2	1.2	1.1
C3	4	2	1.1		1	1	1.3	4	1.1	1.1	2	1.1	1.1	4	1.1	1	1.1	1.3	1.4	1.1
C4	6	3	1.2				1.4	6	1.1	1.2	3	1.1	1.2	6	1.2	1	1.2	1.4	1.5	1.1
C5	8	4	1.3				1.6	8	1.2	1.3	4	1.2	1.3	8	1.3	1.2	1.3	1.6	1.6	1.2

Fig. 3.23 Waste coefficient (W_c) for the sample processes relative to shape classification category.

2 Using the final (finished) volume – If the amount of waste material is not known, then the final component volume may be used. In this case, use the waste coefficient, W_c, which takes into account the waste material consumed by a particular process. The formulation for 'V' for this method is:

$$V = V_f W_c \qquad [3.7]$$

where V_f is the finished volume of the component.

Waste coefficient, W_c, for the sample processes can be found in Figure 3.23, relative to shape classifications provided in Figure 3.9b. While in many cases the values quoted can be used with confidence, estimation of the input volume to the process is the approach preferred (method 1 above). In many applications, when calculating the volume of a component, it is not always necessary to go into great detail. Approximate methods are often satisfactory when comparing designs, and it can be helpful if a design is broken down into simple shape elements allowing the quick calculation of a volume. Before looking at the industrial applications of the design costing methodology it should be noted that material and process selection need to be considered together, they should not be viewed in isolation. The analysis presented here does not in any way take into account physical properties such as strength, weight, conductivity, etc.

Note that for Chemical Milling (CM2.5 and CM5), $W_c = 1$ as the penalty is taken account of in the formulation of the basic processing cost, P_c.

3.2.5 Model validation

In order to validate the approach, a number of companies were consulted, covering a wide range of manufacturing technology and products. Understandably, companies were often reluctant to discuss cost information, even admitting that they had no systematic process or structure to the way new jobs were priced, relying almost exclusively on the knowledge and expertise of one or two senior estimators. However, a number of companies were able to provide both estimated and actual cost data for a sufficient range of components to perform some meaningful validation.

Figure 3.24(a) illustrates the results of a validation exercise in a company producing plastic molded components. The analysis was performed on a number of products at random, and the estimated costs predicted by the evaluation, M_i, have been plotted against the actual manufacturing costs provided by the company. Figure 3.24(b) illustrates another plot, this time

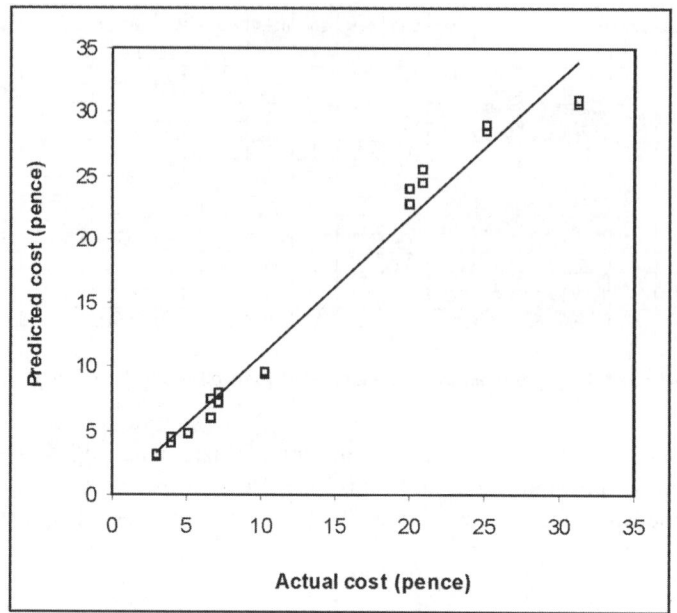

a) Cost Validation Results for Injection Moulded Components

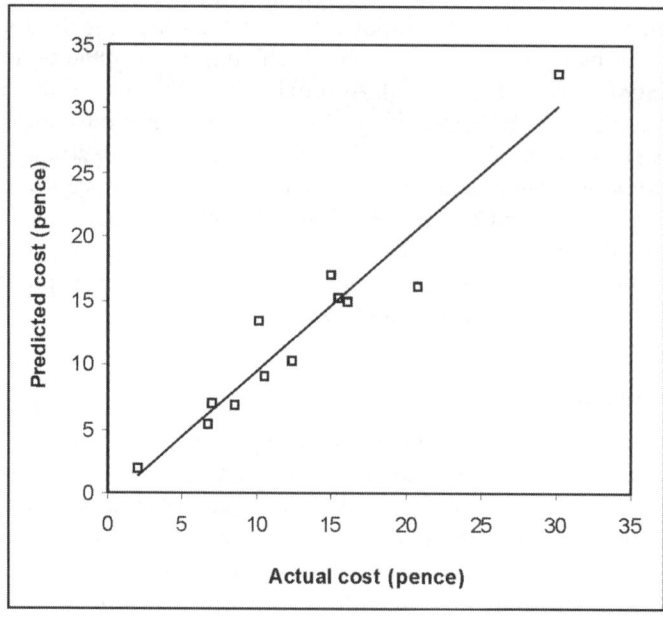

b) Cost Validation Results for Pressed Steel Components

Fig. 3.24 Costing methodology validation results.

from a company producing pressed sheet metal parts. Figure 3.25 illustrates some of the components included in the validation studies.

Validation exercises on a range of component types which was carried out by 22 individuals in industry (mechanical, electrical and manufacturing engineers) showed that the main variability encountered was in the calculation of component volume and in the assignment of the shape complexity index (3.9). While the determination of component volume is mechanistic, it is recognized that the determination of the most appropriate shape complexity classification requires judgmental skills and experience in the application of the methodology. These problems were largely eliminated when the analysis was carried out in a team environment, where highly consistent and reliable results were produced. In addition, training in the application of the methodology yields considerable improvements in the quality and consistency of the results produced proving capable of predicting the cost of manufacture of a component to within 16 per cent. Customizing the data to a particular business would significantly enhance the accuracy of the predicted costs obtained from the analysis.

3.2.6 Component costing case studies

One of the primary goals of the technique is to enable a product team to anticipate the cost of manufacture associated with alternative component design solutions, resulting from the activities of DFA. The technique is currently used to augment the DFA method exploited commercially by CSC Manufacturing in the form of DFA consulting projects and as part of the simultaneous engineering tools and techniques software 'TeamSET' (3.10). As mentioned earlier, one of the main objectives of DFA is the reduction of component numbers in a product to minimize assembly cost. This tends to generate product design solutions that contain fewer but sometimes more complex components embodying a number of functions. Such an approach is often criticized as being sub-optimal; therefore it is important to know the consequences of such moves on component manufacturing costs. Note that a blank component costing table is provided in Appendix C.

An illustration of how the design costing analysis can be used in DFA is given in Figures 3.26 and 3.27. Figure 3.26(a) shows the original design of a trim screw assembly and Figure 3.26(b), the replacement design. The DFA analyses can also be seen in Figure 3.26(a) and (b) respectively. Notice that these figures include data on manufacturing cost and provide the assembly sequence diagram for each design using the standard 'TeamSET' notation. A break-down of the cost analysis for the two components in the new design of the trim screw is given in Figure 3.27. Each component has been assigned a manufacturing index which is represen-tative of the cost in pence. Figure 3.26(c) provides a summary of the resulting measures of performance for each design. Again manufacturing cost values have been included. It can be seen from this that it is possible to fully assess the production cost consequences of each design in terms of both component manufacturer and assembly. Note that the total component manufacturing costs associated with the new design resulting from DFA are less than in the original: this turns out to be the case in many of the DFA studies examined to date by the authors.

A simple illustration of a case where the situation is not quite so clear cut is given in Figure 3.28. The DFA approach drives consideration of the assembly design proposal shown in design 'B'. An investigation of the two designs using the cost analysis suggests that from a component manufacturing point of view design 'A' represents a cost saving. In this example,

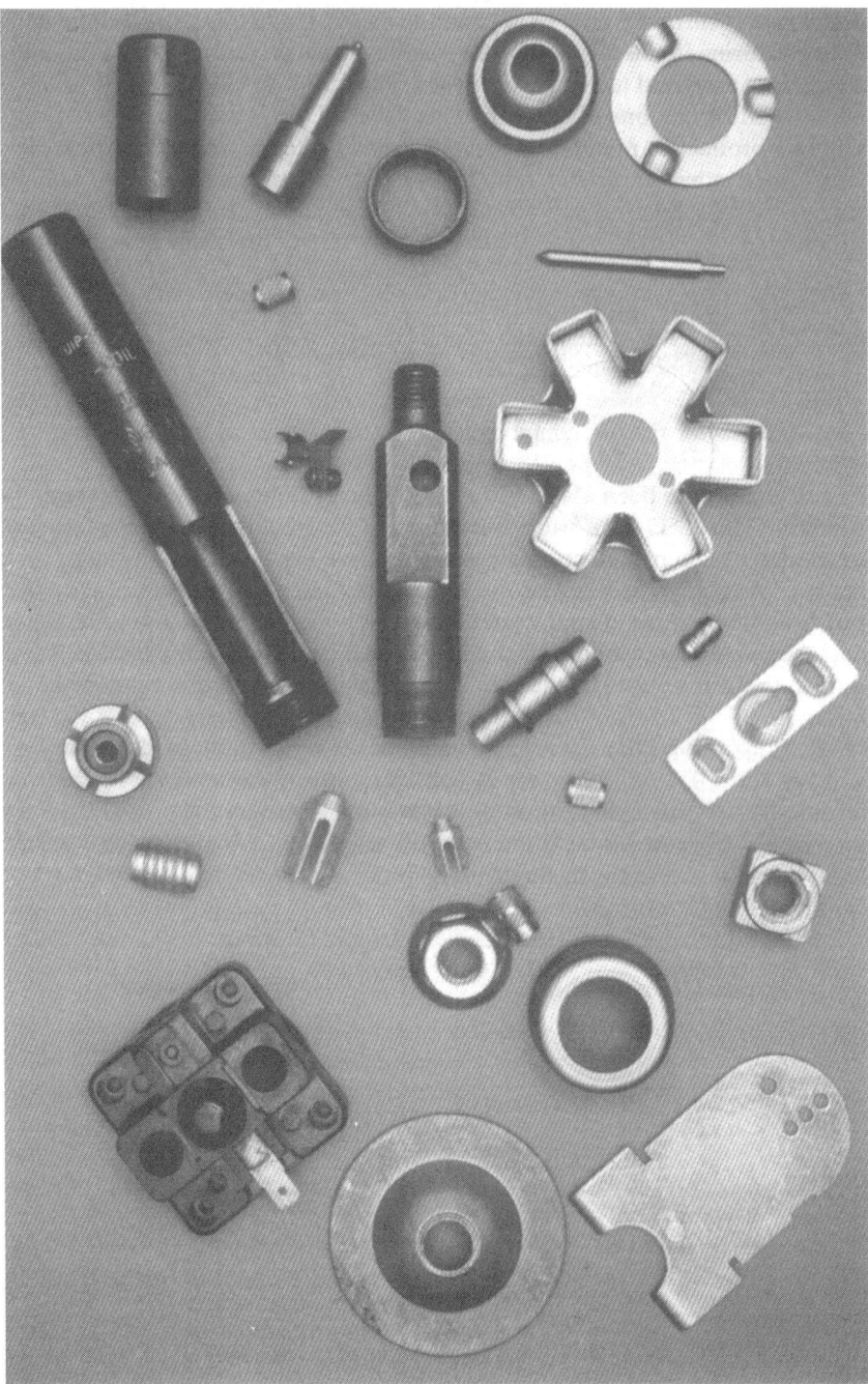

Fig. 3.25 Example components used in the validation exercises.

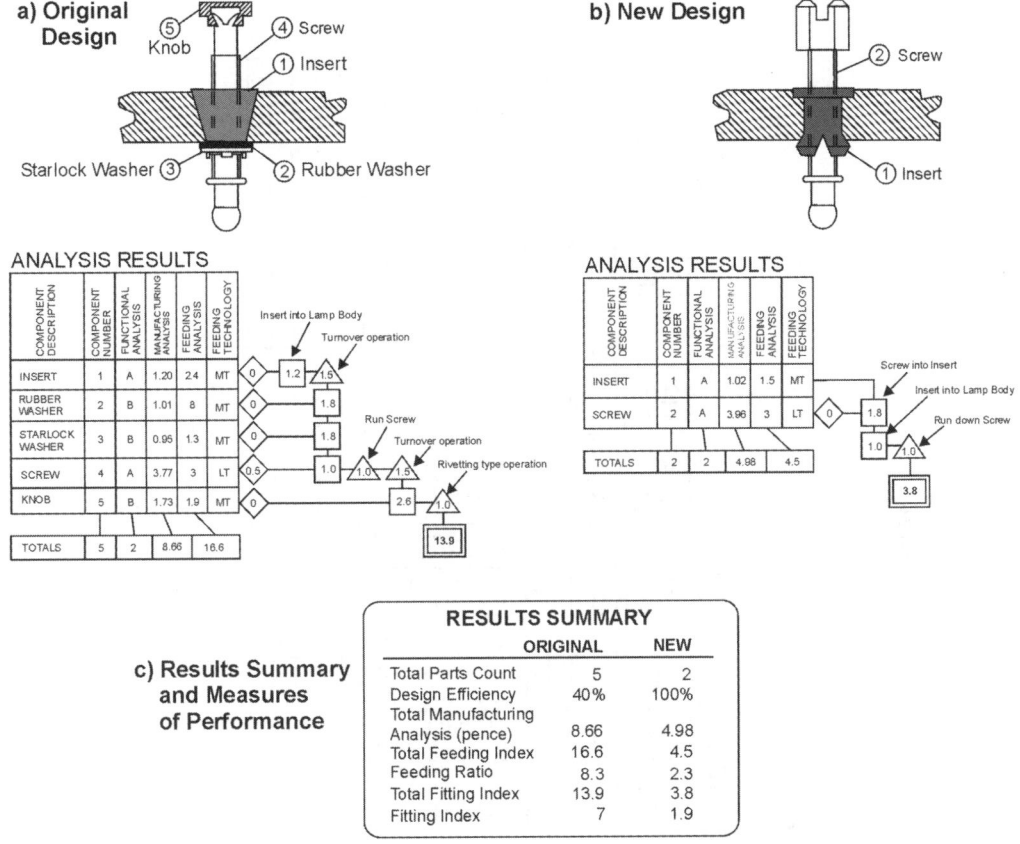

Fig. 3.26 Before and after analysis of a headlight trim screw design.

the same manufacturing process (automatic machining) is used for both pin designs, and the difference in cost results from the different initial material volume requirements. (The values of $P_c = 3$ and $R_c = 2.75$ are the same in each case.) Supplier cost data is used in the case of the standard clip fasteners. Hence, selection on the basis of cost demands a trade-off between assembly and manufacturing cost. Both design solutions are commonly seen in products from various business sectors and product groups.

Comparison of alternative processing routes is illustrated in Figure 3.29. The cold forming and automatic machining processing routes for the plug body design and production quantity requirements show significant manufacturing cost variations. The figure presents the detail of the cost analysis, giving the values obtained from P_c and the individual elements involved in the calculation of R_c, together in the table with details of the design. The benefits of the high material utilization associated with cold forming mean a large cost saving at the annual production quantity of one million components. (The input volume for the machined component is almost five times that required for cold forming.) However, as the annual production requirement reduces, the processing cost moves more in favor of machining, and at 30 000 *per annum* the sample data predicts little difference in cost between the two methods of production (see lower part of Figure 3.29).

PRODUCT NAME   TRIMSCREW
PRODUCT CODE / ID   NEW DESIGN
PRODUCT QUANTITY   1,000,000 pa

COMPONENT DETAILS					Mc = V x Cmt x [Wc]			Mc	Pc	Rc = Cc x Cmp x Cs x [Cft]									Rc	(B) (Pc x Rc)	(A)+(B) Mi (cost in pence)
PART No. ID	PART DESCRIPTION	MATERIAL	PRIMARY PROCESS	SHAPE COMP.	Volume (mm³)	Cmt	Wc	Mc	Pc	Cc	Cmp	Section (mm)	Cs	Toler- ance (mm)	Ct	Surface Finish (μm Ra)	Cf	Cft	Rc	(Pc x Rc)	Mi (cost in pence)
1	INSERT	THERMOPLASTIC	INJ. MOULDING	A1	626	0.00018	1.1	0.12	0.9	1	1	1	1	0.1	1	0.8	1	1	1	0.9	1.02
2	SCREW	LOW CARB. STEEL	AUTO. MACHINING	A2	4262	0.00041	N/A	1.75	1.2	1.2	1.4	2	1	0.1	1.1	0.8	1	1.1	1.84	2.21	3.96
				SHAPE CLASSIFICATION (FIGURE 3.9)		MATERIAL COST (FIGURE 3.22)	WASTE COEFFICIENT (FIGURE 3.23)			BASIC PROCESSING COST (FIGURES 3.1 & 3.3 RESPECTIVELY)	SHAPE COMPLEXITY COEFFICIENT (FIGURE 3.10)	MATERIAL TO PROCESS SUITABILITY COEFFICIENT (FIGURE 3.7)		MINIMUM SECTION COEFFICIENT (FIGURE 3.15)		TOLERANCE COEFFICIENT (FIGURE 3.18)		SURFACE FINISH COEFFICIENT (FIGURE 3.21)			TOTAL	4.98

Fig. 3.27 Cost analysis for the manufacture of the components in the new headlight trim screw design.

PRODUCT NAME: PIVOT PIN
PRODUCT CODE/ID: DESIGNS A AND B
PRODUCT QUANTITY: 30,000 pa

DESIGN A — CLIPS - A2, A3 — PIN - A1

DESIGN B — CLIP - B2 — PIN - B1

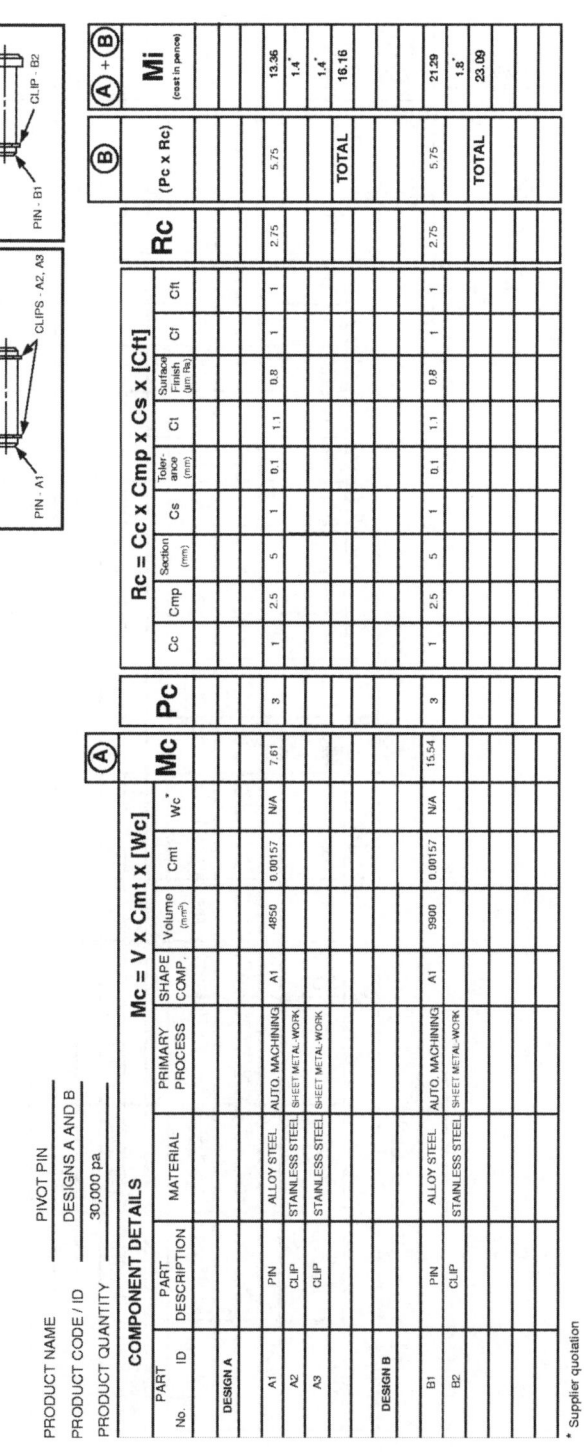

COMPONENT DETAILS

PART No. ID	PART DESCRIPTION	MATERIAL	PRIMARY PROCESS	SHAPE COMP.	Volume (mm³)	Cmt	Wc*	Mc* (A)	Pc	Cc	Cmp	Section (mm)	Cs	Tolerance (mm)	Ct	Surface Finish (μm Ra)	Cf	Cft	Rc	(B) (Pc × Rc)	Mi (A)+(B) (cost in pence)
								Mc = V x Cmt x [Wc]	**Pc**			Rc = Cc x Cmp x Cs x [Cft]							**Rc**	**B**	**Mi**
DESIGN A																					
A1	PIN	ALLOY STEEL	AUTO. MACHINING	A1	4850	0.00157	N/A	7.61	3	1	2.5	5	1	0.1	1.1	0.8	1	1	2.75	5.75	13.36
A2	CLIP	STAINLESS STEEL	SHEET METAL WORK																		1.4*
A3	CLIP	STAINLESS STEEL	SHEET METAL WORK																		1.4*
																				TOTAL	16.16
DESIGN B																					
B1	PIN	ALLOY STEEL	AUTO. MACHINING	A1	9900	0.00157	N/A	15.54	3	1	2.5	5	1	0.1	1.1	0.8	1	1	2.75	5.75	21.29
B2	CLIP	STAINLESS STEEL	SHEET METAL WORK																		1.8*
																				TOTAL	23.09

* Supplier quotation

Fig. 3.28 Estimated costs for alternative designs of pivot pin components.

PRODUCT NAME: PLUG BODY
PRODUCT CODE / ID: -
PRODUCT QUANTITY: 1,000,000 pa

PLUG BODY

						Mc = V x Cmt x [Wc]			Ⓐ Mc	Pc	Rc = Cc x Cmp x Cs x [Cft]									Rc	Ⓑ (Pc x Rc)	Ⓐ+Ⓑ Mi (cost in pence)
COMPONENT DETAILS											Cc	Cmp	Section (mm)	Cs	Tolerance (mm)	Ct	Surface Finish (µm Ra)	Cf	Cft			
PART No.	ID	PART DESCRIPTION	MATERIAL	PRIMARY PROCESS	SHAPE COMP.	Volume (mm³)	Cmt	Wc														
AT 1,000,000 pa																						
1		PLUG BODY	LOW CARB STEEL	AUTO. MACHINING	A2	22100	0.00041	N/A	9.06	1.2	1.2	1.4	1.5	1	0.2	1	1.6	1	1	1.68	2.02	11.08
		PLUG BODY	LOW CARB STEEL	COLD FORMING	A2	3860	0.00041	1	1.58	0.8	1	1.3	1.5	1.2	0.2	1	1.6	1	1	1.56	1.25	2.83
AT 30,000 pa																						
1		PLUG BODY	LOW CARB STEEL	AUTO. MACHINING	A2	22100	0.00041	N/A	9.06	3	1.2	1.4	1.5	1	0.2	1	1.5	1	1	1.68	5.04	14.10
		PLUG BODY	LOW CARB STEEL	COLD FORMING	A2	3860	0.00041	1	1.58	7	1	1.3	1.5	1.2	0.2	1	1.6	1	1	1.56	10.92	12.50

Fig. 3.29 Comparison of automatic machining and cold forming processes for the manufacture of a plug body.

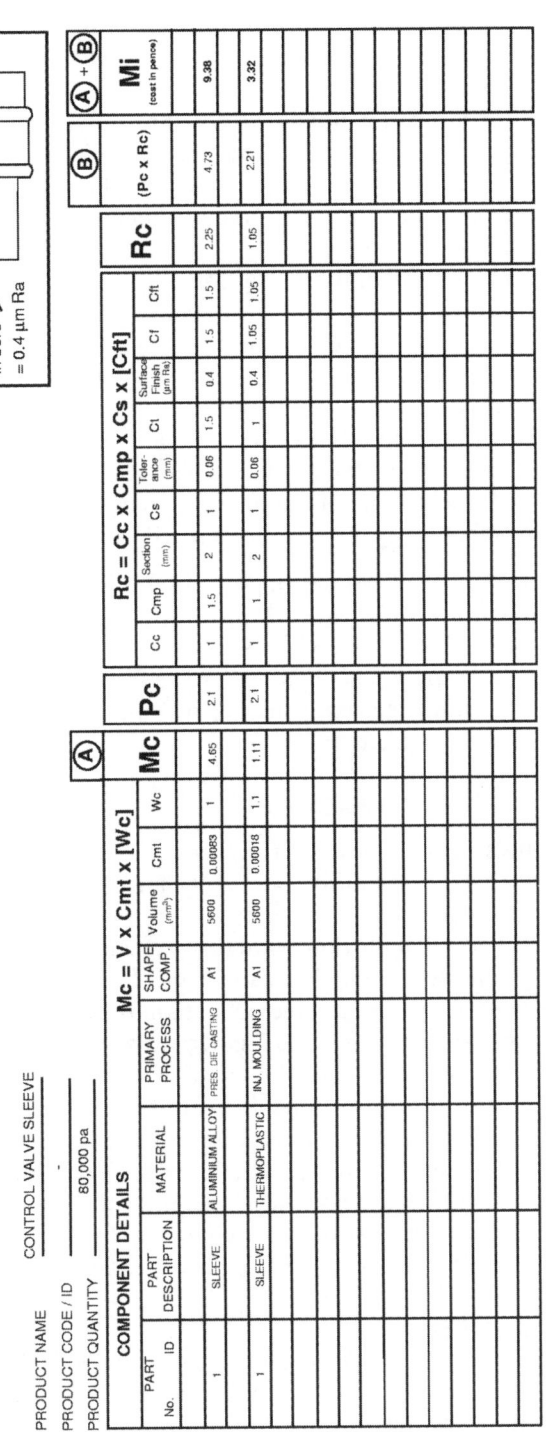

CONTROL VALVE SLEEVE

Surface finish in bore = 0.4 μm Ra

PRODUCT NAME: CONTROL VALVE SLEEVE
PRODUCT CODE / ID: –
PRODUCT QUANTITY: 80,000 pa

COMPONENT DETAILS

$$Mc = V \times Cmt \times [Wc] \quad \text{(A)}$$
$$Rc = Cc \times Cmp \times Cs \times [Cft] \quad \text{(B)}$$
$$Mi = \text{(A)} + \text{(B)}$$

PART No.	ID	PART DESCRIPTION	MATERIAL	PRIMARY PROCESS	SHAPE COMP.	Volume (mm³)	Cmt	Wc	**Mc** (A)	**Pc**	Cc	Cmp	Section (mm)	Cs	Tolerance (mm)	Ct	Surface Finish (μm Ra)	Cf	Cft	**Rc**	**(Pc × Rc)** (B)	**Mi** (A)+(B) (cost in pence)
1	1	SLEEVE	ALUMINIUM ALLOY	PRES. DIE CASTING	A1	5600	0.00083	1	4.65	2.1	1	1.5	2	1	0.06	1.5	0.4	1.5	1.5	2.25	4.73	9.38
1	1	SLEEVE	THERMOPLASTIC	INJ. MOULDING	A1	5600	0.00018	1.1	1.11	2.1	1	1	2	1	0.06	1	0.4	1.05	1.05	1.05	2.21	3.32

Fig. 3.30 Comparison of pressure die casting and injection molding processes for the manufacture of a critical surface finish.

A case where a material and process change eliminates the need for secondary processing is shown in Figure 3.30. An aluminum pressure die casting is initially considered for the sleeve shown, but secondary processing may be needed to ensure conformance to surface finish requirements as the achievement of $0.4\,\mu m$ Ra is on the boundary of technical feasibility. An optional design uses injection molded Polysulfone (PSU). The sample data does not differentiate between plastic injection molding and pressure die casting in terms of basic processing cost. The savings indicated by the cost analysis result from lower material costs, and surface finish capability of the injection molding process reflected in C_{ft} reduced from 1.5 to 1.05. Adopting injection molding here removes additional machining and minimizes the complexity of the manufacturing layout.

The technique can be helpful in producing cost estimates, where design solutions involve a significant amount of sub-contract work. The estimates produced provide support to the make versus buy analysis and the technique can be useful in calibrating supplier quotations. Variations of more than 30 per cent in quotations from sub-contractors against identical specifications are common across the range of manufacturing processes. This has been noted by a number of researchers (3.11). In this way benefits can be gained whether the methodology is applied as a stand-alone tool during product design/redesign or, more globally, as part of a company's integrated application of simultaneous engineering tools and techniques. The applications of the methodology may be summarized as:

- Determination of component cost in support of DFA
- Competitor analysis
- Assistance with make versus buy decisions
- Cost estimating in concept design with low levels of component detail
- Support for simultaneous engineering and team-work
- Training in design for manufacture

3.2.7 Bespoke costing development

Given the wide ranges of processes and their variants, and the problems of producing cost estimates from generic data that businesses can believe in, it is necessary to explore how we might go about getting companies to enter their own process knowledge into the component costing methodology presented previously. In this way, an organization can take ownership of the process costing knowledge and its maintenance. The development of this process of 'calibration' will enable a business to tune the data in the system to known component costs and take into account problems of varying material and processing cost in different parts of the world. However, the problem of enabling the user to add new processes to the methodology is rather complex. The main difficulties are associated with the need to collect and represent process knowledge for the calculation of basic processing cost, P_c and the design dependent relative cost coefficient, R_c. The adding of new material costs, M_c and any necessary waste coefficients, W_c is not considered to be a significant problem. The objective of these notes is to outline a process for the addition of costing information for new processes to the data-base to facilitate the costing of designs in early stages of the design process.

Basic processing cost (P_c)

In order to determine the basic processing cost, P_c of a simple or ideal design, it is necessary to understand the production factors on which it depends. These are equipment costs including installation, operating costs (labor, supervision and overheads), processing times, tooling costs and component demand. The above variables are taken account of in the calculation of, P_c, using the following equation:

$$P_c = AT + B/N \qquad [3.8]$$

where $A =$ total average cost of setting up and operating a specific process, including plant, labor, supervision and overheads, per second in the chosen country, $T =$ average time in seconds for the processing of an ideal design for the process, $B =$ average annual cost of tooling for processing an ideal component, including maintenance and $N =$ total production quantity *per annum*.

The above values of A, B and T are based on processing a simple or ideal design well suited to the process in terms of both material and geometry. They are experience-based quantities and should be based where possible on established standards and expertise in companies specializing in the process under consideration.

Addition of P_c data for a new manufacturing process

The steps proposed are as follows:

1 Select a manufacturing process that is currently covered in Part III of the book, and that is nearest to the new process to be added. For example, consider the adding of reaction injection molding to the system. A similar process would be injection molding.
2 Examine the data used for the quantity 'A' for the surrogate process and determine if this can be used as it stands. If not, decide by how much should it be changed. In the first instance, this should be checked with sources including published material (manufacturing books, manuals), manufacturing experts and specialist suppliers. The average operating cost of an injection molding facility in the UK is taken as 'X'. Obtain a view on a comparative value for reaction injection molding.
3 Repeat process in (2) above for the determination of the value for 'T'. The average operating time for a simple design of component in injection molding is 'Y'. Obtain a view on a comparative figure for reaction injection molding.
4 Repeat process in (2) above for the value of 'B'. The average total tooling cost for injection molding a simple design in the UK is 'Z'. Obtain a view on a comparative figure for reaction injection molding.
5 The values obtained above are used to calculate 'P_c' for a range of values for 'N'. Produce a plot for reaction injection molding and compare and discuss.
6 Add the pilot data to the system and represent as such. Add reaction injection molding data and make as pilot data only.
7 Check the data against known costs for components well suited to the process and calibrate accordingly. Calibrate the new process to known reaction injection molding case studies.
8 Add data to main database, coded as a new process. The user should be informed that reaction injection molding cost estimates are based on new data.
9 Once the data is proven, code as a standard process. The user should be informed as such.

Relative cost coefficient (R_c)

The relative cost coefficient is used to determine how much more expensive it will be to produce a component with more demanding characteristics than the 'ideal' design. In order to determine this quantity, it is necessary to consider the effects of design-dependent criteria.

These are material to process compatibility, geometry to process suitability, including complexity, size and thickness, tolerance requirements and surface finish.

The equation used for the calculation of the relative cost coefficient, R_c is as follows:

$$R_c = C_{mp}C_cC_sC_{ft} \qquad [3.9]$$

where C_{mp} = relative cost associated with material-process compatibility when compared with an ideal material process combination, C_c = relative cost associated with producing different geometries from the ideal for the process under consideration, C_s = relative cost associated with achieving a section reduction/thickness or size outside the envelope of the ideal design and $C_{ft} = C_t$ or C_f (whichever is greater),

where C_t = relative cost associated with obtaining a specified tolerance and C_f = relative cost associated with obtaining a specified surface finish.

The combination of C_t and C_f into C_{ft} is based on the assumption that when a fine surface finish is being produced, fine tolerances can be produced for the same amount of cost and vice versa. This method of comparison and accumulation of costs, based on the product of the above variables, is analogous to the methodologies used by experts in the field of cost engineering and cost estimating. Note that when engineering the above elements going to make up R_c (C_{mp}, C_c, C_s, C_t, and C_f), they need to take account of all the secondary processing required to achieve the specified reductions, tolerances and finish, etc. for the component design. For a simple or ideal design of component each of the relative cost coefficients is unity, but as a component design moves away from that state, the coefficients tend to increase in magnitude thus increasing the processing cost.

If R_c data is not obtained, any estimate produced will be a lower bound only, the quality of the estimate will improve as more information is represented regarding the effect of the design-dependent factors.

Addition of R_c data for a new manufacturing process

The steps proposed are as follows:

1 Following on from the procedure for P_c, select the process in the database nearest to the new process to be added. Again, let us consider adding of reaction injection molding to the system. A similar process would be injection molding.
2 Examine the data used for the variable 'C_{mp}' for the surrogate process and determine if this can be used directly as it stands, if not decide by how much should it be changed. In the first instance, this should be checked with sources including published material (manufacturing books and manuals), manufacturing experts and specialist suppliers. Obtain comparative figures for the materials to be considered and tabulate the values.
3 Repeat process in (2) above for the determination of the value for 'C_c'. Obtain comparative figures against the respective shape categories and plot or tabulate the results. Refer to shape classification charts.
4 Repeat process in (2) above for the value of 'C_s'. Obtain comparative figures taking account of section reductions/thickness and size. Tabulate results.
5 Repeat process in (2) above for the value of 'C_t'. Obtain comparative figures taking account of tolerance requirements. Tabulate results.
6 Repeat process in (2) above for the value of 'C_f'. Obtain comparative figures taking account of finish requirements. Tabulate results.

7 Add the pilot data to the system and represent as such. Add reaction injection molding data and make as pilot data only.
8 Check the data against known costs for components well suited to the process and calibrate accordingly. Calibrate new process to known case studies.
9 Add data to main database, coded as a new process. The user should be informed that cost estimates are based on new data. Once the data is proven, code as a standard process.

3.3 Manual assembly costing

Many designs are created with complex assembly sequences and fitting and handling operations involving complex and restricted motions, poor stability, difficult orientation and alignment and simultaneous multiple insertions. The overall effect is reduced assemblability resulting in increased assembly times and cost. To improve the assemblability of a design, each operation needs to be carefully considered. Since something like 50 per cent of all labor in the mechanical and electrical industries is involved in assembly, fitting and handling processes must be addressed in proactive DFA. The development of suitable insertion ports and handling features is essential for cost-effective assembly operations. In the present DFA methodologies, the fitting and handling analyses are used to evaluate insertion processes, which are ranked quantitatively depending on the difficulty of the task. The higher the score the more inefficient the assembly operation (fitting or handling) is assumed to be, with 1.5 as a threshold value for unacceptable design of an individual operation.

Although the fitting and handling analyses are both well-established means of assessing assembly operations, they are highly judgmental, require training in their application and have no provision for design advice. Within a more proactive DFA methodology, such information needs to be provided to the designer in a transparent and intuitive manner. The data should enable the designer to consider the effects of component and assembly port design on the cost of product assembly. The capability of individual handling and alignment features with respect to their ability to help (or hinder) the assembly operation needs to be presented to the designer.

In order to make progress it is intended to allow the designer to view the data at different levels of detail, ranging from direct comparisons to detailed elements of specific features. The use of different representations will be investigated to make the information user-friendly. One way in which this may be possible is to take a more fundamental approach to the cost/time of component fitting and to use graphical representations of the effects of design geometry to allow for easy comparison at a glance, rather than sorting through tabulated data. In the following, we shall consider manual assembly processes only. Manual assembly is by far the most common assembly system used in industry, in spite of the advent of more dedicated, automatic and programmable systems, mainly due to the inherent flexibility of manual or human operations.

3.3.1 Assembly costing model

The total cost of manual assembly comprises the sum of the total handling and fitting times multiplied by the labor rate (includes tooling cost, equipment costs, direct labor, supervision and overheads) in pence per second. The handling analysis below returns a Component Handling Index, H, related to a time factor for handling. Similarly, the time associated with the fitting of components in assemblies is represented by a Component Fitting Index,

F, through a straightforward analysis of a component's fitting characteristics. Therefore, the total cost of manual assembly, C_{ma}, is:

$$C_{ma} = C_1(F + H) \tag{3.10}$$

where H = component handling index (seconds), F = component fitting index (seconds) and C_1 = labor rate (pence per second).

In order to calculate an assembly cost, two further assumptions must be made:

1 The ideal assembly time for a combined handling and fitting operation is between 2 and 3 seconds. The exact time is dependent on factors such as workplace layout, environment and worker relaxation. In the case where an ideal time of 2 s is assumed, then the indices H and F can be taken as values in seconds. If 3 s is assumed, it is necessary to multiply the indices by 1.5 to obtain an estimate for the assembly time in seconds.
2 The labor rate, C_1, is calculated based on an annual salary of £15 000 (plus 40 per cent overheads for a worker in the UK), for a 250 working day year (5 day week minus statutory holidays), and a 7.5 h working day. This gives the cost of manual labor per second, $C_1 = 0.31$ pence.

Component handling analysis

The component handling index, H, can be defined as:

$$H = A_h + \left[\sum_{i=1}^{n} P_{o_i} + \sum_{i=1}^{n} P_{g_i}\right] \tag{3.11}$$

where A_h is the basic handling index for an ideal design using a given handling process, P_o is the orientation penalty for the component design and P_g is the general handling property penalty.

Basic Component Handling Indices (A_h) (select one only) The basic handling indices, A_h, for a selection of common component handling characteristics are shown in Figure 3.31.

Component Handling Characteristic	Index (A_h)
One hand only	1
Very small (aids/tools)	1.5
Large and/or heavy (two hands/tools)	1.5
Very large and/or very heavy (two people/hoist)	3

Fig. 3.31 Basic handling index (A_h) for a selection of component handling characteristics.

We now go on to consider the determination of the design-dependent, time-related, penalty indices associated with the geometry and characteristics of the design.

Orientation Penalties (P_o) *(select both from Figure 3.32)*

General Handling Penalties (P_g) *(select as appropriate)*   The general handling indices, P_g, for a selection of common situations are shown in Figure 3.33.

End to End Orientation (along axis of insertion)

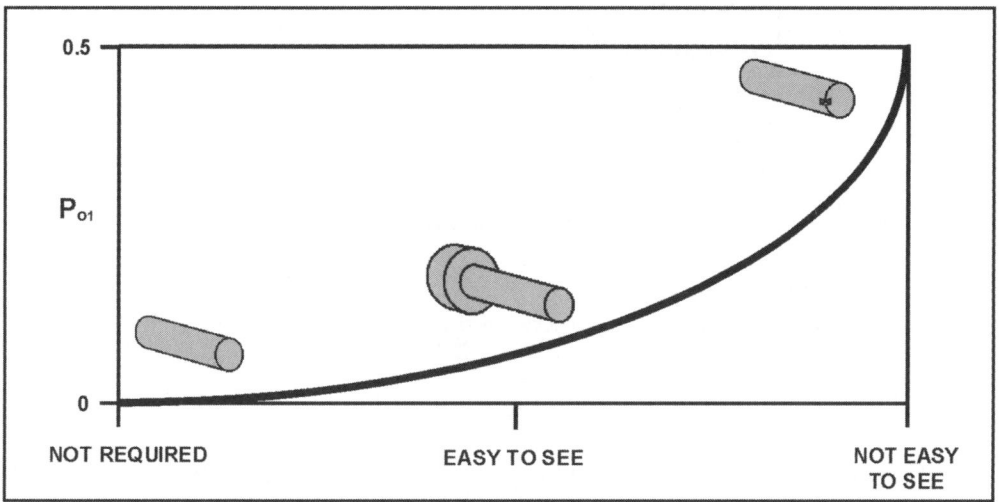

Rotational Orientation (about axis of insertion)

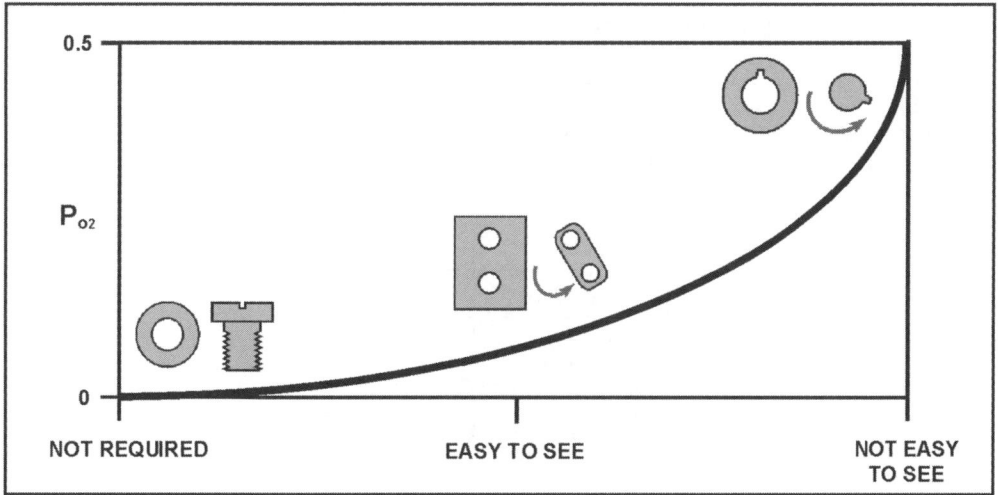

Fig. 3.32 Orientation penalties (P_o).

Component Handling Sensitivity	Index (P_g)
Fragile	0.4
Flexible	0.6
Adherent	0.5
Tangle/severely tangle	0.8/1.5
Severely nest	0.7
Sharp/abrasive	0.3
Hot/contaminated	0.5
Thin (gripping problem)	0.2
None of the above	0

Fig. 3.33 Handling sensitivity index (P_g) for a selection of component handling sensitivities.

Component fitting analysis

The Fitting Index, F, for a particular process in the sequence of assembly is defined as:

$$F = A_f + \left[\sum_{i=1}^{n} P_{f_i} + \sum_{i=1}^{n} P_{a_i} \right] \qquad [3.12]$$

where A_f is the basic fitting index for an ideal design using a given assembly process, P_f is the insertion penalty for the component design and P_a is the penalty for additional assembly processes on parts in place.

Basic Component Fitting Index (A_f) (select one only) Fitting indices for a selection of common processes is shown in Figure 3.34.

Assembly Process	Index (A_f)
Insertion only	1
Snap fit	1.3
Screw fastener	4
Rivet fastener	2.5
Clip fastener (plastic bending)	3
Placement in work holder (P_f and P_a usually not required)	1

Fig. 3.34 Fitting indices (A_f) for a number of common assembly processes.

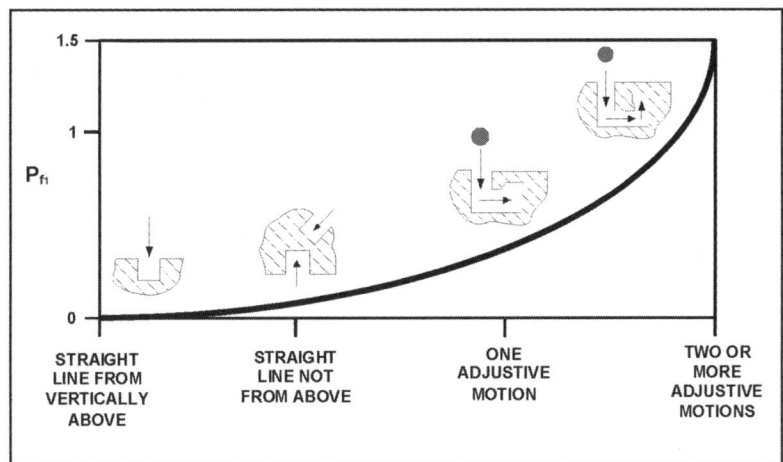

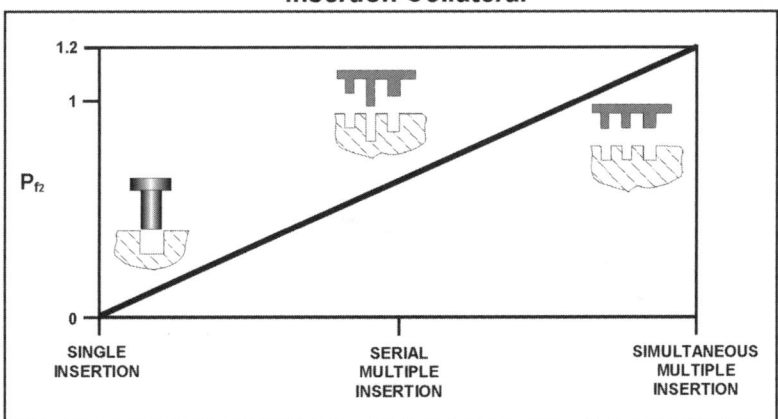

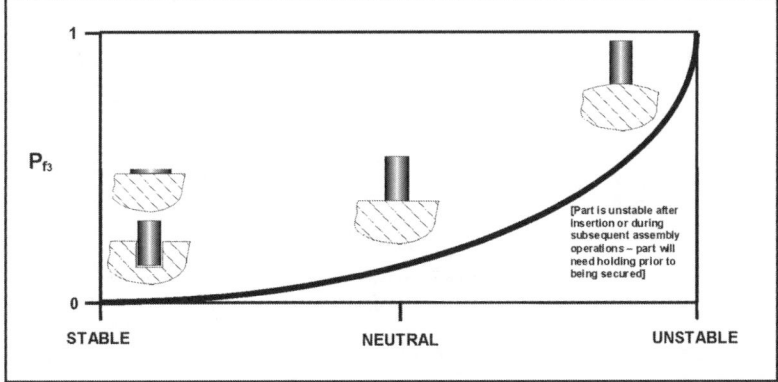

Fig. 3.35 (a) Component insertion penalties (P_{fi}).

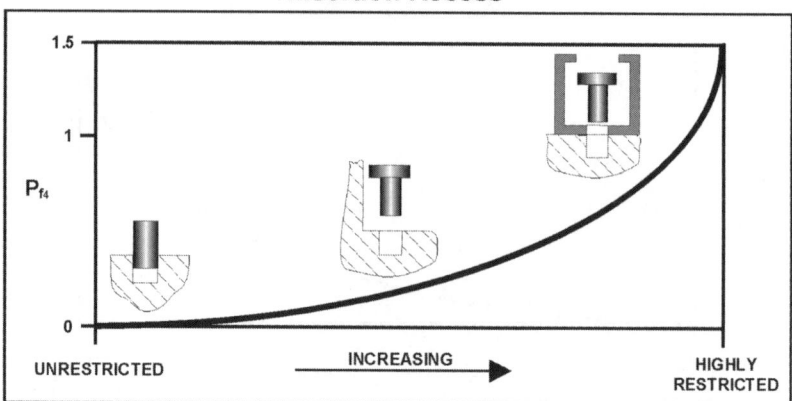

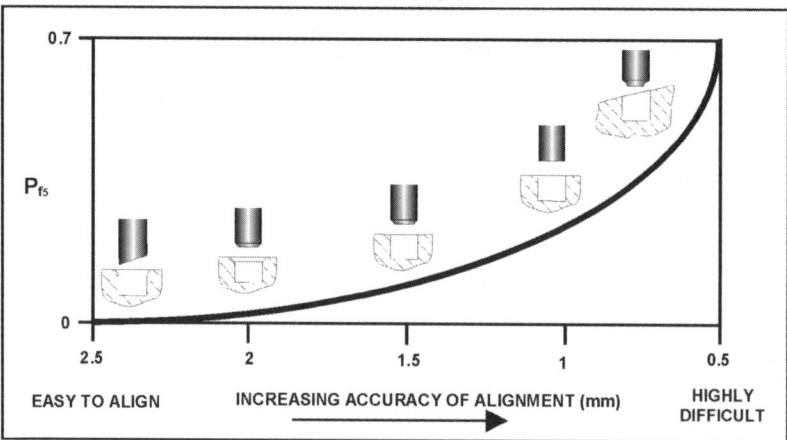

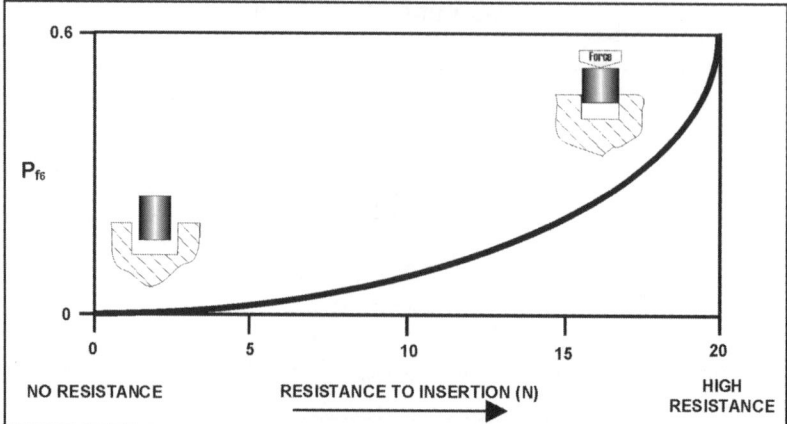

Fig. 3.35 (b) Component insertion penalties (P_{fi}) (contd).

Additional Assembly Process	Index (P_a)
Additional screw running	4
Later plastic deformation	3
Soldering/brazing/gas welding	6
Adhesive bonding/spot welding	5
Reorientation	1.5
Liquid/gas fill or empty	5
Set/test/measure/other [easy/difficult]	1.5/7.5

Fig. 3.36 Additional assembly index (P_a) for a number of common assembly processes.

We shall now go on to consider the determination of the design-dependent, time-related, penalty indices associated with the geometry and characteristics of assembly port designs.

Insertion Penalties (P_f) (select all from Figures 3.35 (a) and (b))

Additional Assembly Processes (P_a) (select as appropriate) Figure 3.36 gives the additional assembly process index, P_a, or a number of assembly processes carried out on components already positioned in the assembly build.

3.3.2 Assembly structure diagram

To facilitate a full assembly costing analysis, it is essential to understand the structure of the proposed product, and an assembly structure diagram is useful in this respect. Through its use, components in an assembly are logically mapped and in essence, represent the product's disassembly sequence from left to right. Constructing this diagram is seen as a beneficial exercise, as it supports an assembly perspective upon the design and compels the designer to focus on each component in the assembly. Included in the diagram are individual component costs, M_i, the manual assembly cost for each component, C_{ma}, total M_i and C_{ma} for the product and sub-assembly, and component identification labels. An example is shown in Figure 3.37. Note that the inclusion of M_i in the assembly structure diagram is optional. A blank manual assembly costing table is provided in Appendix D to support the costing methodology.

3.3.3 Manual assembly costing case studies

The design of a staple remover is shown in Figure 3.38. It is required to find the total production cost of the staple remover, including the cost of manufacturing the components. Figure 3.39 shows the assembly structure diagram for the assembled product, and the assembly costing analysis to support the assembly cost figures for each operation is provided in Figure 3.40. The component cost, M_i, has already been determined from the methodology provided earlier. The total cost of the stapler per unit is found to be approximately £0.23. Of course, a profit margin (typically between 15 and 25 per cent) would be added to this cost, as this is the cost to the company to manufacture and assemble the product. Packaging, shipping and storage could also increase this cost substantially.

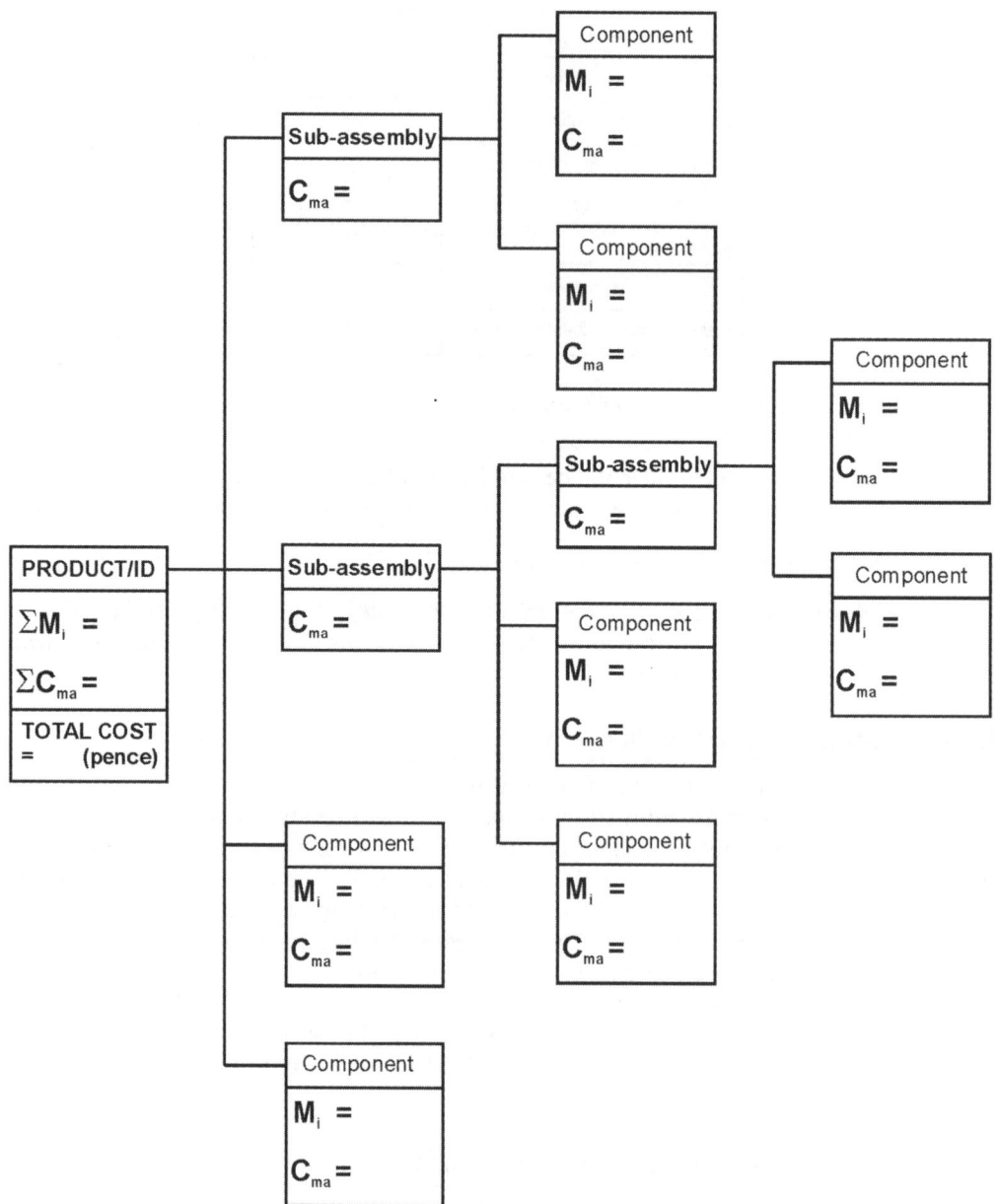

Fig. 3.37 Example format of an assembly structure diagram.

Figure 3.41 shows a possible redesign for the staple remover using just a single pressed sheet metal component made from spring steel. This design eliminates the need for any assembly operations, although the cost of the material and complexity of the press tooling will only be justified if a large volume is produced, in order to be competitive.

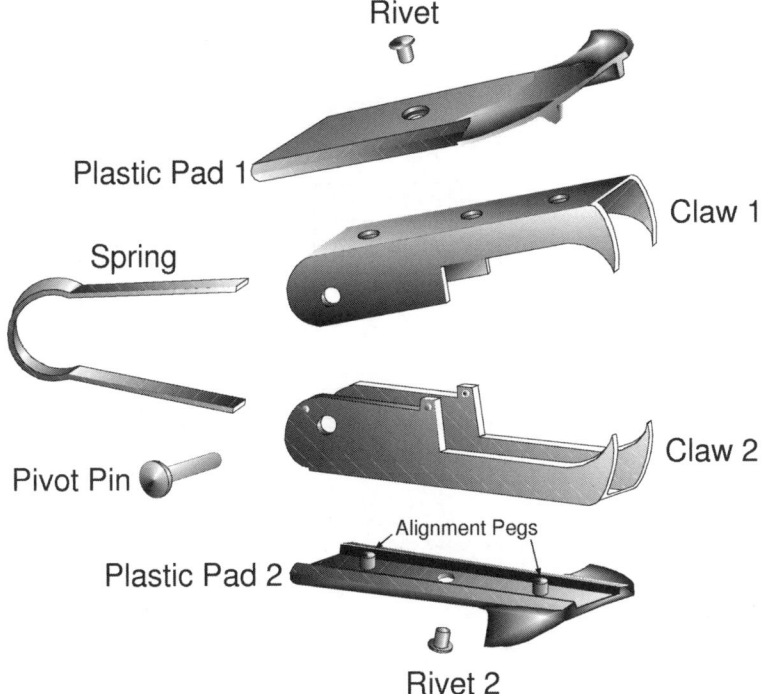

Rivet

Plastic Pad 1

Claw 1

Spring

Claw 2

Pivot Pin

Alignment Pegs

Plastic Pad 2

Rivet 2

Fig. 3.38 Staple remover exploded view.

This second case study is concerned with just the assembly time and cost of a 1.44 Mb floppy disk for use with a personal computer. Figure 3.42 shows the component parts. The results are shown together with the assembly structure diagram in Figure 3.43, and a full assembly costing analysis is provided in Figure 3.44. The total assembly cost of the floppy disk per unit is found to be approximately £0.16, and the calculated assembly time is approximately 52 s. Note that a relaxation is not taken into account and the fact that the operator would be working in a clean environment room wearing protective clothing to stop contamination.

The time contribution of each assembly operation compared to the overall assembly time is shown as a percentage in Figure 3.45. A Pareto Chart format is used with the greatest contribution to the total assembly time to the left. As highlighted, locating the front case sub-assembly on to the back case sub-assembly, whilst the spring is in position, is a difficult and time consuming assembly task. Screen placement and spring fitting are two other operations of a time consuming nature. In order to improve the assemblability of a particular concept design and reduce assembly costs, the use of the metrics in this manner can help identify potentially problematic areas and give guidance on redesign through reference to the charts provided.

3.4 Concluding remarks

The need to provide the concept design and development stages of the product introduction process with carefully structured knowledge about process characteristics and capabilities, together with cost estimating methods has been highlighted. PRIMAs of a standard form and

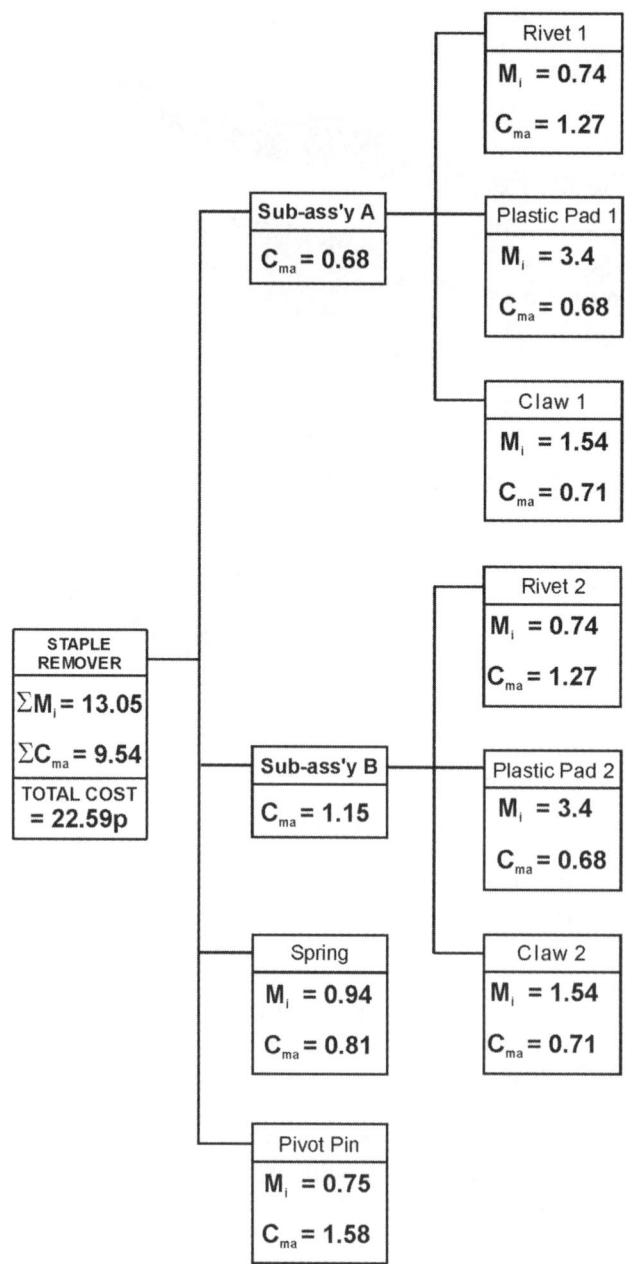

Fig. 3.39 Staple remover assembly structure.

similar level of detail for each manufacturing process have been presented. Simple methods based on economic and technical requirements have been designed to enable the user to focus attention on the most relevant process quickly. The application of the data provided in the PRIMAs as a means of selecting candidate processes has also been illustrated.

PRODUCT NAME Staple Remover

PRODUCT CODE / ID -

LABOUR RATE (Cl) (pence per second) 0.31

CI[H + F]

PART/SUB-ASSEMBLY No.	ID	PART DESCRIPTION	ASSEMBLY PROCESS	Ah	Po₁	Po₂	ΣPo	ΣPg	H	Af	Pf₁	Pf₂	Pf₃	Pf₄	Pf₅	Pf₆	ΣPf	ΣPa	F	Line Total	Cma (cost in pence)
1		RIVET 1	HAND./FIT.	1.5	0.1	0	0.1	0	1.6	2.5	0	0	0	0	0	0	0	0	2.5	4.1	1.27
2		PLASTIC PAD 1	HAND./FIT.	1	0.1	0.1	0.2	0	1.2	1	0	0	0	0	0	0	0	0	1	2.2	0.68
3		CLAW 1	HAND./FIT.	1	0.1	0.1	0.2	0	1.2	1	0	0	0	0	0.1	0	0.1	0	1.1	2.3	0.71
4		RIVET 2	HAND./FIT.	1.5	0.1	0	0.1	0	1.6	2.5	0	0	0	0	0	0	0	0	2.5	4.1	1.27
5		PLASTIC PAD 2	HAND./FIT.	1	0.1	0.1	0.2	0	1.2	1	0	0	0	0	0	0	0	0	1	2.2	0.68
6		CLAW 2	HAND./FIT.	1	0.1	0.1	0.2	0	1.2	1	0	0	0	0	0.1	0	0.1	0	1.1	2.3	0.71
SUB-ASS'Y A			HAND./FIT.	1	0.1	0.1	0.2	0	1.2	1	0	0	0	0	0	0	0	0	1	2.2	0.68
SUB-ASS'Y B			HAND./FIT.	1	0.1	0.1	0.2	0	1.2	1	0.1	0	1	0	0.3	0.1	1.5	0	2.5	3.7	1.15
7		SPRING	HAND./FIT.	1	0.1	0	0.1	0	1.1	1	0.1	0	0.2	0.2	0	0	0.5	0	1.5	2.6	0.81
8		PIVOT PIN	HAND./FIT.	1	0.1	0	0.1	0	1.1	1	0	0	0	0	0	0	0	3	4	5.1	1.58
																				30.8s	TOTAL = 9.54p

Fig. 3.40 Staple remover assembly costing analysis.

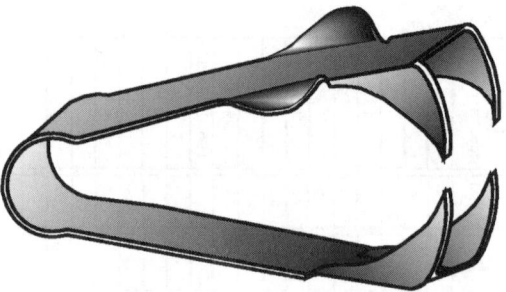

Fig. 3.41 Staple remover redesign.

A method for costing of designs, that can be used from concept to detail, has been introduced. The novelty of the approach is the calculation of processing costs, based on the notion of design-specific relative cost coefficients giving costs for processing idealized designs. Results of validation trails have indicated that the cost analysis can be used to predict component manufacturing costs, across a number of processes, to within 16 per cent of actual values, using average process and material cost data. The performance of the analysis may be much improved through the use of company specific data.

To support assembly-orientated design, it is essential to understand the cost implications of the components designed on the assembly systems used. Various component features and operations are known to exhibit higher assembly times than alternative combinations, and this provides a basis for relatively comparing a number of concepts and calculating the assembly time, and therefore the cost. The case studies have demonstrated the use of the methodology for manual assembly.

The use of the PRIMAs and the costing analyses with DFA provides a more holistic means of evaluating product designs and generating improved design solutions. In this way, the wider application of DFA in industry is encouraged. In addition, the approach presented provides for the carrying out of structured competitor analysis and yields a means for investigating make versus buy decisions. There are opportunities for the development of computer software to enhance the application of the process data and costing analysis. Potential benefits worth noting in this connection include: removal of error prone manual calculation and reference to maps and tables; consistency of results and standardized presentation; adherence to procedure; time saving; ease of editing and 'what if' exploration; people's operation and improved version control.

Integration of computer-based process selection with other concurrent engineering software tools, such as DFA, also offers potential benefits. The machine facilitates improved management of information flow between the applications, and provides for common data entry, a shared database, reuse and control of data and traceability of decisions. The CAD workstation provides additional scope for the application and integration of simultaneous engineering software tools within the design process. Design information from application of the tools supplies useful input to the product modeling process.

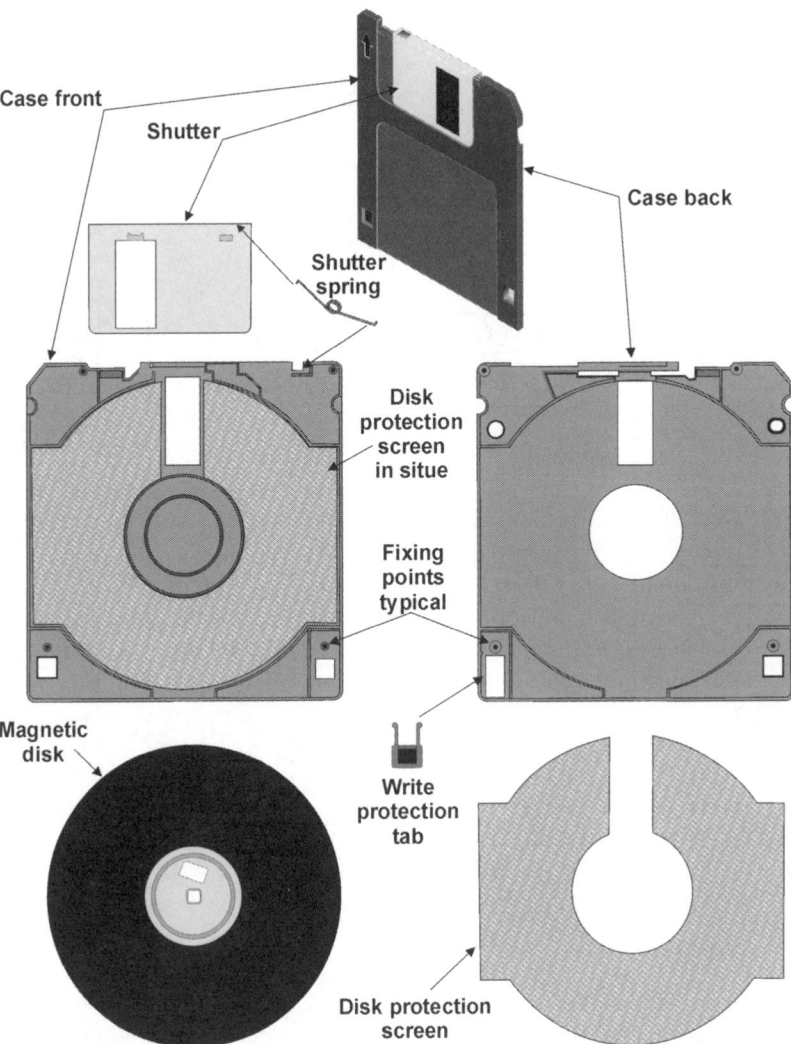

Case front

Shutter

Case back

Shutter spring

Disk protection screen in situe

Fixing points typical

Magnetic disk

Write protection tab

Disk protection screen

Fig. 3.42 Floppy disk component parts.

The development of new and advanced materials and the continuous search for improved capability and lower processing costs means that process development is an important research issue in manufacturing engineering circles. Consequently, the process selection problem is something of a moving target. PRIMA development for standard processes is not currently included, and catering for new processes as they emerge is an activity where research effort is being placed. Also, feedback from users applying the work on new product

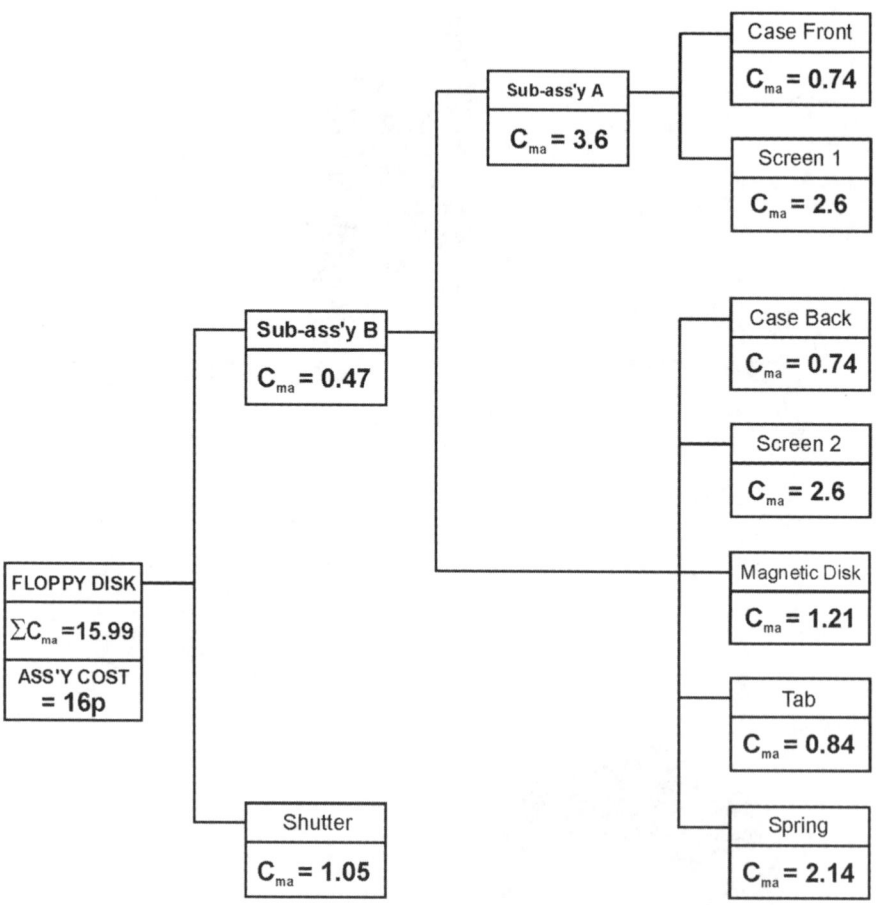

Fig. 3.43 Assembly structure for a floppy disk.

development projects, including views on what additional data they would like to see included in the PRIMAs will provide a useful source of information for PRIMA development. Similarly, user experience is being collected associated with application of the design costing analysis. Investigating the employment of business specific data in place of that provided for the sample set of processes included, and the incorporation by companies of data on methods not in the set are other areas of research. In this way, much more will be understood about ways of improving the analysis and its data, and the confidence that can be put on the resulting cost estimates.

Before leaving the topic of design costing it is worth saying that when costing designs, the costs of non-conformance must always be considered. There is little point in saving a few pence or so on a component if attendant variability means rework, order exchange, warranty claims, etc. The costs of failure can totally swamp any savings on manufacturing cost. The intention behind the material presented here is to encourage the generation of capable design solutions and facilitate the exploration of their likely cost implications. Selection must not be based only on a minimum cost strategy. A 'quality first' strategy must be adopted.

PRODUCT NAME   Floppy Disk

PRODUCT CODE / ID   -

LABOUR RATE (Cl) (pence per second)   0.31

CI[H + F]

COMPONENT/SUB-ASSEMBLY DETAILS			$H = Ah + [\Sigma Po + \Sigma Pg]$						$F = Af + [\Sigma Pf + \Sigma Pa]$											Line Total	**Cma**
PART/SUB-ASSEMBLY No.	ID PART DESCRIPTION	ASSEMBLY PROCESS	Ah	Po_1	Po_2	ΣPo	ΣPg	**H**	Af	Pf_1	Pf_2	Pf_3	Pf_4	Pf_5	Pf_6	ΣPf	ΣPa	**F**			(cost in pence)
1	CASE FRONT	HAND./FIT.	1	0.1	0.1	0.2	0.2	1.4	1	0	0	0	0	0	0	0	0	1	2.4	0.74
2	SCREEN 1	HAND./FIT.	1	0.1	0	0.1	1.2	2.3	1	0	0	0	0	0.1	0	0.1	5	6.1	8.4	2.60
3	CASE BACK	HAND./FIT.	1	0.1	0.1	0.2	0.2	1.4	1	0	0	0	0	0	0	0	0	1	2.4	0.74
4	SCREEN 2	HAND./FIT.	1	0.1	0	0.1	1.2	2.3	1	0	0	0	0	0.1	0	0.1	5	6.1	8.4	2.60
5	MAGNETIC DISK*	HAND./FIT.	1	0.1	0	0.1	1.7	2.9	1	0	0	0	0	0	0	0	0	1	3.9	1.21
6	TAB	HAND./FIT.	1	0.1	0.1	0.2	0.2	1.4	1	0	0	0	0	0.3	0	0.3	0	1.3	2.7	0.84
7	SPRING	HAND./FIT.	1.5	0.5	0.1	0.6	0.8	2.9	1	1	0.5	1	0	0.5	0	3	0	4	6.9	2.14
SUB-ASS'Y A	CASE FRONT	HAND./FIT.	1	0.1	0.1	0.2	0.2	1.4	1	0	1	1	1.5	0.7	0	4.2	5	10.2	11.6	3.60
SUB-ASS'Y B	CASE FRONT + BACK	RE-ORIENT.															1.5	1.5	1.5	0.47
8	SHUTTER	HAND./FIT.	1	0.1	0.1	0.2	0.4	1.6	1	0.3	0	0	0	0.5	0	0.8	0	1.8	3.4	1.05
																			51.6s	TOTAL = 15.99p

* The magnetic disk and central driver are bought-in from a supplier as a single item.

Fig. 3.44 Floppy disk assembly costing analysis.

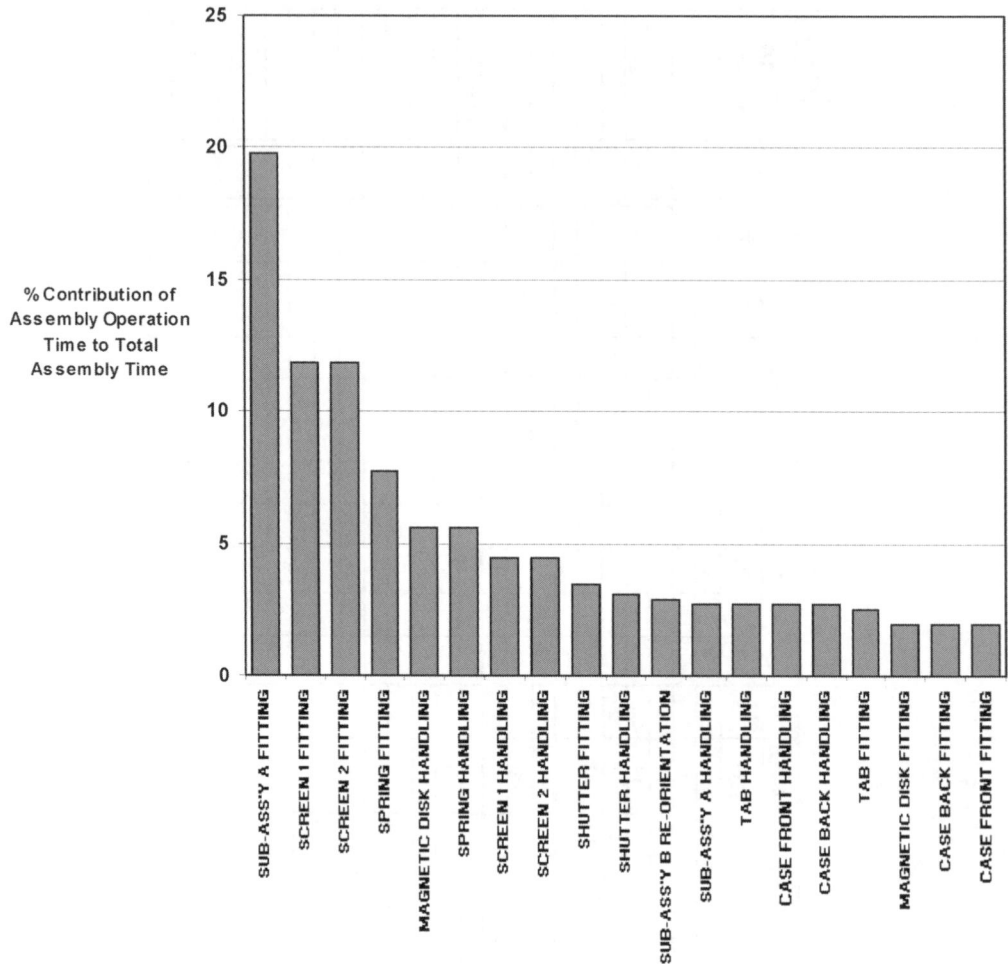

Fig. 3.45 Pareto chart of the assembly operation times for the floppy disk.

Sample questions for students

The sample questions listed below provide some elemental ideas for examination questions and studies for students of engineering and business.

1. In a business concerned with product design and manufacture, why is it worth giving consideration to manufacturing process selection in the early stages of the design process?

2. What are the important criteria that influence process selection in a business? Consider both technological and economic issues. State which of the criteria defined, set limits on what can be achieved by the application of best practice in manufacturing operations.

3. Define a product introduction process and explain how it should be engineered to support the creation of products that are economic to manufacture.

4. Why have businesses implemented formal product introduction process models and how do these differ from the well-established design process models?

5. Present an outline classification of engineering materials indicating the main categories and their subdivisions.

6. Define an outline classification of manufacturing processes indicating the main categories and their subdivisions.

7. Where does process selection fit in a methodology concerned with design for manufacture and assembly? Illustrate your answer with a simple flow chart.

8. Propose candidate material to process combinations for the following engineering components, and justify any decisions made:

   (a) Cylinder head for an internal combustion engine
   (b) Spark plug body
   (c) Radar dish
   (d) 13 Amp power plug body.

9. Select three candidate methods for the manufacture of a low carbon steel tube, 20 mm diameter, 30 mm long with a uniform wall thickness of 2 mm. Rank each candidate for an annual production quantity of 10 000.

10. Why are zinc alloys commonly used for the manufacture of die cast components and give some typical examples?

11. The component shown in Figure Q.1 is to be manufactured by cold forming from a solid cylindrical slug of cold forming steel. Describe the main steps involved in manufacturing the part and comment on how the tooling would need to be proportioned to facilitate metal flow. Illustrate your answer with a sketch.

12. A mezzanine floor is to be fabricated from 1 m square, 5 mm thick low carbon steel panels. Propose methods for cutting the plate to size, preparing the edges, and welding the joints.

13. Compare injection molding and pressure die casting for the manufacture of a small lightly loaded timer gear from a domestic appliance controller in terms of production rates and economics.

14. Contrast the manufacture of toothpaste tubes from aluminum and polymeric material.

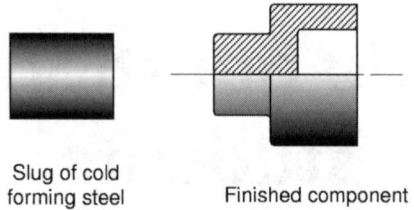

Slug of cold
forming steel

Finished component

Fig. Q.1 Cold formed plug body.

15. Suggest suitable polymeric material and process combinations for the manufacture of the following components, and justify any decisions made:

    (a) Cylindrical bottle (1 l) for vegetable oil
    (b) Automobile handbrake lever
    (c) Computer casing
    (d) Automobile bumper.

16. Compare the processing of metals and plastic by continuous extrusion and explain the differences involved.

17. Contrast the application of adhesive bonding and spot welding for the assembly of pressed steel body panels in automobile manufacture.

18. Suggest suitable composite or ceramic material and process combinations for the manufacture of the following components, and justify any decisions made:

    (a) Golf club heads and shafts
    (b) Aeroplane propeller blades
    (c) Metal cutting tool tips
    (d) High performance hydraulic pistons.

19. In writing a guide for advising the designer regarding injection and compression molding, what design rules would you include and why?

20. Contrast the manufacture of piercing and blanking press tool dies by conventional machining and grinding, with electrical discharge machining.

21. Compare the production of machine tool stands or beds by fabrication techniques and sand casting in terms of economic and technical considerations.

22. The component illustrated in Figure Q.2 is to be manufactured by injection molding unfilled PBT. Given that dimension 'A' is a customer critical characteristic to be maintained at

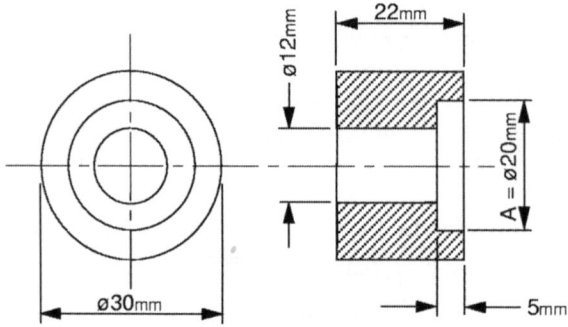

Fig. Q.2 Injection molded bush.

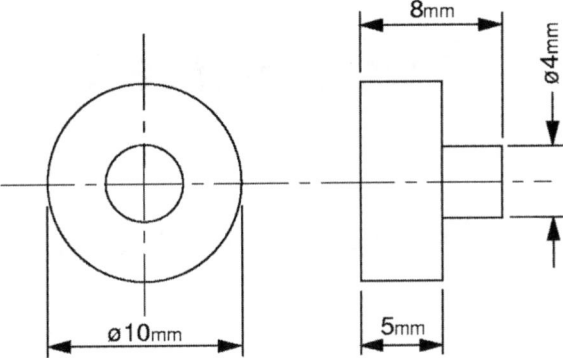

Fig. Q.3 Aluminium alloy button.

$C_{pk} = 1.33$, estimate the cost of manufacture based on a production rate of 20 000 per annum. (Answer: 8.3 pence)

23. Suggest suitable methods for joining the components in the following assemblies, and justify any decisions made:

    (a) A glass lens to the plastic molded automobile headlamp
    (b) Alloy steel bicycle frame tube assembly
    (c) Heavy duty chain links for lifting equipment
    (d) Terminal posts and electronic components in printed circuit boards.

24. Compare the production of phosphor bronze plain bearings by machining and powder metallurgy in terms of manufacturing economics and quality of conformance.

25. Contrast manually operated engine lathes, automatic lathes and CNC lathes in terms of manufacturing economics and technical capability.

26. The small aluminum alloy button shown in Figure Q.3 is currently produced by machining from solid bar at an annual production quantity of 60 000. Would an annual cost saving be possible if the part were to be made by pressure die casting?
    (Answers: Machined = 2.7 pence, Pressure die cast = 3.1 pence)

27. Construct PRIMAs for the following processes:

    (a) Stereolithography
    (b) Water jet machining
    (c) Flux cored arc welding
    (d) Upset forging.

28. Collate and present component costing data that can be used with the costing analysis in Part III of this book for the following manufacturing processes:

    (a) Plaster mold casting
    (b) Rotational molding
    (c) Tungsten inert-gas welding
    (d) Electrical discharge machining.

29. Explain how you would use the process capability charts presented with the PRIMAs in the tolerancing of component assemblies, and in liaison with suppliers.

30. What are the main criteria that influence the cost of a manufactured component? State which of the criteria are predetermined during the design process.

Appendices

Appendix A – Guidelines for assembly-orientated design

Product Level Guidelines

→ Optimise or reduce part-count (and types) by consolidation and integration of features/parts (see Figure A2)
→ Reduce number of fasteners to a minimum
→ Reduce variants and modularise the design
→ Design for an optimum assembly sequence
→ Provide a base for assembly to act as a fixture or work carrier
→ Design the assembly process in a layered fashion (from above)
→ Keep centre of gravity low
→ Use gravity to aid assembly operations
→ Minimise overall product weight
→ Design parts for multi-functional uses where possible
→ Eliminate unnecessary joining processes
→ Strive to eliminate adjustments (especially blind adjustments and shimming)
→ Ensure adequate access and unrestricted vision
→ Use common, efficient fastening systems (when they must be used!)

Component Level Guidelines

→ Use standard components where possible
→ Maximise part symmetry
→ Design parts that cannot be installed incorrectly
→ Minimise handling and re-orientation of parts
→ Design parts for ease of handling from bulk (avoid nesting, tangling)
→ Design parts to be stiff and rigid, not brittle or fragile
→ Design parts to be self-aligning and self-locating (tapers, chamfers, radii)
→ Use good detail design for assembly
→ Avoid burrs and flash on component parts

Fig. A1 Design for assembly guidelines and objectives.

Three basic questions should be asked to evaluate the feasibility of combining parts:

1. When the product is being used, does one part move with respect to mating parts?
2. Must adjacent parts be made of different materials?
3. If parts were combined, would manufacture, assembly of other parts or servicing be made more difficult or impossible?

If the answer is:

YES TO ANY = PROBABLY NOT FEASIBLE
NO TO ALL   = GOOD CANDIDATE TO BE COMBINED

The important principles in combining parts are:
→ Incorporate hinges
→ Use integral springs
→ Use snap fits
→ Incorporate guides, bearings, covers, etc
→ Put electrical and electronic components in one location
→ Use moulded-in labels.

Where fasteners must be used:
→ Use bent tabs or crimped sections
→ Use combined fasteners (integral washers, etc)
→ Use integral locators, hooks or lips
→ Use press fits to replace threaded fasteners.

Fig. A2 Guidelines for part-count reduction (after 2.11).

Appendix B – Weld joint configurations

LAP JOINTS

Single

Double

Offset

CORNER JOINTS

Fillet

Fillets welded both sides

Butt single bevel with root face

Fillets welded both sides (partially overlapped corner)

BUTT JOINTS
(EQUAL THICKNESS)

Compound full penetration

Single-V with backing strip (optional)

Single-V with root faces

Double-V with root faces

Raised edges

BUTT JOINTS
(UNEQUAL THICKNESS)

Fillets welded both sides

Fillet and single-V with deep root faces

Double-V with root faces

Single bevel with root faces (thicker plate tapered)

T-JOINTS

Fillet welded from both sides

Single bevel with root faces

Double bevel with wide root faces welded from both sides

Appendix C – Blank component costing table

PRODUCT NAME _____

PRODUCT CODE / ID _____

PRODUCT QUANTITY _____

COMPONENT DETAILS					Mc = V x Cmt x [Wc]				Ⓐ		Rc = Cc x Cmp x Cs x [Cft]										Rc	Ⓑ	Ⓐ+Ⓑ
PART No. ID	PART DESCRIPTION	MATERIAL	PRIMARY PROCESS	SHAPE COMP.	Volume (mm³)	Cmt	Wc	Mc	Pc	Cc	Cmp	Section (mm)	Cs	Toler-ance (mm)	Ct	Surface Finish (μm Ra)	Cf	Cft	Rc	(Pc x Rc)	Mi (cost in pence)

Appendix D – Blank assembly costing table

PRODUCT NAME _____

PRODUCT CODE / ID _____

LABOUR RATE (Cl) (pence per second) _____

COMPONENT/SUB-ASSEMBLY DETAILS			$H = Ah + [\Sigma Po + \Sigma Pg]$						$F = Af + [\Sigma Pf + \Sigma Pa]$											Line Total	$Cl[H + F]$ Cma (cost in pence)
PART/SUB-ASSEMBLY No. / ID	PART DESCRIPTION	ASSEMBLY PROCESS	Ah	Po_1	Po_2	ΣPo	ΣPg	H	Af	Pf_1	Pf_2	Pf_3	Pf_4	Pf_5	Pf_6	ΣPf	ΣPa	F		

References

Part I

1.1 Parker, A. (1997) Engineering is not enough. *Manufacturing Engineer*, December, 267–271.

1.2 Kehoe, D. F. (1996) *The Fundamentals of Quality Management*, Chapman & Hall, London.

1.3 Mørup, M. (1993) Design for quality. Ph.D. Thesis, Institute for Engineering Design, Technical University of Denmark, Lyngby.

1.4 Fabrycky, W. J. (1994) Modeling and indirect experimentation in system design and evaluation. *Journal of NCOSE*, **1** (1), 133–144.

1.5 Degarmo, E. P., Black, J. T. and Kohser, R. A. (1997) *Materials and Processes in Manufacturing*, 8th Edition. Wiley, New York.

1.6 Kalpakyian, S. and Schmid, S. R. (2001) *Manufacturing Engineering and Technology*, 4th edition. Prentice-Hall, New York.

1.7 Schey, J. (2000) *Introduction to Manufacturing Processes*, McGraw-Hill, New York.

1.8 Groover, M. P. (2002) *Fundamentals of Modern Manufacturing*, 2nd Edition. Wiley, New York.

1.9 Walker, J. M. (1996) (Ed.) *Handbook of Manufacturing Engineering*, Marcel Dekker, New York.

1.10 Waterman, N. A. and Ashby, M. F. (Eds) (1991) *Elsevier Materials Selector*, Elsevier Science Publishers, Essex.

1.11 Shea, C., Reynolds, C. and Dewhurst, P. (1989) Computer aided material and process selection. *Proc. 4th Int. Conf. on Product Design for Manufacture and Assembly*, Rhode Island, USA, June.

1.12 Zenger, D. C. and Boothroyd, G. (1989) Selection of manufacturing processes and materials for component parts. *Proc. 4th Int. Conf. on Product Design for Manufacture and Assembly*, Rhode Island, USA, June.

1.13 Fume, A. and Knight, W. A. (1989) Computer based early cost estimating for sintered powder metal parts. *Proc. 4th Int. Conf. on Product Design for Manufacture and Assembly*, Rhode Island, USA, June.

1.14 Pighini, U., Long, W. and Todaro, F. (1989) Methodical design for manufacture. *Proc. ICED'89*, Harrogate, UK, August.

1.15 Woodward, J. A. and Corbett, J. (1989) An expert system to assist the design for manufacture of die cast components. *Proc. ICED'89*, Harrogate, UK, August.

1.16 Poli, C., Sunderland, J. E. and Fredette, L. (1990) Trimming the cost of die castings. *Machine Design*, 8 March.

1.17 Allen, A. J. and Swift, K. G. (1990) Manufacturing process selection and costing. *Proceedings of Institution of Mechanical Engineers, Part B*, **204** (2), 143–148.

1.18 Schreve, K., Schuster, H. R. and Basson, A. H. (1998) Manufacturing cost estimation during design of fabricated parts. *Proc. EDC'98*, Brunel University, UK, July, 437–444.

1.19 Van Vliet *et al.* (1999) State of the art report on design for manufacturing. *Proc. ASME Design Engineering Technical Conference*, Las Vegas, Nevada, USA, 12–15 Sept.

1.20 Esawi, A. M. K. and Ashby, M. F. (1999) Cost estimation for process selection. *Proc. ASME Design for Manufacture Conference*, Las Vegas, Nevada, USA, 12–15 Sept.

1.21 Esawi, A. M. K. (1994) Systematic process selection in mechanical design. Ph.D. Thesis, Cambridge University Engineering Department, UK.

1.22 Esawi, A. M. K. and Ashby, M. F. (1998) Cost-based ranking for manufacturing process selection. *Proc. IDMME'98*, Compienge, France, 27–28 May, **4**, 1001–1008.

1.23 Liebl, P., Hundal, M. and Hoehne, G. (1999) Cost calculation with a feature-based CAD system using modules for calculation, comparison and forecast. *Journal of Engineering Design*, **10** (1), 93–102.

1.24 Shercliff, H. R. and Lovatt, A. M. (2001) Selection of manufacturing processes in design and the role of process modelling. *Progress in Material Science*, **46**, 429–459.

1.25 Miles, B. L. and Swift, K. G. (1992) Working together. *Manufacturing Breakthrough*, March/April 1992.

1.26 TRW (2002) Product Introduction Management (PIM).

1.27 Parnaby, J. (1995) Design of the new product introduction process to achieve world class benchmarks. *Proc. IEE Sci. Meas. Tech.*, **142** (5), 338–344.

1.28 Miles, B. L. and Swift, K. G. (1992) Design for manufacture and assembly. *Proc. 24th FISITA Congress, Institution of Mechanical Engineers*, London.

1.29 Chrysler Corporation, Ford Motors, General Motors Corporation (1995) *Potential Failure Mode and Effects Analysis (FMEA) – Reference Manual*, 2nd Edition. February.

1.30 Clausing, D. (1994) *Total Quality Development*, ASME Process, New York.

1.31 Kapur, K. (1993) Quality engineering and robust design. Kusiak, A. (Ed), *Concurrent Engineering: Automation, Tools and Techniques*, Wiley, New York.

1.32 Booker, J. D., Raines, M. and Swift, K. G. (2001) *Designing Capable and Reliable Products*, Butterworth–Heinemann, Oxford.

1.33 Miles, B. and Swift, K. G. (1998) Design for manufacture and assembly. *Manufacturing Engineer*, October, 221–224.

1.34 Boothroyd, G. and Dewhurst, P. (1988) Product design for manufacture and assembly. *Manufacturing Engineering*, April, 42–46.

1.35 Boothroyd, G., Dewhurst, P. and Knight, W. (1994) *Product design for manufacture and assembly*, Marcel Dekker, New York.

1.36 DFA Practitioners Manual (Version 12) (2002). CSC Manufacturing, Solihull, UK.

1.37 Shimada, J., Mikakawa, S. and Ohashi, T. (1992) Design for manufacture, tools and methods: the assemblability evaluation method (AEM). *Proc. FISITA'92 Congress*, London, No. C389/460.

1.38 Andreasen, M., Kahler, S. and Lund, T. (1988) *Design for Assembly*, 2nd Edition. IFS Publications/Springer-Verlag, Berlin.

1.39 Ashby, M. F. (1999) *Material Selection in Mechanical Design*, Butterworth–Heinemann, Oxford.

1.40 ASM Handbook (1997) *Materials Selection and Design – Volume 20*, 10th Edition. ASM International, Ohio.

1.41 Miles, B. L. (1989) Design for assembly – a key element within design for manufacture. *Proceedings of Institution of Mechanical Engineers.*, **203**, 29–38.

1.42 Corbett, J. (Ed) (1991) *Design for Manufacture: Strategies, Principles and Techniques*, Addison-Wesley, Bath.

1.43 Edwards, L. and Endean, M. (1990) *Manufacturing with Materials*, Butterworth–Heinemann, Oxford.

1.44 Bakerjian, R. (Ed) (1992) *Design for Manufacturability – Tool and Manufacturing Engineers Handbook Volume 6*, 4th Edition. Society of Manufacturing Engineers, Dearborn, Michigan.

1.45 Huang, G. Q. (Ed) (1996) *Design for X – Concurrent Engineering Imperatives*, Chapman & Hall, London.

1.46 Plunkett, J. J. and Dale, B. G. (1991) *Quality Costing*, Chapman & Hall, London.

▨ Part II

2.1 Kehoe, D. F. (1996) *The Fundamentals of Quality Management*, Chapman & Hall, London.

2.2 Kotz, S. and Lovelace, C. R. (1998) *Process Capability Indices in Theory and Practice*, Arnold, London.

2.3 Redford, A. (1994) Design for assembly. *European Designer*, Sept/Oct, 12–14.

2.4   Swift, K. G., Raines, M. and Booker, J. D. (1997) Design capability and the costs of failure. *Proceedings of Institution of Mechanical Engineers, Part B*, **211**, 409–423.

2.5   Sporovieri, J. (1998) Justifying Automation (www.assemblymag.com).

2.6   Boothroyd, G. (1999) Quality and Automation will Shape Design (www.assemblymag.com).

2.7   Shtub, A. and Dar-El, E. M. (1989) A methodology for the selection of assembly systems. *International Journal of Production Research*, **27** (1), 175–186.

2.8   Khan, A. and Day, A. J. (2002) A knowledge based design methodology for manufacturing assembly lines. *Computers and Industrial Engineering*, **41**, 441–467.

2.9   Dewhurst, P. and Boothroyd, G. (1984) Design for assembly. *Robots Machine Design*, 23 February.

2.10 Noori, H. and Radford, R. (1995) *Production and Operations Management: Total Quality and Responsiveness*, McGraw-Hill, New York.

2.11 Bralla, J. G. (Ed) (1998) *Design for Manufacturability Handbook*, 2nd Edition. McGraw-Hill, New York.

2.12 Wang, J. and Trolio, M. (1998) Predicting quality in early product development. *Proc. 3rd Annual International Conference on Industrial Engineering Theories, Applications and Practice*, 28–31 December, Hong Kong, 1–9.

2.13 Lees, W. A. (1984) *Adhesives in engineering design*, Design Council, London.

2.14 http://209.64.216.70/Adhesive

2.15 Darwish, S. M., Al Tamimi, A. and Al-Habdan, S. (1997) A knowledge base for metal welding process selection. *International Journal of Machine Tools and Manufacture*, **37** (7), 1007–1023.

2.16 Beitz, Q. and Küttner, K.-H. (1994) *Dubbel Handbook of Mechanical Engineering*, Springer-Verlag, London.

2.17 http://www.manufacturingtalk.com/indexes/categorybrowseg.html.

▧ Part III

3.1   Pugh, S. (1977) Cost information for the designer. *First National Design Conference*, London.

3.2   Pahl, G. and Beitz, W. (1996) *Engineering Design: A Systematic Approach*, 2nd Edition. Design Council, London.

3.3   Farag, M. M. (1997) *Materials Selection for Engineering Design*, Prentice-Hall, Hemel Hempstead.

3.4   Hundal, M. S. (1993) Cost models for product design. *Proc. ICED'93*, The Hague, 17–19 August, 1115–1122.

3.5   Lenau, T. and Haudrum, J. (1994) Cost evaluation of alternative production methods. *Materials and Design Journal*, **15** (4), August 1994, 235–247.

3.6   Allen, A. J. and Swift, K. G. (1990) Manufacturing process selection and costing. *Proceedings of Institution of Mechanical Engineers*, **203**, 143–148.

3.7   Allen, A. J. *et al.* (1991) Development of a manufacturing analysis and costing system. *International Journal Advanced Manufacturing Technology*, **6**, 205–215.

3.8   Mazzilli, A. Electroplating Costs Calculation (http://www.ipt.dtu.dk/~ap/ingpro/surface/elecomk.htm).

3.9   Hird, G. (1994) *Personal Communication*, Lucas Industries Ltd.

3.10 TeamSET Software Manual (1996). CSC Manufacturing, Solihull, UK.

3.11 Sealy, M., Berriman, P. and Marti, Y.-M. (1992) A practical solution for implementing sustainable improvements in product introduction performance. *Proc. FISITA'92 Congress*, London.

▧ Bibliography

- Black, R. (1996) *Design and Manufacture: An Integrated Approach*, Macmillan, Basingstoke.
- Bolz, R. W. (Ed) (1981) *Production Processes: The Productivity Handbook*, 5th Edition. Industrial Press, New York.

- Bralla, J. G. (Ed) (1998) *Design for Manufacturability Handbook*, 2nd Edition. McGraw-Hill, New York.
- Dieter, G. E. (2000) *Engineering Design: A Materials and Processing Approach*, 3rd Edition. McGraw-Hill, New York.
- Koshal, D. (1993) *Manufacturing Engineer's Reference Book*, Butterworth–Heinemann, Oxford.
- Mucci, P. (1994) *Handbook for Engineering Design: Using Standard Material and Components*, 4th Edition. BSI, Milton Keynes.
- Oberg, O. E. (Ed) (2000), *Machinery's Handbook*, 26th Edition. Industrial Press, New York.
- Parmley, R. O. (Ed) (1996) *The Standard Handbook of Fastening and Joining*, McGraw-Hill, New York.
- Todd, R. H., Allen, D. K. and Alting, L. (1994) *Manufacturing Processes Reference Guide*, Industrial Press, New York.
- Ullman, D. G. (2002) *The Mechanical Design Process*, 3rd Edition. McGraw-Hill, New York.
- Wick, C. and Veilleux, R. F. (1987) *Quality Control and Assembly – Tool and Manufacturing Engineers Handbook Volume 4*, Society of Manufacturing Engineers, Dearborn, MI.

Relevant British Standards (http://bsonline.techindex.co.uk)

- BS 1134 (1990) Assessment of Surface Texture – Part 2: Guidance and General Information.
- BS 4114 (1984) Specification for Dimensional and Quantity Tolerances for Steel Drop and Press Forgings and for Upset Forgings Made on Horizontal Forging Machines.
- BS 6615 (1985) Dimensional Tolerances for Metal and Metal Alloy Castings.
- BS 7010 (1988) A System of Tolerances for the Dimensions of Plastic Mouldings.
- PD 6470 (1981) The Management of Design for Economic Manufacture.

Useful web sites

- http://www.adept.com/main/solutions
- http://www.assemblymag.com
- http://claymore.engineer.gvsu.edu/~jackh/eod/index.html
- http://www.eFunda.com
- http://www.npd-solutions.com
- http://www.teamset.com
- http://www.manufacturingtalk.com/indexes/categorybrowseg.html
- http://www.dfma.com

Index